3495
8/9/95

Technimanagement

Prentice Hall International Series
in Industrial and Systems Engineering

W. J. Fabrycky and J. H. Mize, Editors

Amos and Sarchet   *Industrial Safety and Health Management, 2/E*
Amrine, Ritchey, Moodie, and Kmec   *Manufacturing Organization and Management, 6/E*
Asfahl   *Safety and Health Management, 2/E*
Babcock   *Managing Engineering and Technology*
Badiru   *Expert Systems Applications in Engineering and Manufacturing*
Banks and Carson   *Discrete-Event System Simulation*
Blanchard   *Logistics Engineering and Management, 4/E*
Blanchard and Fabrycky   *Systems Engineering and Analysis, 2/E*
Brown   *Technimanagement: The Human Side of the Technical Organization*
Bussey and Eschenbach   *The Economic Analysis of Industrial Projects, 2/E*
Buzacott and Shanthikumar   *Stochastic Models of Manufacturing Systems*
Canada and Sullivan   *Economic and Multi-Attribute Evaluation of Advanced Manufacturing Systems*
Chang and Wysk   *An Introduction to Automated Process Planning Systems*
Chang, Wysk, and Wang   *Computer Aided Manufacturing*
Eberts   *User Interface Design*
Elsayed and Boucher   *Analysis and Control of Production Systems, 2/E*
Fabrycky and Blanchard   *Life-Cycle Cost and Economic Analysis*
Fabrycky and Thuesen   *Economic Decision Analysis, 2/E*
Francis, McGinnis, and White   *Facility Layout and Location: An Analytical Approach, 2/E*
Gibson   *Modern Management of the High-Technology Enterprise*
Hall   *Queuing Methods: For Services and Manufacturing*
Hammer   *Occupational Safety Management and Engineering, 4/E*
Hutchinson   *An Integrated Approach to Logistics Management*
Ignizio   *Linear Programming in Single- and Multiple-Objective Systems*
Ignizio and Vavalier   *Linear Programming*
Kroemer, Kroemer, Kroemer-Elbert   *Ergonomics: How to Design for Ease and Efficiency*
Kusiak   *Intelligent Manufacturing Systems*
Lamb   *Availability Engineering and Management for Manufacturing Plant Performance*
Landers, Brown, Fant, Malstrom, Schmitt   *Electronics Manufacturing Processes*
Mundel and Danner   *Motion and Time Study: Improving Productivity, 7/E*
Ostwald   *Engineering Cost Estimating, 3/E*
Pinedo   *Scheduling: Theory, Algorithms, and Systems*
Pulat   *Fundamentals of Industrial Ergonomics*
Shtub, Bard, Globerson   *Project Management: Engineering Technology and Implementation*
Taha   *Simulation Modeling and SIMNET*
Thuesen and Fabrycky   *Engineering Economy, 8/E*
Turner, Mize, Case, and Nazemetz   *Introduction to Industrial and Systems Engineering, 3/E*
Wolff   *Stochastic Modeling and the Theory of Queues*

Technimanagement

The Human Side of the Technical Organization

David B. Brown
The University of Alabama

Prentice Hall P T R, Englewood Cliffs, New Jersey 07632

Library of Congress Cataloging-in-Publication Data

Brown, David B.
   Technimanagement: the human side of the technical organization / David B. Brown
     p.   cm.
   Includes bibliographical references and index.
   ISBN 0-13-180811-7
   1. Industrial management.   2. Professional employees.
3. Organizational change. I. Title
HD31.87654   1995   94-39933
658.3'14—dc20   CIP

Editorial production: bookworks
Acquisitions editor: Bernard Goodwin
Manufacturing manager: Alexis R. Heydt
Cover design:Wanda Espane

© 1995 Prentice Hall P T R
Prentice-Hall, Inc.
A Simon & Schuster Company
Englewood Cliffs, NJ 07632

All rights reserved. No part of this book may be
reproduced, in any form or by any means,
without permission in writing from the publisher.

The publisher offers discounts on this book when ordered in bulk quantities.
For more information, contact:

   Corporate Sales Department
   Prentice Hall P T R
   113 Sylvan Avenue
   Englewood Cliffs, NJ 07632
   Phone: 800-382-3419 or 201-592-2498
   FAX: 201-592-2249
   E-mail: dan_rush@prenhall.com

Printed in the United States of America

10 9 8 7 6 5 4 3 2 1

ISBN 0-13-180811-7

Prentice-Hall International (UK) Limited, *London*
Prentice-Hall of Australia Pty. Limited, *Sydney*
Prentice-Hall of Canada, Inc., *Toronto*
Prentice-Hall Hispanoamericana S.A., *Mexico*
Prentice-Hall of India Private Limited, *New Delhi*
Prentice-Hall of Japan, Inc., *Tokyo*
Simon & Schuster Asia Pte. Ltd., *Singapore*
Editora Prentice-Hall do Brasil, Ltda., *Rio de Janeiro*

To Carolyn,
Lesley and Jonathan

Contents

PREFACE

1 WHY TECHNIMANAGEMENT?   1

    1.1 First: What Is It? 1
    1.2 Why Do We Need It? 2
    1.3 Where Do We Go from Here? 6

2 ELEMENTS OF CONTROL   8

    2.1 Why Control? 8
    2.2 Analysis of Control 9
    2.3 Goals and Objectives 11
    2.4 Measurement and Evaluation 13
    2.5 Correction 14
    2.6 Control of Professionals? 15

3 THEORY OF REWARDS   18

    3.1 Basic Theory 18
    3.2 Rewards versus Motives 20
    3.3 Hierarchy of Needs 23
    3.4 Protestant Ethic and Social Darwinism 26

3.5 Motivational Conflict and Frustration 29
3.6 Defense Mechanisms 32
3.7 Productivity through Goal Integration 34
3.8 Motivational Maintenance Theory 36
3.9 Consideration of the Human Factor 38
3.10 Hawthorne Effect 40
3.11 Industrial Humanism 42
3.12 Human Internal Complexity: Mind/Emotions 44
3.13 Rewards and the Meta-control System 46

4 McGREGOR'S THEORY X AND THEORY Y 49

4.1 Introduction 49
4.2 Theory X 50
4.3 Theory Y 52
4.4 What Happened to Theory Y? 53
4.5 Technimanagement and Theory Y 57

5 DECISION MAKING AND THE PRINCIPLE OF AUTHORITY 59

5.1 Definitions: Authority and Stability 59
5.2 Authority and Organizational Evolution 61
5.3 Authority versus Leadership 63
5.4 Responsibility without Authority 65
5.5 Decision-Making Process 69
    5.5.1 Who Makes the Decision? 69
    5.5.2 Decision-Making Mechanism 71

6 MANAGEMENT BY EXCEPTION 74

6.1 Definitions 74
6.2 MBE under Technimanagement: MBNE 75

6.3 MBNE Deterrent: Resentment 78
6.4 Toward a Solution 80

7 MANAGEMENT BY OBJECTIVES 83

7.1 Definitions 83
7.2 MBO and Technimanagement 84
7.3 MBO Complications 87

8 OPTIMIZATION, EQUIFINALITY, AND SYSTEMS VIEW 90

8.1 Definition of the Problem 90
8.2 Optimization 95
8.3 Equifinality 100
8.4 Systems View 101

9 PARKINSON'S LAW 103

9.1 General Applicability 103
9.2 Compression Corollary 106
9.3 Applications to Time Management 108

10 THE PETER PRINCIPLE 110

10.1 Definitions 110
10.2 Validity of the Peter Principle 112
10.3 Shortcomings of the Peter Principle 117
10.4 Solutions to the "Peter" Problem 119
    10.4.1 Individual Countermeasures 120
    10.4.2 Organizational Countermeasures 122

11 THE PARADIGM PROBLEM 128

11.1 Definition of the Problem 128
11.2 Validity of the Paradigm Problem 130
11.3 Solving the Paradigm Problem 134

12 INFORMAL LEADERSHIP AND ORGANIZATIONS 138

- 12.1 **Definitions, Methods, and Problems** 138
  - 12.1.1 Identification Methods 139
  - 12.1.2 Problems with Current Approaches 141
- 12.2 **Benefits of Informal Organizations** 145
  - 12.2.1 Filling the Decision-Making Gaps 145
  - 12.2.2 Personnel Development 148
  - 12.2.3 Useful Channels of Employee Communication 148
  - 12.2.4 Improving Managers and Management 149
  - 12.2.5 Improving Worker Morale 150
  - 12.2.6 Downside of Informal Organizations 152
- 12.3 **Empowerment of Informal Organizations** 153
  - 12.3.1 Effective Empowerment 156
  - 12.3.2 Empowerment Mechanisms 158
- 12.4 **Informal Communication** 162

13 ORGANIZATIONAL ENTROPY 165

- 13.1 **Definitions** 165
- 13.2 **Principle of Organizational Entropy** 166
- 13.3 **Spontaneous Group Formation** 170
- 13.4 **Management Response to Group Formation** 173
- 13.5 **Spontaneous Group Decomposition** 174

14 GROUP DYNAMICS 178

- 14.1 **Definitions** 178
- 14.2 **Group Empowerment** 181
- 14.3 **Group Selection** 184
- 14.4 **Group Sizing Theory** 189
  - 14.4.1 Project Group Size 189
  - 14.4.2 Sizing of Organizational Support Groups 192

Contents   xi

14.5 Group Harmony 193
    14.5.1 Group Synergism 194
    14.5.2 Group Membership Selection Revisited 197
    14.5.3 Maintaining Good Morale 199

14.6 Consensus-Building Techniques 200
    14.6.1 Meeting Conduct 201
    14.6.2 Brainstorming 204
    14.6.3 Prioritization Techniques 205

15 TOTAL QUALITY MANAGEMENT 210

15.1 Introduction 210
15.2 General Characteristics of TQM 212
15.3 Cost of Short-Term Gains 214
15.4 Deming's 14 Obligations of Management 215
    15.4.1 Constancy of Purpose 216
    15.4.2 Major Paradigm Change 217
    15.4.3 Process Improvement, Not Inspection 218
    15.4.4 Price Tag Independence 219
    15.4.5 Constant Improvement Forever 220
    15.4.6 Immediate Training and Education 221
    15.4.7 Supportive Management 223
    15.4.8 Reversal of Motivation Mechanisms 224
    15.4.9 Elimination of Departmental Barriers 228
    15.4.10 Honest Maturation 230
    15.4.11 More Effective Measurement 230
    15.4.12 Pride of Workmanship 231
    15.4.13 Long-Term Education and Retraining 231
    15.4.14 Teams 232
    15.4.15 Closure on Deming's Fourteen Obligations 232

15.5 Quality Management Approaches 233
    15.5.1 Quality Function Deployment 233
    15.5.2 Statistical Process Control 234
    15.5.3 Total Employee Involvement 235

15.6 TQM Infrastructure: Customer Orientation 236
    15.6.1 Defining the Customer 236
    15.6.2 Prioritizing the Customers 237

15.6.3 Involving the Customers 238
15.6.4 Orienting the Organization to the Customer 238
15.7 TQM and Technimanagement 239

16 THE CLINICAL APPROACH 242

16.1 Definition 242
16.2 Perception of Current Reality 244
16.3 Differences in Opinion 246
16.4 What Is Said and What Is Meant 246
16.5 Perception of Future Reality 248
16.6 Informal Organizations 249
16.7 Ego Growth and Ego Protection 249
16.8 Objectivity and the Control System 251

17 PRINCIPLES OF FUNCTIONAL CONFLICT 252

17.1 Definitions and Causes 252
17.2 Benefits of Functional Conflicts 255
17.3 Resolution of Functional Conflicts 258

18 RESOLUTION OF PERSONAL CONFLICTS 262

18.1 Statement of the Problem 262
18.2 Preventing Personal Conflicts 263
18.3 Personal Conflicts involving Your Subordinates 264
18.4 Personal Complaints against You 272
18.5 Role Conflict 276

19 INDIVIDUAL ENTROPY AND PERSONAL POWER 279

  19.1 Definitions 279
  19.2 Second Law of Thermodynamics 280
  19.3 Sources of Power 285
     19.3.1 Time Value of Power 286
     19.3.2 Influence Spectrum 290
     19.3.3 Winning and Losing 292
     19.3.4 Informal Sources and Uses of Power 294
  19.4 Power Problems 297

20 THE NEGATIVE FEEDBACK PRINCIPLE 300

  20.1 Introduction 300
  20.2 Killing the Messenger 301
  20.3 Reaction to Criticism 302
     20.3.1 Internal Reaction to Criticism 302
     20.3.2 External Reaction to Criticism 303
  20.4 Inspiring Criticism 308
  20.5 Inspiring Acceptance in Others 309

21 LEADERSHIP 311

  21.1 Counterintuitive Notion 311
  21.2 "Walk behind Them" 311
  21.3 Ego of the Leader 313
  21.4 Traits of Successful Leaders 314
  21.5 Traits of Successful Managers 317
  21.6 Leadership or Management? 319
  21.7 Leadership Styles 322

22 LIMITATIONS OF EXPERTS 325

    22.1 A Notion Revisited 325
    22.2 The Problem 326
    22.3 Dealing with Expert Limitations 330

23 PRINCIPLES OF COMMUNICATION 335

    23.1 Definition and Objective of Successful Communication 335
    23.2 Deterrents to Successful Communication 338
        23.2.1 Stereotyping 338
        23.2.2 Poor Symbology 340
        23.2.3 Selective Perception 341
        23.2.4 Breakdown of Effective Informal Communications 342
        23.2.5 Failure to Modularize 343
    23.3 General Conclusions 344

24 PERFORMING THE TRANSITION 345

    24.1 Essential Requirement 345
    24.2 Evolution of a Customer Reorientation 348
    24.3 The Transition 351
        24.3.1 Top-Down 352
        24.3.2 Bottom-Up 355
        24.3.3 Middle-Out 367
    24.4 Postscript on Training 371
    24.5 Where Will All This Take Us? 372

GLOSSARY 374
REFERENCES 389
INDEX 393

Preface

Criticism from employers of engineers and scientists rarely focuses on technical deficiencies. Invariably it is the lack of "human skills" that causes their concern. *Technimanagement* provides scientists, engineers, and other professionals with an understanding of basic principles of human nature as they apply to managing organizations of professionals. It is directed primarily toward those who have been out in the workplace for a few years and are facing their first management challenges. However, the principles apply generally, and specific approaches are given for applying these principles at middle and upper management levels as well.

Technically trained people are usually the ones chosen to manage technically trained people. Since these managers have spent most of their academic and work careers dealing with technical as opposed to people problems, they are ill-equipped to handle the interpersonal relationships of management. Thus it would seem that a compilation of principles with regard to human skills in management would provide the solution—*or so I thought*.

But as I synthesized the principles of management that are currently bearing fruit in our nation's "quality movement," I recognized that the simple synergism which enables a beehive to produce honey is woefully lacking in most contemporary organizations. Its emergence requires much more than the indoctrination of a few of our organizational "bees." The problem is a holistic one, and it cannot be addressed merely by training (or retraining) a subset of the organization's members. For, once they are absorbed back into the organization they regress to organizational consistency.

This led to a major premise that defines the goals of technimanagement: *The increased productivity of technical organizations requires a major transition wherein the entire management structure itself undergoes constant and adaptive improvement.* This premise has not been accepted by traditional management because it requires protracted meta-control efforts over a long period of time. It cannot be brought about merely by sending everyone off to a three-day training course.

The progress made in our transportation, power, communication, and computer systems has taken decades to evolve. Can we expect the human aspects of our organizations to change overnight? Should we throw up our hands because there is no quick fix?

There is hope! The reason that our physical systems have evolved so effectively is the *immediate* benefit that was produced as each new component was integrated into the rest of the system. Similarly, there are immediate benefits that will support and nurture the transition of our management approaches—the *immediate* productivity increases that come from the empowered organization.

Empowerment can occur only when a manager possesses the courage to act on an *accurate perception of reality* as it applies to people and their interactions within organizations. This can be attained independently at the lowest levels even within the most traditional of organizations. However, our most progressive managers are recognizing and applying the principles throughout their organizations now!

It is here that the greatest hope lies. For when the productivity of these organizations becomes clear, there will be no turning back to traditional management approaches. This will not be tolerated by managers who are reaping the rewards of increased productivity. (Listen to them!) But, more important, turning back will not be tolerated by the other members of the organization, who are having so much fun while being so productive.

David B. Brown

1

Why Technimanagement?

1.1   First: What Is It?

Technimanagement is concerned with applying time-proven principles of human interrelationships to improvement in the productivity of technical organizations. In this context, *technical* refers to organizations of scientists, engineers, or other professionals who are performing creative design and development activities. This can be contrasted with *traditional* production-oriented organizations which have a well-developed and tested set of products and whose major objective is to produce them at a competitive quality and cost. Although many principles apply equally to both, technimanagement is as different from the traditional labor—management interactions of industrial organizations as the products that they produce.

Since very few organizations are purely technical or traditional, we are talking degrees of application rather than absolutes. Nevertheless, we will use the term *technical organization* to refer to an organization that is composed primarily of technically trained and/or experienced professionals. Similarly, we will generally refer to nontechnical organizations, and the corresponding management practices which they have evolved, as *traditional organizations*. Finally, the word *technimanagement* will be used throughout to refer to the application of those principles that have been proven over time to apply to the effective management of technical organizations. This brings us to a point where we can begin to define the problem that technimanagement is designed to solve.

1.2 Why Do We Need It?

One of the major initiatives of our efforts to improve curricula in the engineering and science disciplines is based on ongoing feedback from the employers of our students. Rarely do we hear of technical shortcomings. The two deficiencies in science and engineering students most often cited are communication and human skills. These are highly interrelated, since deficiencies in one can cause difficulties in the other. It seems that as soon as technical graduates get to a point where they are given authority over others, they run into problems for which neither their education nor their experience has prepared them. In today's downsized job environment, this opportunity comes quickly, often within the first year or two. Even in larger companies, the first management assignment (e.g., to project group leader) carries with it a tremendously magnified potential to either create or destroy initiative.

Insight into the source of the problem can be obtained by evaluating the criticisms, which generally go something like this: "He was a computer whiz but just could not relate to his fellow workers." "She was our best engineer, so we made her the team leader, and then everything seemed to go down the tubes." "The design project got bogged down with endless debates over the system requirements; we finally had to reassign several people and virtually start over." The variations are endless, but they all have the common thread—the lack of management and human interpersonal skills on the part of otherwise valuable and competent technically trained people.

Like most educators, the author has had the tendency to believe that education can solve this problem. However, both the complaints and this proposed solution are not of recent origin, as evidenced by the following quotation:

> We have failed to train students in the study of social situations; we have thought that first-class technical training was sufficient in a modern and mechanical age. As a consequence we are technically competent as no other age in history has been; and we combine this with utter social incompetence. This defect of education and administration has of recent years become a menace to the whole future of civilization. (MAYO45,120)

This was made by Elton Mayo in 1945, a time when many of our readers may not have been born and which few now perceive to be a time of technological sophistication. Nevertheless, this problem was clearly recognized nearly two generations ago. The fact (1) that a failure "to train students" was the assigned cause and (2) that academia has had adequate time to respond leads us to wonder if our education system will ever produce a solution. In fact, the display below indicates that the current approach of technical education might contribute heavily to the problem.

> **Quandaries of Technical Education**
>
> *Its restricted nature.* All technical subjects are getting more and more complex. There is little space in technical curricula for nontechnical subjects, and accreditation agencies all but specify most of these.
>
> *Objectivity of technical subjects.* On tests there is one right answer. Push all the right buttons and everything works right. Get the right formula, plug in the numbers, and the answer is always the same. This establishes a thought paradigm into which people just do not fit.
>
> *The student's instilled perception of methods for bringing about change.* Having the methodology for determining an optimal decision should be enough, but it's not. Very little emphasis is given to the necessity for *selling* the decision or, more important, the organizational involvement that would make such selling unnecessary.
>
> *The educators themselves.* Despite claims of liberal views, academicians tend to be biased toward their own disciplines. This is often defensive in light of internal competition, but more important, it is the result of an honest belief in one's own chosen field (instilled, at least initially, by the very same educational system). A true appreciation for diversity by professional educators is a critical first step.
>
> *Temporal irrelevance.* Even when technical students take management courses, it is at a time when few of them can visualize themselves as managers of other technical people. It is very difficult to conceptualize these principles in the abstract without actually "being there."
>
> *Positional irrelevance.* Most management courses have the perspective of the highest level of management, which is quite different from the requirements imposed on the first-line or even the middle manager. Such irrelevance can be counterproductive in producing a disregard for all management principles.
>
> *Irrelevance of application.* Many management courses fail to distinguish between traditional management and the management of technical professionals.

McGregor reinforced the potential cause of "training" when describing the technically trained staff person:

> His training, as a matter of fact, tends to make him unusually vulnerable. He has been taught to find "the best answer." He has great confidence in the objectivity of his scientific techniques. His natural expectation is that the solutions resulting from his application of these techniques will be immediately and gratefully accepted by the organization. Management by direction and control seems perfectly reasonable, with the modification that "the authority of knowledge" is for him final. No reasonable human being would challenge it. (MCGRE60,154)

The time-proven words of McGregor and Mayo help to demonstrate that we dare not wait for our academic institutions to lead us out of this quagmire.

The fact that many engineers and scientists, especially some who start their own companies, become great managers is ample proof that technically educated people *can* acquire the skills necessary for technical management. A counterexample comes from the other extreme: Some business students schooled in management and human interrelationships still fall flat on their faces when trying to manage technical people.

The major premise of this book is that the problem is a behavioral one primarily on the part of managers, not on the part of the people they manage. A symptom of an even deeper underlying problem is that these technical managers rarely realize this. Although they see the problem as behavioral, they usually perceive the deficiency to be on the part of their subordinates or their management. It is only human to point the finger at others rather than taking the blame ourselves. Whether we point the finger at academic institutions, our subordinates, or our management, we are falling into the same trap of personal complacency that prevents any form of self-improvement.

Although changes in behavior cannot be accomplished by education alone, there is no hope for correcting personal behavior in the absence of an improved perception of reality. Change for the sake of change has as much chance of exacerbating as of ameliorating the problem. Changes based on plans that conform to sound principles generally produce the desired results.

Each of the principles of technimanagement will define and confront one element of the underlying problem. Consistent application of the principles involved will solve a particular part of the problem, and the integrated application of all these principles will solve a large portion of the problem.

We would expect the following question to arise: Given (1) that technimanagement will solve the management problem, and (2) that most of the principles which it encapsulates have been around for dozens of years, why are these principles not currently being applied? We will try to address this question on a principle-by-principle basis, since each might be somewhat different. However, so that we will not have to be overly repetitive, we offer the following general observations.

First, they *have* been applied to some extent; and to the extent that they have, they are bringing about great gains in productivity and satisfaction. Compare the principles given with the advice in many relatively recent and popular books to validate this (from Robert Townsend's *Up the Organization* of the 1970s to DeMarco and Lister's quite recent *Peopleware*). There are a number of great success stories that could be cited in favor of the fact that application of these principles does produce positive results.

On the other hand, inconsistent or self-serving manipulative applications of the same principles are destined to fail. Failures have been due to either (1) a lack of knowledge of the principles, or (2) a failure to apply them properly and consistently. Obviously, a principle cannot be applied correctly unless first known and then understood to the point where it can be applied properly. While the first part of the problem must be addressed by education, the second involves the character of the manager.

We will consider both throughout our discussions. As an example of the second, we have already mentioned the natural human tendency to blame others for our own shortcomings, which is a character flaw fatal to self-improvement. A second example is the perception by management that these are merely techniques for the manipulation of others, and this will be given considerable attention in the chapters that follow.

Another independent reason that these principles have not received proper application is that they were tried on nontechnical organizations and then discarded because they did not apply fully. We will not be concerned with the applicability of the principles of technimanagement to what we have defined as *traditional* organizations. Although many of them certainly apply, others can reach fruition only with considerable change in the attitudes of both the work force and management. Change is still required in technical organizations but to a much smaller degree. Thus significant progress is quite feasible in the near future.

Even in technical organizations, the lack of faith in counterintuitive principles either by management or by those managed could result in serious setbacks. If lower-level managers are not rewarded properly for their efforts despite these setbacks, it is only reasonable that the principles would be abandoned in favor of upper-management directives.

But we are getting ahead of ourselves. To clarify our objectives further, we recognize that most management books present definitions and models in an attempt to bring a degree of order to the confusion of human organizations. Although some of this is essential if we are going to discuss and understand the nature of these organizations, we believe that the primary need of the technically trained is just the opposite. Generally, the preponderance of their training and experience has instilled an intuitive sense that the world should be a very orderly place. The attempt to transfer this paradigm of order to the management process is one of the primary causes for misunderstanding the demands of technical management. This aspect of the problem might well lead us to wonder where we go from here.

1.3 Where Do We Go From Here?

The problem statements given above imply that technical managers need to establish paradigms for human interaction which are considerably different from those which their disciplines have instilled in them. Technimanagement synthesizes many of the principles of human nature and human interaction necessary for this. This strikes at the true source of the problem, and most technical people will have little problem in learning the other organizational aspects of management (e.g., personal time management, planning, accounting, etc.).

We begin with the most basic concepts, which are presented in Chapters 2 through 7. Essential to an understanding of organizational behavior is a general understanding of the elements of control, which we present in Chapter 2. Since these elements are the same as those necessary for the control of all physical systems, we use this context to begin to distinguish between the mechanical and human systems. It will be demonstrated that within professional organizations, managers must recognize a higher-order function, that of *meta-control*, which will be defined.

A second component of our foundation is laid in Chapter 3 with a discussion of reward theory. This chapter is lengthy, to cover the relevant aspects of motivation and perception as they apply to complex organizations of professionals. This enables us to understand McGregor's theories of management (Chapter 4), which begin to formulate the image of a blurred road map from the current state of organizational structures to the ideal target structures of the future.

To begin to focus this image requires a proper perception of decision making and authority, which is covered in Chapter 5. Then in Chapters 6 and 7 we critique two popular recent management techniques. Together these three chapters clearly demonstrate the counterintuitive requirements of technimanagement. In turn, they illustrate that many of the approaches of traditional management are counterproductive when applied to technical organizations.

A pivotal chapter (Chapter 8) follows, on optimization, equifinality, and the systems view, which has the goal of transforming the thought process from the limited perspective that is so inherent in working with physical devices to the broader outlook required to deal effectively with people and organizations. In Chapters 9 through 11, which deal with Parkinson's law, the Peter principle, and the paradigm problem, we test both this new perspective and the counterintuitive nature of technimanagement.

Having digested the milk, we get to the meat of the book in Chapters 12 through 15. We begin with a discussion of a subject that most managers ignore at their grave peril: informal leadership and spontaneously generated informal organizations. This provides the theoretical basis for the concepts of natural organizational evolution covered in Chapter 13. To reverse the natural trend, we suggest some basic principles of group dynamics. Finally, to address the current wave within the "quality movement," we provide a brief review of total quality management, which encapsulates a subset of the principles of technimanagement.

The advanced topics are developed in Chapters 14 through 23—advanced in the sense that they require a certain degree of personal maturity for their acceptance. These include the clinical approach, principles of functional conflict, resolution of conflicts, power, negative feedback, and leadership. Although principles of communication are utilized throughout the book, some of the most important are formalized in the last conceptual chapter.

We end in Chapter 24 with the process of transition through a suggested procedural application of the principles expressed in the earlier chapters. It stresses what should be apparent throughout the book: *that technimanagement is not a procedure or technique that can be implemented merely by following a set of rules or implementing a methodology.* It is a way of thinking about people on the personal level, assessing each one as a person with a different but priceless potential to contribute to the organization. And it places the responsibility for unleashing this potential squarely on the shoulders of you, the manager.

2

Elements of Control

2.1 Why Control?

All management strives for control. But over what? And to what end? In this chapter we show that these questions are not trivial. It is essential to understand the nature of control and its ramifications when attempting to "control" an organization.

The technically trained understand control of physical devices quite well. Placing a bomb precisely down the ventilation shaft of a high-rise building from an altitude of several miles, as was done in the Gulf War, is an amazing display of control. Guiding a spacecraft to the moon and back is nothing short of spectacular. The mechanism by which we can harness and control modern computers mystifies even the most technical of minds. This raises a question: With such an understanding of control mechanisms, why should there be such difficulty in controlling the behavior of an organization?

Quite simply: Maintenance of control within the human organization is an order of magnitude more complex than the control of any physical device or system. This is because most physical devices behave in relatively deterministic and repeatable ways. Organizations rarely respond identically to identical stimuli. Most people are different in their desires and motivations, and hence they differ considerably in the way they react to a set of circumstances. To complicate things further, individuals and their interrelationships change over time, and sometimes they change quite dramatically.

This is not to say that the basic principles of control do not apply. Indeed, as we get into the many other principles of human behavior, the basic concept of control is the one to which we will refer most often. It forms the foundation for our understanding of both human nature and human organizations. This is the reason that we start with a discussion of it. Some of the concepts presented in this chapter are detailed further in an excellent treatise by Kazmier (KAZMI69).

2.2 Analysis of Control

Control is an abstract concept. Much like the concepts of quality and reliability, there are degrees to which control exists in any given situation, mechanical device, or organization. Thus we cannot assess the degree of control without a criterion against which it can be measured. We may never be able to state unequivocally that a system is "in control," but we can usually identify when it is not by the obvious consequences—but by that time the damage has already been done.

The term *control* gains substance when we analyze it into its essential elements. Figure 2.1 introduces the three essential elements of control. This model is obviously a simplification, but we will flesh it out in the remaining sections of this chapter. However, the *goal–measurement– correction* paradigm is an extremely useful one for practical applications.

Element	**Purpose**
Goal	To generate specific *objectives* for which degrees of attainment can be definitively measured
Measurement	To assess the degree of progressive realization of objectives
Correction	To restore the system when the *evaluation* of measurements shows unacceptable deviation from objectives

Figure 2.1 Role of the Three Essential Elements of Control

The elements essential to control are totally independent of the system being controlled. Consider such systems as a vehicle on the roadway, an aircraft in the sky, or the temperature in a room. Identify the goal, measurement, and correction elements that must exist to maintain control. Now, mentally remove any one of the three elements. Can the system stay in control? This leads to a principle for which we believe there is no counterexample:

> Without the presence of all three elements of control, control cannot possibly be maintained.

Thus the most logical thing to do when it is clear that control has been lost is to determine which element(s) of the control system are missing and go about the task of creating or restoring them. However, the natural tendency is to treat the symptoms of lost control as the problem, in which case the actions taken may have no impact on the control system itself.

We stated above that there was no way to determine if a system is in control but that it is relatively easy to determine if it is out of control. This is because *the absence of out-of-control symptoms does not necessarily imply the presence of control*. Measurement systems are not perfect, and a false sense of reality disguises subtle harbingers of disaster. Indeed, one of the greatest threats to an organization is to judge that it is under control when in fact it is not. It is easy to assess that control has been lost after the symptoms clearly reveal that some undesirable event has occurred. This is usually followed by a knee-jerk reaction under the guise of correction, which in all probability will neither restore control nor effectively heal the control system.

This leads us to the following proposition:

> For control to exist, the measurement elements must be able to detect the increasing probability of undesirable events sufficiently before they happen so that corrective action can be taken to prevent their occurrence.

In other words, the presence of goals, measurements, and corrections alone is not sufficient. These elements must be of sufficient quality themselves to provide control. This is clear in the control of physical devices; it is just as applicable to human organizations.

This helps us discern the difference between operation of the control system and its establishment and maintenance. One primary shortcoming of management in general is that we get so absorbed with operations that we virtually disregard the construction and maintenance of the control systems themselves. We call this critical function *meta-control* (i.e., the control of control systems), and its effective performance is the essence of technimanagement. Physical devices do not stay in control unless their control systems are designed, developed, and maintained (i.e., meta-controlled). Similarly, meta-control is essential to the progress and effectiveness of all human organizations.

One other concept will be useful as we apply control concepts further; this will require introduction of the term *countercontrol*. Analogous to the meaning of the word *counterproductive*, which we use quite frequently, it goes beyond the idea of being out of control to the point where attempts at correction actually cause the system to deviate further from the goal. Common examples of this include (1) mistaking the hot water valve for the cold water in the shower, and (2) hitting the gas rather than the brake in an emergency-stopping situation. Countercontrol leads to the worst consequences of having a misperception of reality.

The integration of control theory into the maintenance of human organizations is a recurring theme in this book, so all the practical ramifications will not be made here. (Additional principles are needed before this can be done.) However, it is important to recognize that the model given above is overly simplistic, and the objective of the following sections is to begin to augment this model to make it more useful. This process will be continued in subsequent chapters.

2.3 Goals and Objectives

In the context of human organizations, *goals* are the articulation of the highest-level ideals of the organization. These are qualitative statements that may not be directly measurable. Some organizations have effective mission statements, but most do not. Needless to say, if the individual members of an organization do not have a clear understanding of its goals, they can hardly be expected to attain them.

On the other hand, the potential of collective human effort is incalculable when a well-defined goal statement is understood and endorsed by an organization's membership. Consider a task that was unthinkable in the 1950s: putting a human being on the moon. How did the United States go from its third-rate space program to a preeminent leader in less than a decade? A large part of the answer lies in this clear goal statement by President Kennedy: "I therefore ask the Congress . . . to provide the funds . . . to achieve the goal, before this decade is out, of landing a man on the moon and returning him safely to the earth" (John F. Kennedy, 1962). That goal could not have been better defined: Go out on most clear nights and take a look at it. It not only enabled us to visualize its accomplishment but also provided a time limit for its realization. Most important, its accomplishment had the support of an entire nation.

We have been interchanging the singular and plural: *goal* and *goals*. It might seem preferable to have just a single goal, as in the moon mission.

Since the accomplishment of the moon landing, however, NASA and those who dictate policy (i.e., Congress) have essentially lost their way and at this writing are still attempting to find it. One reason for this is a multiplicity of goals imposed by an ever-changing "leadership." Generalizing this example, we might state that a singular goal is better than a plurality of goals. There is also a more theoretical foundation for this statement, and it involves the reduction of *functional conflicts*, as we discuss further in Chapter 17. However, there *are* multiple competing goals within most organizations. This is a reality that we cannot sweep under the rug merely by stating that a single goal is preferable, and we will address this shortly.

The existence of even a very clear-cut goal might not be sufficient to provide an effective *goal* element for the control system essential to its accomplishment. This was certainly true with regard to getting to the moon. There were at least three different ways proposed for accomplishing this unique goal: (1) the "science fiction" method of taking a single spacecraft from earth to the moon and returning it, (2) the space station approach of constructing a platform in orbit and then building and launching the mooncraft from that platform, and (3) the idea of escaping Earth's gravitational pull with a large rocket and then discarding pieces of it along the way until only the bare minimum remained for the journey back. Given the time limit, the third proved to be the most effective approach despite its most counterintuitive nature (witness the fact that no science fiction writer ever appears to have imagined it). The point here is that it was necessary to *refine* the goal by proposing methods for its accomplishment before the (refined) goal could become an effective element of the control system. This process of *goal refinement* must be recognized as an organizational function which itself must be controlled (more meta-control).

To detail our control model further, we introduce the notion of objectives. An *objective* is the articulation of one activity (usually among many) in the accomplishment of a goal. Ideally, each goal is analyzed into a number of objectives, and these objectives are used as the first elements of control in the child control systems that are generated. To be effective for control purposes, objectives must be measurable on a fairly continuous basis. While an organization might have competing alternative goals that vie for its dedication, a goal can have many alternative objectives by which it might be accomplished. It is important to resolve *goal conflicts* during the goal refinement process before establishing objectives. The meta-control function of establishing optimal objectives is a primary goal of technimanagement that we revisit in subsequent chapters.

Considerable methodology could be developed for the purpose of defining goals and objectives, but this is a sufficient introduction for the technical reader. In following sections we introduce the other two elements of control so that they can be used to help communicate other principles of technimanagement.

2.4 Measurement and Evaluation

Let us assume that a set of goals has been specified that reflect the mission of the organization. Further, suppose that these goals have been decomposed into sets of objectives. It is essential that the realization of these objectives be measurable so that action can be taken before an obvious out-of-control situation results. With a car-on-the-road example, some of the measurements consist of a comparison of distance of the vehicle from the side of the roadway. To the experienced driver this is virtually subconscious. When the vehicle gets too close to one or the other side of the lane, a correction is made before the system goes out of control. A loss of control is certain if the driver cannot see (perform measurement), perhaps due to extremely heavy rain or loss of lights at night.

The transfer of this theory to human organizations is not simple. While objectives are usually obtained directly by analyzing the organization's goals, the development of measurements that can adequately determine if these objectives are being realized can be quite difficult. For example, in the area of software development, programmers often struggle for weeks on the development of a given module. Can we tell if this time is spent productively? How do we measure professionals working on design requirements? Some attempts at this are extremely counterproductive. For example, U.S. Department of Defense projects usually require reams of documentation, which becomes an end in itself, and the documentation may be produced by personnel other than those engaged in the actual software development. The measurement of number of lines of code produced might be of interest, but it is not of great use in maintaining control. Many such metrics essentially indicate that the car is functioning properly on the road, but they fail to determine whether it is on the right road. In most cases *only the individual professional or close team member can make this assessment.*

The fact that this is not an easy problem to solve does not mean that we should throw up our hands and give up—or, what is worse, ignore it, get involved in another project, and come back in a few weeks to attempt to see what is happening. This will have the same effect as that of the driver who dozes off for a couple of minutes and then wonders why he drove off the road. Between the actual production of deliverables, measurement might consist of close observation, periodic progress reports, or other written or oral formal or informal feedback. Of overriding importance, however, is the integration of subordinates into the control process, for without their cooperation no effective measurement is possible. This is probably one of the most challenging aspects of managing the creative process, because *the very imposition of measurement into the process can kill true creative productivity* quicker than anything else. (But we are getting ahead of ourselves; this will be taken up in Section 2.6.)

The primary function of control metrics is to guide the correction process. It is essential that these metrics be designed to give data that can be translated into useful information for correction. This process of translating control measurements into useful information for correction is called *evaluation*. Evaluation could be broken out as a separate element of control, but we prefer to embed it in the measurement component. The concept of data for the sake of data is counterproductive and is the grist for the *bureaucratic* mill. (Since we never use the terms *bureaucratic* or *bureaucracy* with anything but a negative connotation, we feel we owe the reader our definition. We define a *bureaucracy* to be an organization that has evolved to a point where it is *rich in procedure but virtually devoid of principle*. It is defined to be everything that technimanagement is not, and our wildest fantasy is to see the two words established as opposites in the English language. *Bureaucratic* simply means "of or pertaining to a bureaucracy.")

We postpone further discussion of metrics against objectives until other principles have been introduced. The formulation and communication of objectives are discussed much more specifically in Chapters 7 and 23.

2.5 Correction

When control measurements indicate that the process is deviating significantly from the ideal, an evaluation of this information should lead to correction (i.e., restoring the system to an improved course). The process of correction is critical to technimanagement. When a problem is identified it must be addressed in some way. We hear the admonition: *Deal with it*! But there are a large number of ways that any given problem can be "dealt with," and these will vary greatly with the situation and the particular people involved. The imposition of a rule is usually the first attempt of the technically trained in the correction process. However, this is the pavement of the road leading to bureaucracy. Inevitably, this rule will not apply to all cases, and the process of correction only succeeds in creating further deviations from the intended goal in the future. (We recognize that no organization can function without some rules. However, they should be developed only after clear patterns are established, not as the reactive functioning of the control system.)

Recognize that there are several control systems being implemented concurrently in all productive organizations. It is essential that the correction be applied at the right level. The very first impulse of most managers is to find out who created the problem and deal directly and definitively with that person so that the problem does not recur. Sounds good, doesn't it? If this is the natural impulse, why doesn't it work?

Simple answer: The manager is the cause of most problems. In most cases had the right control mechanisms been in place, the problem would never have occurred (and meta-control is the manager's responsibility). Correction must begin at this level. So get a mirror, keep it in your desk, and whenever something goes wrong in your organizational component, look into it and you will generally be looking at the right person to blame. This is not to say that other people in your organization do not make mistakes. But we will learn by the principles presented in subsequent chapters that this is generally not their intent. (We recognize the obvious exceptions, but these should be addressed in radically different ways than mere correction.) The point that we are emphasizing here is that if you had been smart enough to anticipate people's shortcomings, you would have taken steps to prevent them from causing the problem in the first place. (If not, why not?)

Given that correction starts with better self-control and self-evaluation (and it must!), the next step is to project this attitude of correction onto the organization. This is how correction makes its maximum impact on the functioning of the organization. Although managers are an integral part of this correction process, the ideal is to enable the organization to correct itself without the imposition of action on your part. This is a major goal of technimanagement, and the objectives for its realization as well as the measurement and correction processes form the underlying theme of this book. In essence, technimanagement is effective meta-control.

2.6 Control of Professionals?

Scenario: You are thrust into a situation that is obviously out of control. What do you do? Let's explore a few things that you *don't* do in terms of the elements of control. First, you do not impose a system of goals, objectives, measurements, and corrections on the system (i.e., the people). That works fine for physical objects and is fairly good for some human organizations where there is a division in labor and management skills. But we are dealing with professionals, and they do not like being "controlled." Thus this "solution" will only create morale problems (the worst kind).

Recognize, however, that it is the *imposition of*, not the *necessity for*, a control system that is at issue. Few would disagree that control is necessary, and more morale problems accrue from lack of control than from anything else. In fact, imposition is a symptom of lack of control. Recognizing this and the obvious downside of imposition, many managers opt for the end run. That is, sneak up on your organizational component with the idea that you (or they) have a solution; impose a control system without anyone even knowing about it. This is what we earlier called *manipulation*, but now that we have an understanding of the elements of control, we can give it a formal definition:

> Manipulation is the attempt to implement a control system on people without their understanding it fully.

We will use this term frequently, so it is important that its meaning be clear.

Two alternatives to effective control are manipulation and no control at all; both bring chaos and ultimate disaster. The question is not whether or not a control system must be in place; the question is: Who puts it in place—or more precisely: How does it come into being? We will state a principle at this point that will readily become apparent as the additional principles of technimanagement are revealed:

> For control systems to work effectively, they must be devised and implemented by the same people and/or groups that they are to control.

Critical to this concept is the absence of the "they must *buy in*" mindset, which constitutes the bywords of common management practice. The manager does not figure it out first and then promote it through clever salesmanship. Neither is this the traditional committee concept in which the committee formulates a recommendation and then submits it to the manager for approval or modification. Both of these approaches are extremely counterproductive in that they do not utilize fully the shackled potential that lays dormant within the members of an organization. To avoid manipulation, *managers must really believe in the power of their people.*

This is not to say that managers should be totally passive in the process. The degree of influence they exert in this regard is totally dependent on the situation and the characteristics of the people involved. It will range and evolve initially from rather forceful recommendation, to being an equal member of a team, to exercising very little or no influence at all (this is the transition process, discussed in Chapter 24). Although no fixed rule will serve all circumstances, an understanding of the nature of a technical organization will go a long way toward resolving the optimal level of involvement for each situation as it arises.

To summarize this section, let us propose a simplistic model that we will expand on later. We have stated that one goal of the technical manager is to establish a relatively low-level presence in the operation of the technical organization's control system. However, the next higher level (i.e., the meta-control system) has the goal of assuring that effective control systems are in place. This is the manager's primary responsibility. This does not necessarily mean that managers need to participate directly in their establishment or maintenance. In fact, nonparticipation can greatly facilitate the process, and it certainly promotes subordinate involvement, which can promote morale. At the same time, the manager cannot engage in manipulation in bringing all of this about. If this seems complicated, please be patient— we will add substance to it as we proceed.

As a postscript to this chapter, we broach one other major component of technimanagement. We hope that there has been an implicit introduction to a notion of *leadership* through this meta-control paradigm. We define this formally later, but for now, recognize the part that leadership played in establishing the human control systems that ultimately got us to the moon. Consider the following: "We choose to go to the moon in this decade . . . and do the other things not because they are easy but because they are hard . . . because that goal will serve to organize and measure the best of our energies and skills . . . "(John F. Kennedy, 1962). Notice that this was not a raw display of authority, it was an appeal to professionalism. He had already laid his reputation on the line with a definitive measurable goal and time limit, whether followed or not. Now he was arguing for support, not out of costs and benefits (although he could have made that case), but out of a sense of what was the right thing to do. Don't argue with this model: It worked. Most technological failures can be traced to the absence of leadership, which, in turn, results in the lack of effective control systems for the organization.

Having whetted your appetite for further principles by which organizational control can be established, the next step is to begin to satisfy it. We begin by demonstrating the principles of human motivation to provide an understanding of how effective control systems can evolve within the organization. The accomplishment of measurable objectives provides the personal and organizational satisfaction called *success*. In the next chapter, these are called *rewards*.

3

Theory of Rewards

3.1 The Basic Theory

What we call the *theory of rewards* is a well-established principle of experimental psychology:

> Behavior that is rewarded tends to be repeated and emulated; behavior that is punished tends to be avoided and not emulated.

We call this a theory because although it has been validated by endless experimentation with virtually all types of animals, it is not without exception in its application to human beings. Although seemingly an obvious and simple principle, in the sections that follow we demonstrate that it is not practiced consistently either by managers or by the general public (from which our managers come).

Both the theory and its failure to be applied can be illustrated by the behavior of most small children. It is interesting to observe that, in general, children are much more expert in controlling their parents than the parents are in controlling them. The next time you see a child misbehave in public, see if that child is not rewarded for this behavior. For example, the child screams at the candy counter, the parent gets the child some candy. Chances are that if similar behavior had not been rewarded in the past, it would not be occurring in the present. We are not advocating child abuse. The point is that if parents (the managers of families) would express their displeasure merely by consistently *not rewarding* improper behavior, the punishment would not have to be very extreme. In fact, the withholding of a reward might be ample punishment to discourage misbehavior in the

Sec. 3.1 The Basic Theory

future. Persistence in denying rewards for improper behavior will *ultimately* (remember that word) bring that behavior to an end. (We recognize that children's undesirable behavior is often symptomatic of other causes, such as a wet diaper or illness. In these cases we hope that *undesirable* is not mistaken for *improper* behavior, for this too would be a violation of the principle of rewards. Most parents can tell by the tone of the complaint whether the child is having a real problem or is fussing needlessly.)

There is also the other side of the coin: Children's ability to control their parents is a direct application of the administration of rewards and punishments. The reward is quietness (good baby); the punishment is fussing, pouting, crying, screaming bloody murder! Children know this almost instinctively from birth as they cry out for attention, and it is reinforced as they obtain it. So they learn the theory of rewards by the success that they attain in practicing it on their parents. This is good in that it assures survival of the species. However, it is up to parents to see that it does not go too far.

This is all food for thought. First, why do so many parents fail to practice consistently the seemingly obvious theory of rewards? There are two possible explanations: (1) they do not know the true power of the theory of rewards, or (2) they seek short-term relief over long-term solutions. In fact, to some extent both of these are true: *They do not realize how perfectly true it is and how disastrous the long-term consequences will be if they violate it.* So they violate it whenever it appears that the short-term benefits outweigh the long-term consequences (which they often even fail to imagine). So although persistence in denying rewards for improper behavior will *ultimately* bring that behavior to an end, there is another pressing problem right now. (Sound familiar? The problem of short-term thinking is such a problem that we will revisit if often, especially in Chapter 19.)

We cannot leave this example without emphasizing how much rationality can be thrown to the wind due to the personal relationships of the people involved. The theory of rewards is known and applicable almost universally, but that does not mean that those who know it understand how to apply it properly. For example, we seem to know all too well how to discipline everyone else's children but fail miserably when it comes to our own. Such things as dedication, loyalty, and love (or sometimes their ugliest counterparts) completely change the nature of its proper and effective application. And if our old nemesis *manipulation* creeps into the picture, even a consistent application of the principle will be counterproductive.

It is safe to assume that the theory of rewards works and that it is generally known to work. The problems discussed above indicate, however, that it is not a simple matter for most people to apply this theory even to a single person over whose rewards they have virtual control (i.e., their own child). The difficulty in applying it to groups and organizations should be clear, and some of the more apparent problems are summarized in Figure 3.1. This sets the stage for the next six sections in which we address each of these problems.

Cause	Subject (Section)
Unclear linkage between rewards and motivation	Rewards versus Motives (3.2)
Changing nature of individual needs and desires over time	Hierarchy of needs (3.3)
Influences of intangible aspects of motivation	Protestant ethic and social Darwinism (3.4)
Conflicting perceptions of motivations and rewards	Motivational conflict and frustration (3.5)
Irrational behavior (however mild)	Defense mechanisms (3.6)
Deficiencies of monetary rewards	Goal integration (3.7)

Figure 3.1 Causes of Misapplication of Reward Theory

The effective application of reward theory requires that we go well beyond addressing misapplications. In Sections 3.8 and 3.9 we discuss the effective administration of rewards in general through motivational maintenance and increased consideration of the human factor. In Sections 3.10 and 3.11, where we discuss the Hawthorne effect and industrial humanism, we provide some summary concepts that tend to unify the principles presented in earlier sections. The final two sections can be classified as advanced principles: In Section 3.12 we discuss the complexity of the human personality, and in Section 3.13 we integrate reward theory into our meta-control model. We will see that the major goal of the meta-control system is to create a resonant condition of reinforcement within the reward process.

3.2 Rewards versus Motives

In the ideal situation the accomplishment of organizational goals produces rewards realized by its members. Although it is always true that the accomplishment of a personal goal produces a reward (at least in the short run), this may not transfer to organizational goals. In establishing principles relating organizational goals to personal goals, the differences between rewards and motives must be clarified. A contrast between rewards and motives, suggested by Kazmier (KAZMI69), is given in Figure 3.2. This defines *motives* for this context, and it clarifies the differentiation in function between rewards and motives.

Motives	Rewards
Internal to the person	Associated with the work environment
True cause of behavior	Change behavior only when they affect motives
Complex origination and formation	Alter the direction or intensity of motives but rarely create them
Vary considerably from person to person	Limited by organizational and societal constraints

Figure 3.2 Distinction between Motives and Rewards

Rewards are formulated by management in an attempt to make personal goals consistent with organizational goals. However, specific types of rewards are not universally applicable; instead, their effectiveness depends on the motives of the people to whom they are applied. Managers often make a major mistake by assuming that their subordinates' personal goals are (or even, ought to be) synonymous with their organizational component's goals. The reason that they make this error is that they are judging their subordinates' motives on the basis of their own. (*Note*: We use the term *organizational component* to refer to the manager's span of authority and responsibility, as opposed to department, division, and so on, which might not be generally applicable.)

In most cases even lower-level managers will not have goals totally synonymous with those of the organization. However, being a part of management, first-line managers' personal goals tend to be much closer to organizational goals than do those of their subordinates. Further, the degree to which personal and organizational goals naturally coincide is directly proportional to the level of management. Thus, at the very top, personal and organizational goals become virtually synonymous. (We are speaking from a perspective of the traditional organization; a goal of technimanagement is to reduce this disparity to nearly zero.)

Let us start with the lowest-level managers to see why their motives might not be consistent with organizational goals. Quite simply, each manager is competing with every other manager at a comparable level for organizational resources. At times key staff members are assigned to two teams under two different managers, creating competition between managers for their subordinates' services. There is competition for the "best" people, competition for more people, and competition for more dollars and space to perform whatever assignment the organizational component has been given. Further, the success of the manager is often measured by upper management on a relative basis in comparison with the other managers at comparable rank. The net effect is that the personal motives of first-level managers are very focused on their own group's functions, almost to the exclusion of the rest of the organization.

Now consider this question: If managers' goals are not automatically consistent with organizational goals, why should we assume that subordinates' goals are naturally consistent with those of the organizational component? We stated above that managers make this assumption because they assume that their subordinates have the same perception of organizational goals that they have. A recognition of the disparity between our own personal goals and those of the organization is the first step toward understanding the same condition in our subordinates. In turn, we cannot expect our subordinates to acquire the goals of the organization, or even the organizational component, when our own personal goals are strictly localized.

Although this is a difficult concept to accept, it is fundamental to most of the remainder of this chapter. It does not matter what position you occupy in management—your subordinates essentially have the same problem of personal and professional resource competition with their peers as you have with yours. It affects their salary, equipment, software, technical assistance, and so on—it all comes out of the same finite pot. The only difference between levels is the specific pot, which has little to do with the motivating forces involved. Thus there is no more reason to expect your subordinate to have an organizational motivation than there is to expect it of yourself.

Given that we can understand this, the distinctions (Figure 3.2) and conclusions derived from Kazmier tend to acquire an entirely different meaning. First, we can all but exclude salary as a motivator. The base level is already given (an accomplished goal), and the chances of significant increase without a major promotion are generally restricted by the organizational structure. True, the threat of loss of salary is ample motivation to provide a level of performance that will maintain employment, but it is not sufficient motivation to unleash the person's full potential. For this it will be necessary to unite personal goals with those of the organizational component.

This process, called *goal integration*, is one of the most challenging aspects of management. It begins by performing a transition on yourself by which you integrate the goals of the organization with your own personal goals; the realization of this process enables you to lead your subordinates to a similar integration at the level of your organizational component. However, this is not as simple as it sounds, and we need to define a few more principles related to personal motivations (needs) before we come back to this.

3.3 Hierarchy of Needs

The hierarchy of needs principle is well established in management literature, and it relates very heavily to the proper selection and administration of rewards (see Maslow MASLO43, MASLO54). We define a *need* for this context to be anything that a person perceives at the moment to be essential to his or her well-being or happiness. Hence needs are not limited to the classical essentials of food, shelter, and clothing. Neither are they totally synonymous with either *goals* or *motives* as these were defined in Section 3.2. However, there is considerable overlap, which will become clear as we proceed.

It has been determined by a wide array of social and management scientists that needs are highly dependent on a person's current state. Once one need is satisfied, it is no longer perceived to be a need, and the fact that it continues to be attained does not provide the same satisfaction (the intent of a reward) as it did when first attained. Further, it is postulated that these needs can be arranged in a hierarchy, according to the priority in which most people seem to value their attainment. We say *postulated* both because (1) there are exceptions to this arrangement itself, and (2) there are deterrents to the evolution that this hierarchy might imply.

Figure 3.3 provides a classification of needs in traditionally recognized priority order, with satisfaction generally assumed to proceed from the lowest to the highest. The *first-order basic needs* include food, shelter, clothing, and a degree of comfort in addition to the mental and spiritual stimulation to enable physical and psychological health. The remaining aspirations will be referred to collectively as *higher-order needs*.

Hierarchy	General Categories
Lowest	First-order needs
	Basic physiological needs
	Higher-order needs
	Safety and security
	Belonging and social activity
	Esteem and status
Highest	Self-realization and fulfillment

Figure 3.3 Hierarchy of Needs

To fit this into our meta-control model, recognize that the realization of a need produces a reward. This will, in turn, reinforce the action that was perceived to produce the reward, thus increasing motivation. However, as a person becomes accustomed to the acceptance of the new status (i.e., life with the need fulfilled), the intensity of the reward diminishes. Thus that which once provided increased motivation for a given behavior may now fail to do so. Of course, withdrawal of the need satisfaction may act as a punishment, but the use of such for anything other than major corrective surgery is totally inconsistent with the principles that we are trying to establish.

Most technical organizations do not have to give a tremendous amount of consideration to the *safety* need, since this is covered by occupational safety and health laws [exceptions might occur in some experimental contexts (e.g., medical, chemical, biological, etc.)]. *Security* is another issue. This is especially true of employees who are on probation and others in industries that have considerable variation in workforce size over time (e.g., aerospace). Effective management requires that job security be instilled as a function of individual competence, and in this regard, even if the particular organization loses a major part of its business, the person's skills will still be in high demand by others. More important, the need for security can serve to promote individual acceptance of organizational goals, since the greater the success of the company, the greater the chance for advancement and job security. (Some qualifiers to this statement will be presented when we discuss goal integration.) Of course, security goes beyond job security to cover such things as health insurance and retirement. The absence of these items can so preoccupy the members of an organization as to reduce their productivity. However, the presence of these "security" items will produce little positive incentive, as their realization is largely taken for granted once a person joins an organization.

Ideally, you would love to be a manager of an organization in which everyone has attained all of his or her needs except self-realization and fulfillment. In this ideal organization, all members are pulling together, going out of their way to serve one another, and together attaining the goals which they collectively lay out for themselves in a perfect manner, since their own personal goals no longer conflict in any way with organizational goals. There are some organizations that approach this ideal, but rarely is this because lower-level needs have been fully attained. Usually, it is a characteristic of the organizational component, as opposed to the satisfaction of individuals, which brings this about. Thus in most technical organizations the *belonging and social activity* needs are quite relevant.

This provides an opportunity to broach the subject of individual versus group goals and motivations, which will be an ongoing subtheme as this chapter unfolds. If we learn nothing else from the hierarchy of needs, it is that, theoretically at least, the collective needs of belonging and social activity require satisfaction before esteem, status, self-realization, and fulfillment. When managers provide rewards (e.g., honors, awards, special recognition) that boost the esteem of some at the expense of others, they do so at the risk of violating this principle. Similarly, when status is given to one at the expense of others, the same danger exists. These types of rewards tend to dissolve the glue that forms the cohesive force necessary for a diverse group to act as a unit. We saw that the natural competition which exists between individuals tends to keep people from assuming group goals. Instead of attempting to play off this competition to attain productivity, the effective manager will create a team spirit which has as its sole purpose mutual support for all members of the team.

The hierarchy of needs model is a good one, and has served management scientists well, but it is far from perfect. Its primary underlying assumption is that there is *upward mobility*. By this we mean that there is the perception of *freedom* to enable a person to move to the next-higher need when the next-lower need is satisfied. The effective use of rewards implies further that it is within the prerogative of the rewarder to grant or deny the next level of need. There are a number of circumstances that argue against both of these assumptions, which are discussed in the adjacent display. There may be a number of others, but these are sufficient to illustrate the potential for interrupting the process of moving subordinates to higher levels of needs for purposes of motivation.

> **Complications in Need Attainment**
>
> *Lack of educational opportunities.* Some company policies prevent professionals who have Master's degrees from being managed directly by those who do not. Clearly, everyone is not in the same position to obtain this degree due to financial, family, time, or other constraints.
>
> *Lack of alternative employment opportunities.* Due to the extremely specialized nature of particular past job experiences within specific industries.
>
> *Geographic constraints.* Due to family, friends, other relationships, love for the particular part of the country ("home"), recreational opportunities, and so on.
>
> *Golden chains.* Low house payments, the high cost of moving, retirement incentives, and so on.
>
> *Comfort with current position.* The current job is a known entity, and there is no positive assurance that the person would be better off elsewhere.
>
> *Fear of changing jobs.* Some people are insecure as to their abilities even though they are very competent technically.

Recognize that these deterrents do not set aside the theory. The theoretical laws of physics virtually never hold perfectly on Earth because of such factors as friction and gravity. This does not mean that they are invalid. In the absence of atmosphere, these relationships enable very precise computations that can maneuver a spacecraft within millimeters of its goal. Similarly, the hierarchy of needs assumes that the given deterrents (and possibly others that we have not discovered) are absent. Although this does not give us a perfect prescription for motivation in the real world, it provides a model that we can adjust for these other factors (just as we adjust the theoretical laws of physics).

Before continuing to explore other motivational aspects of management, we need to supplement the hierarchy of needs to account for spiritual needs. This will be exemplified by two major philosophies that have had a major impact on traditional management philosophy.

3.4 Protestant Ethic and Social Darwinism

In dealing with exceptions to the recognized hierarchy of needs, spiritual beliefs are of interest since they are one of the few factors that have been shown historically to be able to motivate people to act consistently against what they perceive to be in their best physical interests. The hierarchy of needs defined above were all consistent with the physical interests of the person in that they were perceived to satisfy a physical or psychological need. It might be argued that religion also satisfies a psychological need, or even that it is consistent with the self-realization and fulfillment need. However, this does not explain the many examples of people who have sacrificed their lower-level needs for their religious beliefs, thus providing another striking counterexample to the hierarchy of needs theory.

We do not wish to get into a classification of what we believe to be *spiritual needs*, which seem to have the capability of displacing any of the other needs in the hierarchy. We only know from direct observation that when leaders want to motivate their followers to go against what is generally perceived to be their own physical self-interest, they most often appeal to an underlying religious motivation. In the case of religious cults, recent history gives us the extremes of Jim Jones and David Koresh as vivid examples. In the political realm few wars have been fought effectively without the warriors being convinced of the ultimate rightness of their actions in terms of the Supreme Being. The recent Gulf War was not an exception on either side, and the wars that continue to rage in Eastern Europe and the Middle East also fit this pattern.

We can hardly expect to use religion for motivation within a technical organization in a society with such diverse religious beliefs. Further, to use religion for any purpose in the secular realm would be extremely manipulative. However, this does not mean that managers can be oblivious to the religious beliefs of their subordinates. A number of sensitive issues could cause severe morale problems if they are assumed to be insignificant. The following is but a partial list of examples: (1) attitudes toward the use of alcoholic beverages, (2) the use of subordinates' time on weekends and religious holidays, (3) dietary requirements at team meals, and (4) acceptability of certain forms of entertainment. Problems arise when managers are not sensitive to the diversity in beliefs among their subordinates. This is an area where an accurate perception of reality can go a long way toward avoiding motivational conflict. To summarize: *Never expect your subordinates to violate their respective religious consciences.* Although it might seem that you are going out of your way for the sake of a given person's conscience, generally there is sufficient organizational flexibility to enable this rule to be followed. The downside of its violation is that you stand to harm the morale of at least one subordinate, and perhaps of others who respect that person's right to hold certain beliefs even if they do not share them. (The reason for respect of other's beliefs should be a matter of principle, however, not an attempt to minimize the downside.)

With this in mind, let us explore the possible effects of the established religious attitudes of the contemporaries of those who developed many of the principles of technimanagement. Consider the following quotation from Kast: "The tenets of Calvinism provided the basic framework for the encouragement of capitalism and set the stage for the *Protestant ethic*. In the new world, puritanism continued the stress on the virtues of hard work, sobriety, and an accumulation of worldly goods as a sign of being in God's grace" (KAST70,33).

To a large extent the "Protestant ethic" is alive and well in modern capitalistic societies. Rewarding hard work with material wealth is a generally accepted ethic. However, any ethical principle can be pressed beyond its original intent. Such is the case of extrapolating the puritanistic emphasis to infer that material success is "a sign of being in God's grace." Perhaps this is the opportunistic interpretation of many, from the drug pusher in the ghetto to the junk-bond pusher on Wall Street. Reality is quite clear: For every person who gets caught, dozens never get detected. Thus material success is not proof of the "rightness" of a person or an organization. However, those attempting to justify their exploitation of others might find the doctrine quite comforting.

This discussion serves to introduce a major irony with regard to the motivating power of religion. That is, we have a very strong tendency to modify our beliefs to conform with the reality of our actions. Thus, rather than using religion to improve our behavior, we modify our religion to fit our behavior. We do not agree with this paradigm: Spiritual values should modify behavior, not vice versa. Further, we believe that professionals are probably very evenly divided on the extent to which religion should influence behavior. (This is merely the result of our own observations; we know of no studies to verify it one way or the other.) Nevertheless, this does not nullify religion's motivating potential; if anything, either view tends to reinforce spiritual values as a motivator of action, since both views tend to lead to a unity of physical and spiritual goals. That is, whether there is a modification of religious belief or a modification of actual behavior, ultimately the two tend to become compatible.

To develop these concepts a bit further, consider the transformation at the end of the nineteenth century away from spiritual precepts to what was accepted by some as inevitable evolution within the social structure. As a formalized extreme example, consider the primary tenets of social Darwinism:

> *Social Darwinism* suggested that the most capable and resourceful of people would rise to the top of the social hierarchy and that this was the natural order of things. Under Social Darwinism it was only natural that there would be poor and rich classes, and any attempt to upset this hierarchical order was considered unnatural and against the best interest of society. Social Darwinism clearly reinforced the Protestant ethic and Adam Smith's concept of laissez faire. It provided the basic ideology for the businessman

in the late nineteenth century and helped justify the accumulation of resources and their use for self-interest. (KAST70,35)

This was certainly a self-serving proposition—we dare not call it a principle. Taken to the extreme, it would have provided Hitler with ample justification for his atrocities. Its primary fault is that it judges "rightness" purely on the basis of (short-term) results and not on the ethics or morality involved in establishing those results. Further, it is clearly an attempt to manipulate one's belief system to provide self-justification as opposed to the striving for everlasting improvement. (We inserted *short-term* because it is our belief that any empire built on a flawed set of principles is destined to destruction, although it actually might take decades to dematerialize.)

But the validity of social Darwinism is not the issue. Our original premise was that *spiritual needs* have the potential to dominate and thus rearrange the priority within the hierarchy of needs. It appears from a study of societies throughout history, however, that there is a need within each of us to seek some basis on which our actions can be considered as either (1) unquestionably correct, or (2) justified (by the circumstances or by forgiveness). We might seek this frame of mind through a concept of a supreme being or by adhering to a philosophy such as social Darwinism. In either case we are responding to something that is beyond physical explanation, which we are calling spiritual needs. It is not our objective to provide an explanation for this phenomenon, only to state that we are denying an essential element of reality by ignoring it.

The fact that human beings have the capacity to rationalize the integration of other self-serving concepts into their spiritual beliefs tends to make this principle of even greater consequence. However, the wide range of both the underlying religious belief and the degree of such rationalization make the resulting motivation difficult to predict. This might account for the fact that it is virtually ignored in most management books. Nevertheless, it would be folly for a manager to fail to take this most important factor into consideration, or, even worse, to assume that the effect of spiritual motivations of one's subordinates is identical to one's own.

Due to their perceived source (whether real or not), spiritual needs might come into conflict with the traditionally recognized hierarchy of needs. Even without the interposition of the spiritual aspect, the needs within the hierarchy itself can conflict with each other. Further, the means of attaining any of these needs can come into conflict, leading to a confused set of motivations, which results in motivational conflict and frustration.

3.5 Motivational Conflict and Frustration

According to Kazmier: "The presence of incompatible motives that are more or less equally influential results in an internal situation that psychologists called *motivational conflict*" (KAZMI69,204). Note that the motives need to be more than just different, or plural. For motivational conflict to occur, the various motives must first be incompatible and then close to the others in their influence on personal decision making. *Incompatibility* here implies that the satisfaction of one will not support the satisfaction of another. For example, the motives of wealth and promotion are generally not at all incompatible, whereas the motives of loyalty to one's fellow workers might be totally incompatible with the motive of promotion. In this second case, if promotion is of much greater personal importance than the collegial social bond, there will be no motivational conflict. However, if both are desired to an indistinguishable degree, motivational conflict will follow. (Additional dollars usually serve as an effective tie-breaker.)

Let us introduce another term from Kazmier (KAZMI69,207): "*Frustration* indicates the failure to satisfy personal motives because of barriers to goal attainment," caused either by motivational conflict or external factors. Those caused by external factors are usually much easier to either resolve or accept, since there is usually a clearly assignable cause. This is not the case with motivational conflict, since given that motivational conflict is understood, the adjustments are purely internal. However, often there is a lack of ability to recognize the internalized nature of this problem (i.e., one believes in having one's cake and eating it too). In this case the problem is further exacerbated by an attempt to blame external factors, and a vicious cycle results from this misperception of reality.

We can all agree that frustration is not a good thing, either psychologically for the individual or in its effect on the organization. Consider some alternative responses to frustration (roughly from best to worst): (1) removal of external barriers, (2) motive alignment, (3) compromise, (4) withdrawal, and (5) aggression. The first two, which are discussed in the adjacent display, are almost equivalent in their potential positive effects. The last three are *defense mechanisms* and almost always result in negative consequences; they are discussed further in Section 3.6

One of the major problems encountered by many managers is the assumption that their subordinates have the same abilities that they have to apply motive alignment and barrier removal to eliminate the frustration that is inherent in most professional jobs. This is a natural assumption since we have ourselves for a laboratory and are constantly learning from our own experiences. However, the very fact that a person has been promoted to a managerial position presupposes that he or she has a greater capacity to resolve motivational conflicts than that of other candidates. The capacity of managers to handle motivational conflicts in a positive manner is itself highly variable.

Sec. 3.5   Motivational Conflict and Frustration

The concepts in this section have been introduced to further our understanding of motivation and the barriers that inevitably crop up which prevent its satisfaction. When this occurs, there can be either positive or negative responses. It is essential that managers not only take positive actions to eliminate their own frustration, but that they are prepared to guide their subordinates toward overcoming their frustrations as well. In all cases we have stressed the individual responsibility for resolving motivational conflict. Effective management is essential to provide an environment in which this can occur. However, without the full participation of the subordinate, these actions reap the same counterproductive results as do other forms of manipulation. We will consider this further after we explore defense mechanisms more deeply in the next section..

Positive Responses to Frustration

Removal of external barriers. This involves identification of the source of frustration and dealing with it directly.

- *Approach:* First, clearly identify the barrier. Then design and implement an effective method for its removal. This is not always easy, and often misperceptions of reality tend to misidentify the true barrier and lay the blame on all but the true cause.
- *Example:* A person is given conflicting orders by two superiors. Potential solution: Meet simultaneously with the two managers who issued the orders and get them to resolve the conflict or prioritize an order of accomplishment.

Motive alignment. This is a process of internalizing a peaceful coexistence between the motives that are at conflict.

- *Approach:* Establish internal priorities between anticipated conflicting demands recognizing that both cannot be fully realized.
- *Example:* Quite often the desire to please one's spouse comes into conflict with the desire to please one's parents. The barriers to mutual satisfaction cannot be removed (by any civilized person), and thus, the only positive way to resolve the conflict is by motive alignment. In this case the person makes an overt decision and internalizes the fact that priority must be given to one or the other. This is followed by an explanation to the involved parties (a type of damage minimization), and at this point "the chips fall where they may." Notice that the motives have not changed: The desire to please both sets of parties remains. However, the motives have now been aligned such that there is not continued frustration every time a potentially conflicting situation arises.

3.6 Defense Mechanisms

The approaches given above for resolving the causes of frustration were stated to be "positive" inasmuch as they tend to result in positive and proactive activity to eliminate the cause. Opposed to the positive, proactive approach is a retreat to what Kazmier (KAZMI69,209) appropriately called *defense mechanisms*. He stated that they "serve to protect the individual's feeling of self-worth in the face of continued motivational frustration." It is essential that each person have a feeling of self-worth. In these cases, a false perception of self-worth replaces the true perception that should exist. The three defense mechanisms discussed by Kazmier are addressed in the adjacent display.

Defense mechanisms are the result of a loss of perception of reality in at least one of two aspects: (1) the basic cause of the frustration, and (2) the most effective method for eliminating the cause. In the most extreme cases the cause becomes irrelevant and the quest for motive satisfaction is replaced by a quest for vengeance. It is hard to imagine that any positive effects can accrue from such counterproductive actions, yet they persist in most organizations. Managers should avoid escape mechanisms themselves and should do everything in their power to provide a *positive environment* in which their subordinates do not resort to these actions: one based on trust and honesty as opposed to manipulation. This involves everyone within the organizational component, and possibly some people outside it.

Negative Responses to Frustration

Compromise. Of the three counterproductive reactions to frustration, compromise is generally the least damaging. It is essential to recognize that we are using the term *compromise* in a technical sense and are not trying to imply that all compromise is a reaction to frustration or motivational conflict. (Compromise can be quite productive in resolving some disagreements between individuals or groups, although it has its downside as well.) Specifically, compromise in this context can take one of two forms: *compensation* or *rationalization*. In the case of compensation, a different goal is substituted for the one that (it is perceived) cannot be reached, and the subject derives satisfaction from attaining this rather than the original goal. In the case of rationalization, the person, in frustration, invents a reason to excuse the failure to satisfy a motive. Although this reason is usually related to some aspect of the problem, it is also usually quite irrelevant to the real cause of the problem (hence the rationalization).

Negative Responses to Frustration (Continued)

Withdrawal. Withdrawal may exhibit itself in the form of *regression* or *emotional isolation*. Regression implies that a level of maturity has been reached but that the person has regressed to an earlier degree of immaturity and/or incompetence. Emotional isolation is an attempt to keep from getting hurt further emotionally, and as a result, many of the communication contacts that are essential to the organization are eliminated. Either of these reactions to frustration can have a devastating effect on the person, and the corresponding effects on the organization may be equally damaging. An additional problem is the likelihood that it may ultimately reveal itself in a radical form of aggression.

Aggression. This is the opposite of withdrawal, but its consequences are generally much more counterproductive. Aggression is a perverted form of barrier removal. However, instead of action being directed toward productive communication and agreement, it is directed against something or, worse, someone. Aggression can be *direct* or *displaced*. The deceptive short-term soothing effects seem to be the same. In the direct case, the frustrated person identifies the cause accurately but does not properly identify the means of resolving the problem. Displaced aggression fails to identify the true cause and thus cannot possibly identify a method for resolution. The innocent victim is classically called a *scapegoat*.

Even though a positive environment is the key to preventing escape mechanisms from motivating behavior on the part of subordinates, the manager will inevitably have to deal with such behavior since no environment persists in its ideal form for very long. Past problems, external interactions, and individual personal problems all take their toll. Once lost, the restoration of a positive environment may be quite complicated, and it will generally take a prolonged period to reestablish. Attempts at communication can be counterproductive since they might be interpreted as manipulation. (Valid or invalid does not matter; for with confidence, perception is reality.) The very fact that the communication is coming from management can destroy its credibility. Unless the manager can in some way influence the *informal organization* (see Chapter 12) without this being interpreted as manipulation, any attempt at communication will be doomed to failure. At this point the only hope for the organizational component might be to bring in an entirely new manager, not necessarily because the first has done anything improper, but because the lack of credibility has resulted from problem circumstances of the past.

A problem that warrants additional consideration is the persistence of defense mechanisms despite the existence of a positive environment for their resolution. This could be due to psychological problems or the presence of malicious intent. It is very tempting to blame most negative behavior on this cause, so we urge managers not to jump to this conclusion until they have exhausted all other possibilities. However, when this has been resolved, every attempt should be made to design a mutually acceptable parting of the ways.

With this background we are ready to consider definitive ways to fashion a positive environment. It must be recognized that this is not an abstract concept; rather, it is something that affects every person in the organization. The remaining sections of this chapter are dedicated to this purpose.

3.7 Productivity Through Goal Integration

Goal integration was introduced above without a formal definition. We correct that deficiency at this point. Consider the following propositions suggested by Kazmier (KAZMI69):

1. Generally, personal goals are different from organizational goals, especially in nonmanagerial positions.
2. This results in a conflict between these goals, in terms of self-fulfillment if not preoccupation.
3. Efforts to satisfy personal goals at the workplace generally increase morale, but they can result in a loss of productivity.
4. The ideal way to increase productivity is the *integration* of organizational and personal goals, which means that the realization of one causes simultaneous realization of the other.

The third premise is in reaction to those who liberalize the work environment to allow the satisfaction of personal goals at the workplace. These might include such things as use of the phone for personal calls, in-house child care centers, and allowing employees to set their own hours and/or work at home. More important, it might include provisions for the satisfaction of sociological needs by a variety of social programs. Kazmier contends that within themselves, these have neither positive nor negative effects on productivity. It might be quite positive for morale to develop a number of these, especially if not offered elsewhere. However, they all have their downside, not only in cost and time, but most important, in the distraction of management from activities that have a much more dramatic impact on productivity: those that focus on the work itself. Thus, while each of these must be evaluated in terms of the costs and benefits for the particular people and circumstances, their effects are negligible compared to the potential of goal integration. (We are not suggesting that any of these should be discontinued, only that they are supplementary to the primary motivating influence.)

The simplest example of goal integration is integration of the personal desire for financial security. As improving performance is rewarded, the goal of the individual is integrated with the goal of the organization to make both more successful. A serious problem occurs, however, when the manager fails to recognize that money is only the first step toward overall organization-individual goal integration. Monetary rewards are limited for several reasons: (1) generally they are not under direct control of the manager since (2) they can only be administered at certain times and (3) within a given range. In fact, these limitations can be quite counterproductive since superior employees might develop skills that are marketable at a higher price elsewhere. Although managers should certainly promote the highest reasonable pay scales for their subordinates, the overall structure of the organization and its need to stay competitive will impose definite restrictions on the use of money as a motivating factor. Further, we might notice that money is concerned primarily with the lowest needs in the hierarchy. Thus, while monetary rewards provide a starting point for understanding goal integration, the technical manager who thinks that employees are going to be satisfied by picking up a paycheck at the end of the pay period is in deep trouble from the outset.

Consideration of the potential for goal integration should begin as early as possible—when interviewing prospective employees. It should be the single most important consideration in the selection among alternative candidates. In certain circumstances it might be more important than technical skills, since those can be acquired given the proper motivation. Once an employee is hired, a control system should be established to measure, evaluate, and correct deficiencies in the goal integration process. Symptoms (measurements) of goal incompatibility include complaints, insubordination (neither of which should be discouraged), and most important, complacency (this is a real problem). These signals should trigger the manager to assist the subordinate to identify those conflicts that exist and recommend approaches to goal integration (i.e., correction). That is, the manager should find or make a pathway by which the subordinate can assure the fulfillment of goals within the context of the organization.

The actual control system itself is best developed and maintained by the subordinates themselves. This starts with an identification of personal and professional goals and continues with the development of metrics for their attainment. Assistance is furnished by the manager through frequent discussions with regard to problems in either personal or professional goal attainment. This maintenance process cannot be performed by default or exception; it must be a concerted proactive effort on the part of all managers with regard to their subordinates. In summary, managers should have a personal concern for the career and personal ambitions of their subordinates, even if these extend beyond the realm of the organization.

We appeal to the theory of rewards and the latitude of the manager for the basis of goal integration. There might be as many alterations of goals and tactics on the part of the organization to bring about goal integration as there is change on the part of the individual. The goal integration process should be seen as both a change on the part of employees toward the organization and the corresponding change in the organization toward the employees. The distributed decision-making concepts of technimanagement lend themselves to a natural evolution in this direction, since as this proceeds, the goals of the organization will be determined by the employees themselves. There is no more effective way to produce goal integration than this. Goal integration cannot be a manipulative process induced by management. It must proceed naturally out of the mechanisms established for moving traditional organizations toward technimanagement. The objective of establishing these mechanisms is to produce an environment in which the satisfactions of personal motives are totally consistent with the satisfaction of organizational goals. The result is a consistency of reinforcement that eliminates conflict and produces the highest possible morale.

Further complicating the process of goal integration is the fact that people change with time. Thus the same approaches that may have resolved conflicts in the past may not work in the future. This is consistent with the hierarchy of needs principle, but it is not totally determined by it. It should also be recognized that the motive resolution process tends to change with the maturity of the organization as well as with the person.

Goal integration is required of both individuals and organizational components, due to functional conflicts. However, we postpone discussion of the resolution of these problems to Chapter 17.

3.8 Motivational Maintenance Theory

The notion that a continuous effort is required on the part of management to preserve a work environment in which goal integration can occur was a main theme of Herzberg (HERZB59), called *motivational maintenance*. He differentiated between what he called the *motivation factors* and the *maintenance factors*. As discussed by Kazmier: "The factors that lead to high job satisfaction and goal-oriented effort—called the *motivational factors*, or *motivators*—are different from the factors that lead to dissatisfaction and discontent—called the *maintenance factors*" (KAZMI69,232). The new principle introduced here is the counterintuitive fact: *High morale is not caused by the absence of the factors that cause job dissatisfaction.* Extreme examples could be cited in many hazardous occupations (e.g., military and paramilitary operations), which are often typified by very high levels of morale.

We talked above of the array of higher-order needs which might be translated into rewards available to all members of an organization. These motivational factors include achievement, recognition, responsibility, advancement, and the sense of accomplishment of the work itself (especially its quality). These are the items that lead to satisfaction. Without them the workplace is at best tolerable. However, it is quite rare to hear direct complaints about the absence of these items (although indirectly we often hear of complaints about dead-end jobs).

Contrasted with motivational factors are maintenance factors, which include company policy and administration, supervision, salary, interpersonal relations, and working conditions. These are the factors about which we most often hear complaints. They are related to environment or job context, and they are often organization-wide in their effect, as opposed to the motivational factors, which are related to job content. The motivational factors tend to be job-centered and hence determined more by the group and the immediate supervisor than by the total organization.

This leads us to reemphasize the necessity of applying reward theory at the group rather than the individual level. Individual rewards tend to create competition, if not outright resentment. This tears at the fabric of team mentality, creating a set of independent persons, each in it for himself or herself. The reward system has to evolve in a way to create the incentives for cooperative effort. This occurs almost automatically in the military, where each person is dependent for survival on every other person in the unit. When asked the reason why people lay their lives on the line, country and idealogy seem to play a relatively minor part. The major response is "I did not want to let my buddies down." Surely if people will risk their lives on this basis, they can be motivated to work together on technical activities by a similar motivation. This is a classical example of where the traditional management practice of competitive rewards tends to destroy a natural motivator.

Another major application of this principle involves the interpretation of complaints. It is important for the manager to recognize that complaints are most often a symptom of a much broader underlying problem. The direct treatment of symptoms tends to bring about a short-term gain which has the downside of covering up the real problem. Most short-term fixes sacrifice a more effective longer-term solution. (Complaints are discussed further in Section 20.3.2.)

By placing the emphasis on higher-order needs, motivational maintenance theory emphasizes *the human factor* above those considerations that typically emanate from bureaucracies. This will provide additional focus for the next section.

3.9 Consideration of the Human Factor

The structure of traditional management is basically one which treats labor as a commodity that can be purchased, much like interchangeable parts. However, in this and the following two sections we demonstrate that this occurred despite, and not because of, the clear results of scientific studies that were conducted on organizational productivity. Begin this consideration by assuming the aspect of your subordinates. This should not be difficult since virtually all of us have been in our subordinates' position. However, for some reason, when we assume a management position we fail to relate to our subordinates as we did in our "previous existence." To assist in this regard, the adjacent display, suggested by Keith (KEITH71,176), presents a viewpoint of needs and expectations from the employee's perspective.

One of the most important and controllable aspects of humanizing the reward process involves criticism and commendation. It is most important that mixed signals not be sent. This is a major flaw in most organization's annual review system. Commendation is mixed with criticism, and the net effect is neither (more accurately, the net effect is whatever the employee's mindset infers from the experience, which puts it beyond the control of management). The control system has run amuck. Countercontrol results, and management can only stand back and watch the effects of their incompetence in pulling the trigger on anything ranging from a motivational firecracker to a destructive A-bomb. Criticism must be administered in very astutely measured and timed quantities. Like leaven in dough, it takes only a very small amount to bring about the desired effect, and in excess (that which brings about humiliation or which is unjustified) it can have long-term ruinous effects. It must be matched in time and greatly exceeded in quantity with commendation when there is a positive response.

> **Questions Demanding Consideration of the Human Factor**
>
> *"What am I within the organization?"* All employees should have a firm grasp of their perceived contributions to the organization. This begins with an understanding of their contribution to their own organizational components. Then the contribution that the component makes to the entire organization must be understood. However, organizational components tend to undergo restructuring over time. This process can be quite disconcerting to employees unless some perception of value within the broader organizational context can be imparted. This requires middle management to communicate an understanding of the plans of each organizational component and how it fits into the fulfillment of overall organizational goals. The irony here is that middle managers often do not have their own goals fully resolved.
>
> *"What status is available; what must I do to attain it?"* One of the most often-cited causes of job dissatisfaction is that the job is a dead end. Employees are more than willing to put up with adversities on the job if there is an expectation that something better is available in the near future. As important is the specification of the bridge over the obstacles which stand in the way of attaining that expectation. An implicit contract evolves between the employee and the organization over time. The clearer that it is perceived that this contract leads to the satisfaction of personal goals, the greater the motivation to construct whatever bridges are necessary to overcome these obstacles. A clear perception of personal status is essential to maintenance of the personal control system, which, in turn, is essential to individual motivation and group morale.

> **Questions Demanding Consideration
>   of the Human Factor (Continued)**
>
> *"Who knows what I do; to what extent is commendation or criticism due?"* One of the most deadly blows to morale is the failure of management to reward productive efforts on the part of the employee. The withholding of commendation and criticism leaves the employee in the state of the unknown, and the fear of the unknown is often far worse than the fear of known reality. Over the longer term, constructive criticism can have a conciliating effect since it demonstrates the importance of the task and the person. The short-term negative morale effects of constructive criticism can be negated quite easily by following up with commendation when there is a response. Most important, the individual employee must have these questions answered so that the following are legitimate and truthful perceptions: *Management knows what I do, they are appreciative of it, and they are helping me to improve my performance.*

A proper appreciation of the human factor results in dealing with subordinates with total candor and truthfulness. This, in turn, requires three things: (1) the intent on the part of managers to convey an accurate sense of reality to their subordinates, (2) a knowledge of this reality on the part of the manager, and (3) the skills and credibility to communicate this to their subordinates. The first and third of these are easier to attain, although the many gimmicky techniques that essentially guide managers through a process of manipulation so often convey just the opposite. To these we submit the following principle: *Nothing appears to be succeeding with such certainty but is so assured of utter failure as a scam.* The second item, a true perception of reality, takes a very long and concerted effort on the part of a manager to attain. However, we cannot reemphasize too often that a thorough understanding of reality is essential to the control process. If it is absent, it is the manager who is out of control, and the process cannot be in much better shape.

One of the most interesting phenomena in the early study of the human factor is discussed next.

3.10 Hawthorne Effect

Most managers have heard of the classic study conducted by Mayo and others beginning around 1927 at the Hawthorne wire works. However, the ramifications of this study are not fully appreciated, and a full understanding of the principles revealed could avert many management fiascoes. Management lore on the study goes something as follows. Mayo and his team of early management scientists attempted to determine if any factors within the work environment could be altered to increase productivity. They chose a large number of factors and performed controlled experiments. For example, they increased the light intensity and found that the number of units produced per hour increased. This seemed to be an important factor related to human perception and performance. However, before drawing conclusions, they lowered the light intensity from the average for another test group to establish the relationship fully. To their amazement, lowering the light intensity had practically the identical positive effect on productivity as did raising the light intensity.

As is true in many scientific experiments, Mayo and his colleagues had serendipitously discovered something far greater than what they were searching for. The immediate conclusion was that the workers were responding to the change—any change; and this simplistic interpretation persists to this day in many management doctrines. However, the real discovery was that of determining the cause of this behavior. Many hypotheses were put forward, most of which were skewed by adherence to the common worker paradigm that existed at the time. For example: (1) they were trying intentionally to confound the studies, or (2) they just liked the attention and were responding to it. We believe that the real underlying cause was best expressed by Blau: "The conclusion finally arrived at was that increased productivity was a function of improved human relations. The entire social situation had been altered in ways that fostered friendly relations among workers. In addition, the supervision of workers had been taken over by the researchers who, in the interest of maintaining worker cooperation in the experiments, were very informal and non-directive in their approach" (BLAU62,90). In short, the workers wanted the productivity to increase, perhaps in general support of the researchers who befriended them. We cannot think of a more powerful motivating force than this to increase productivity.

The principle here is very simple; now for the applications. Let us start with a few misapplications. It has been said that it does not matter what change you make just so you make a change. People will respond and productivity will increase. This totally misses the point! Once again, our urge to manipulate has overcome good sense. Do we not realize that once change becomes the norm, the lack of change becomes a change? Hence in some organizations you will hear that the only thing that stays the same is that nothing stays the same. This absurdity will cause confusion, not increase productivity.

Well, then, how about planning an optimal number of changes just so we don't overdo it? This again is the manipulation trap, and if your subordinates do not see through it, they probably are not capable of very creative thought. Once again, we have taken a very useful scalpel and tried to use it as a hammer.

How, then, do we apply this principle? First, recognize that change had nothing to do with it. Go back and reread the passage by Blau. If by a change you can bring about "improved human relations [and] friendly relations among workers," go ahead and make the change. But make it for the sake of better human relations, not productivity! It is quite paradoxical that the exact reason that we try to accomplish something is the reason that causes it to fail. This is the reason that arbitrary changes are akin to most short-term solutions, which ultimately cause many more problems than they eliminate.

Now, consider this corollary of the principle of the Hawthorne effect:

> Just because a positive effect is noted after you initiate a change does not mean that this effect was caused by the genius of your change.

We have personalized this statement because it is so tied up with the ego of the manager. Who is there among us who does not want to take the credit when we implement a change for the better and the expected benefits follow? But there are two other possibilities: (1) the positive effect would have occurred anyway (i.e., it had nothing to do with your change), or (2) it was in fact caused by the sociological effect of the change, not the substance of the change itself (i.e., it was due to the Hawthorne effect).

The failure to recognize the validity of the above would have destined Mayo to the same fate as that of most of his contemporaries who were doing the same types of experiments. So anxious were they to claim the credit that they jumped to raise the light intensities to blinding levels or perhaps shut off the lights altogether. They missed the point. However, the human factor was not hidden from the early researchers, as is shown in the next section.

3.11 Industrial Humanism

We begin with another finding of Mayo, which sets up the definition of industrial humanism. According to Kast, "Mayo emphasized the necessity for reevaluating the traditional hypothesis of economic theory which considered society to be made up of individuals who were trying to maximize self-interest. He called for modifications in the industrial system to give a greater recognition to human values" (KAST70,89).

Mayo was not proposing an alternative to capitalism; he was only trying to make it work better. Indeed, he tried to "sell" his concepts based on greater productivity, which would certainly appeal to capitalists. It might not appeal to the workers, however, since many plant managers punished increased productivity by laying off workers. (This was quite possible since the unskilled labor market was a buyers' market.) Increased productivity could certainly be rationalized as being beneficial to society as a whole. After all, someone has to purchase and use all the products that are being produced, and the more that is produced, the more people would have the opportunity to acquire the products. But this is of little comfort to workers who are about to lose their jobs. This accounts for much of the social unrest, unionization, and other attempts to overcome these problems that are inherent in our system.

This movement, which was initiated by Mayo and his contemporaries as they determined that a happy workforce is more productive, stalled before it got off the ground. We might look for causes in the interruption of the Great Depression, World War II, and the virtual monopoly that the United States had in industrial power after the war. Compounding this were the problems of division of labor, which led to an ongoing conflict between labor and management. Although this is not directly relevant to technical management, it is important to understand this because the management practices of most technical organizations evolved from these early production organizations.

As opposed to this deeply entrenched paradigm of management is the philosophy of industrial humanism. According to Scott, "basic to the philosophy of *industrial humanism* is the design of the work environment to provide for the restoration of man's dignity" (SCOTT67,43). This would not seem to be too much to ask. This is especially true for technical professional personnel, who are expected to know much more about their area of specialization than their superiors and to make design and development decisions on which our technological society can progress. The following paragraphs provide commentary on some of the key terms within the definition of industrial humanism.

- *Restoration.* Hopefully in the technical organization, human dignity has not suffered as in the traditional workplace, where the proposers of industrial humanism perceived it to be lost. However, to the extent that the traditional management styles have been applied, restoration will be required. This can be performed at the local level despite the fact that the rest of the organization is still in the dark ages. While bottom-up transformation is not optimal, it is feasible (see Chapter 24); and without cooperation and development at the lowest levels, transition is impossible. Lower-level managers must never allow errors of their management to become the standard. Instead, they must prove the validity of sound management principles by the productivity that results from its application within the local organizational component.

- *Dignity.* In the absence of a definition from Scott, we will take the liberty to define this word for our context. Most important to dignity is the *perception of recognized accomplishment.* This is impossible in a situation where employees cannot perceive their own personal unique contribution to the organization. The internalization of this perception is the primary contributor to dignity. Next in priority is recognition from peers, which is more important than recognition from superiors. However, in a fully functional organization, recognition will exist at all these levels.

- *Design of the work environment.* We have used the term *work environment* consistently throughout this chapter to refer to the human relationships conducive to a productive workplace. The idea of design implies forethought on the part of management. This requires their recognition that to a large extent, the work environment is under their control. The loss of dignity on the part of such managers' subordinates is due to their inability to administer rewards properly, which is related directly to their lack of leadership.

In this chapter we have presented a large variety of principles of human behavior. In the final section these will be synthesized into the technimanagement meta-control model. However, in the following section we present a necessary qualifier to keep the model in proper perspective.

3.12 Human Internal Complexity: Mind/Emotions

We can approach an understanding of our own individual minds and emotions, but we can never hope to explain them thoroughly, at least not in the foreseeable future. This concept has been stated in a variety of ways: "If our minds were simple enough for us to understand, we would be too stupid to understand them" (author unknown). (Even this oxymoron is intriguing.) But if we cannot even understand ourselves, how can we hope to understand others totally? We might perceive that the designer of the human being made this an integral part of the design requirements.

All the computer power in the world cannot match the power of one human brain. The best robot in the world cannot navigate as effectively as the common house fly. As much as we talk about artificial intelligence, those who believe that we are even close to it are naive, despite their academic prowess. We need only open our minds' eyes to realize the complexity of human reasoning. Even if we could program a computer to think, we could hardly begin to approach the capability of meta-thought (thought about thought), a human characteristic that we take for granted and have been utilizing throughout this chapter.

We hope that we have not treated technimanagement in this chapter as though it were a simple formula (as are many engineering and science problems). It is just not as simple as all that, and this is the primary reason that the technically trained (and most others) fail so often as managers. Technimanagement cannot be modeled mathematically without more assumptions being made than the model itself would explain. Rather than recognizing it for the complexity that it is, we have a tendency to write it off as being "common sense." We use the term *counterintuitive* throughout this book to discourage simplistic, intuitive reasoning. The point is quite simple: People are not as simple as machines; they are all totally different, and they are each changing constantly. They have *emotions*, which make them act in quite inconsistent ways even when the identical stimulus is applied. Emotions are the basis of human motivation and can lead a person to act either consistently with or counter to what would otherwise be considered the *rational* senses.

We do not believe that the technically trained person should be at a disadvantage here. Consider other physical things. There are few, if any, physical things in this world that we understand totally. Yet we are able to control many, if not most, of them. The approach to controlling human technical organizations lies in applying basic principles that generally hold, while recognizing their limitations. As a preface to the next, concluding section of this chapter, let us state this major principle of differentiation of control:

> The control of an organization is attained by establishing the means and environment by which each person's control system is consistent with organizational goals; it is not attained by controlling every person in the organization.

Indeed, it is not attained by controlling anyone in the organization—except oneself.

In closing this chapter we tie the concept of rewards back to the principles of control systems introduced in Chapter 2.

3.13 Rewards and the Meta-Control System

The major difference between technimanagement and traditional management concepts is primarily one of control. Whereas standard management techniques rely on methods to get others to do what the manager wants (i.e., manipulation), in this chapter we have introduced the concept of the organization responding to the needs of the individual. The organization is seen in a holistic way to have a life of its own with distributed decision-making capabilities, as opposed to a single-headed hierarchy. But the administration of rewards and the establishment of a friendly work environment remain, at least initially, in the hands of management. (This will change as the organization goes through the transition described in Chapter 24.)

In introducing the concepts of control in Chapter 2 we were explicit that they should not become a means of manipulation. As the various reward types and their ramifications were introduced in this chapter, we have tried to reinforce these concepts (i.e., that rewards are not to be used for manipulation). To merge these two broad concepts, it is necessary to see rewards as the means for implementing corrections in the control process, not for controlling the actions of a specific person in the organization.

The application of rewards to produce change is not a new concept. Indeed, most accepted management techniques, such as management by objectives, explicitly recognize the administration of rewards as an integral part of the control process. How, then, is technimanagement different? Ideally, the technical manager does not establish or administer the control system directly. Rather, the primary objective of the technical manager is to provide an environment in which these functions will be performed by the organization itself. This added degree of freedom is a further step toward dignity, as defined in Section 3.11.

Does this mean that the technical manager becomes totally passive and has no responsibilities (as we expect some to conclude at this point)? Indeed, several names have been given to this perception: *laissez-faire, hands-off, free rein,* and others. We will not attempt to define or argue with the definitions of these terms at this point. However, we submit that if you conclude that technimanagement requires you to sit by passively, you have not been reading very carefully. We have described the construction of control systems and the administration of rewards as a very complex and labor-intensive process. But it is work directed at the organization itself and correction of the flawed mechanisms of the traditional organization, not work directed at correcting the people in the organization. Indeed, to attempt to implement technimanagement by means of controlling the members of the organization directly is countercontrol, as defined

Sec. 3.13   Rewards and the Meta-Control System   47

in Chapter 2. This is true because it is the traditional organization's tendency to control people which creates its single greatest deficiency.

Bear with us now; this will not be as difficult as it might seem. We wish to wrap up the preceding two chapters into a single new paradigm for technical management. The control process that the manager is to maintain is indirect. It is, in fact, a meta-control process that seeks as its goal that all employees and work groups have properly established their own control systems. It measures not by ultimate performance (as is the case of individual control systems) but by assessing the realization and functioning of each control system. Finally, corrections are implemented through the indirect reward system of approval, constructive criticism, and the realization of organizational goals.

It should be recognized that the establishment and operation of the meta-control process is much more difficult than the traditional control system in which the manager imposes a goal and then comes back later to determine by some simple metric if it has been accomplished. This is what we have been emphasizing when we state that the manager must establish the environment in which the various reward and control systems discussed above can be established. In the remaining chapters of the book we provide further principles by which the environment for the evolution of technimanagement can take place. (Technimanagement cannot be accused of being the management fad of the year because it takes forever to establish it fully. At the same time, it feeds spontaneously on the good parts of these fads, because its mechanisms are open to anything that might improve organizational effectiveness.)

As a final concept, consider the *principle of resonance.* In a physical system, a resonant condition exists when the forces tend to reinforce the existing conditions. For example, if you push a person on a swing at just the right time, very little force will propel the person to greater and greater heights on each push. On the other hand, improper timing of the same force, when the swing is coming toward you, damps out the momentum that has been established. Resonance occurs at certain speeds of an automobile that has unbalanced tires, and if that exact speed is maintained, it can do damage to the vehicle. Resonance has been known to destroy otherwise well-designed bridges. The point is, resonance within your department can turn your company on its ear. The manager who is properly regulating the meta-control system will administer the "push" at just that right instant to keep each employee in a resonant condition. It does not take a large reward any more than it takes a large force to maintain resonance. However, timing is crucial. In the meta-control system it is not when the contract is landed or when the report is done that the reward should be administered, for by then it is too late. The rewards must be administered as the various control systems are being established and as they are seen to be functioning effectively. Meta-control is effectively meta-management—assuring that effective self-management is in place. This requires considerable self-control on the part of the manager, who at times seems to be losing control over the actions of certain employees. However, the net overall effect is increased control of the organization, not of the person. For if employees are establishing their own goals and controlling their own actions, they will have the highest motivation and from this will result the greatest possible productivity.

4

McGregor's Theory X and Theory Y

4.1 Introduction

Without doubt, some of the most revolutionary work in solidifying the movement toward industrial humanism was first published in the late 1950s by McGregor (MCGRE57, MCGRE60). In retrospect, we can see that McGregor's principles go back at least to Mayo's work. However, they were not given a very effective handle, and thus the result of their attempted introduction may well have stimulated far more discussion than action. It might be fine for college professors or even staff personnel to espouse the doctrines of industrial humanism, but for line managers to practice them would take a major departure from the corporate culture. Without total upper-management support, their efforts would generally be viewed as, at best, weak (at worst, gutless). This is not to say that these concepts were not practiced to some degree "in the closet" by many traditional firms and overtly by some very progressive company owners.

McGregor's use of the names Theory X and Theory Y had several effects, not all of which were good. On the positive side, it enabled managers to understand and contrast these two basic extreme philosophies of management in simple and straightforward terms. This greatly facilitated communication in both industry and the academic world. However, the downside had two facets. First, it tended to oversimplify what even McGregor insisted would require a fairly complicated process of transition. Many believed that the wholesale adoption of Theory Y was a rather simple and straightforward process that could be accomplished by the stroke of a pen or the statement of a high executive. Nothing could be further from reality. This led to the second problem, which was the perversion of Theory Y to the status of a management fad. Indeed, from the 1960s on, professional managers in their right minds would never claim to be oriented toward Theory X, and the proclamation the "we're a Theory Y shop" was the byword of recruiting. Yet we saw no revolution in the way that businesses were managed, although this lip service did have the positive effect of making fashionable attempts by some to distribute management responsibility, if not authority.

Because McGregor's theories were intended to apply to management in general, we divert our focus temporarily from the technical organization in order to understand the context for his introduction of these principles. The principles are of such importance that no technical manager should be ignorant of any of McGregor's work. In this chapter we summarize his major points and interpret them in terms of the principles of control and rewards discussed above.

4.2   Theory X

McGregor observed quite correctly that the predominant paradigm of management of the 1950s was based on the assumption that (1) managers know more than their subordinates and thus can better direct the workers' activities, (2) managers must be proactive in this direction since the employees are not generally motivated by the needs of the organization, and (3) the application of science to management consists of establishing a number of control systems, based on the first two assumptions, to assure that workers do what they are charged to do. McGregor called this approach to management *Theory X*.

According to Kazmier, this was "conventional theory" in 1957. He stated: "In a society with a relatively low subsistence level and shortage of employment opportunities this 'carrot-and-stick' theory of management would tend to work rather well" (KAZMI69,223). Recognize that traditional management styles evolved from the small single-owner businesses that existed prior to the industrial revolution. These were, in turn, based on their agricultural counterparts. Virtually all work at these times was boring, painful, and except for the few animals used, totally labor intensive. It was only natural that rich landowners would watch their poorer employees very carefully to assure that they put in their hours, since time worked and physical effort exerted on the job correlated almost perfectly with productivity. In many cases there was also close supervision to assure that employees did not steal what they might believe was justifiably theirs.

In terms of the control/reward system, the landowner (or industrialist) became a virtual mini-king, lord, or dictator. This power often extended to the point of life and death over the worker's entire family. If the loss of a job is devastating to a person now (and it is!), just think of what it meant in the days prior to government-sponsored safety nets. (It was even worse in the environment of slavery, to which we are not currently referring.)

There were many good land and factory owners; and they were no more to blame for the way the system evolved than are most of today's managers. Some practiced a type of stewardship system, and the wise ones allowed their subordinates to develop themselves to the point of independence. Still, the rule was a pure top-down system in which the control mechanism was imposed by the owner on each person in the organization. Goals were set, measurements were taken and evaluated, and the correction process was administered by the owner. The growth of business required that both the establishment of controls and the administration of rewards be delegated. However, the use of rewards were almost purely for manipulation, and there was no notion of a meta-control system.

McGregor recognized that what may have had to work in the agricultural and early industrial societies of the past was not effective in the 1950s. This is summarized in the critique of Theory X given in the adjacent display. Additional insight into the ramifications of these assumptions when applied to technical organizations is obtain by reflection on another assumption made above, which holds fairly well in nontechnical organizations: that *time worked and physical effort exerted on the job were both almost perfectly correlated with productivity*. The adherence to this assumption leads to the strict regimentation of professionals, which is characteristic of bureaucracy. The ultimate in counterproductivity (i.e., countercontrol) results when the members of the organization spend the major part of their creative energies in determining methods for circumventing the control systems that are established based on these false assumptions. Sometimes these tasks are even formalized by the bureaucracy and given position names, such as *expediter* or *facilitator*.

Critique of Theory X

Managers know more than their subordinates. This is true only for a very limited scope of activities (e.g., interaction with other components of the organization). However, that is often sufficient to reinforce management's belief in this assumption. It is never generally true in either technical or nontechnical organizations. That is, the collective knowledge of subordinates as to methods that can be employed to improve both quality and productivity is always superior to that of the manager. The reason for this is their sheer number and their proximity to the problem. The error in making this assumption is far worse in technical organizations, where many subordinates have specialized skills well beyond those attained by the manager.

> **Critique of Theory X (Continued)**
>
> *The manager can better direct the subordinate's activities than can the subordinate.* It is possible that in some nontechnical organizations the ability to direct one's activities might not correlate directly with direct knowledge of the job. However, in an organization composed almost completely of professionals, this assumption is nothing short of an insult to their intelligence.
>
> *Employees are not motivated by organizational goals.* That this is true for many organizations, both technical and nontechnical, cannot be disputed. However, we submit that its probability of being true is well below 50 percent for technical organizations (even in our current Theory X culture), making it a very precarious base of management behavior. Further, it is the responsibility of management to bring about goal integration (Section 3.7), not to surrender to its absence.
>
> *Control systems must be based on these assumptions.* Because of the invalid assumptions on which they are based, all such control systems will be attempts to modify individual behavior (i.e., manipulation) as opposed to creating an effective meta-control system.

4.3 Theory Y

Although it is clear that there were many who had proposed that management accept a paradigm that was more friendly toward employees, McGregor's Theory Y went beyond this. It was his contention that if just given the chance, each worker had the capacity and the motivation for behavior that was consistent with organizational goals. He blamed Theory X for the problems in management and contended that managers could unleash a bonanza of productivity if they would allow people to direct their own behavior. He did not see management as being responsible for developing this motivation. Instead, management was responsible for developing the environment in which this potential could be realized.

The degree of difference between Theory X and Theory Y cannot be underemphasized. It was not just a softening of attitudes toward workers, or the addition of a series of social activities. According to Kazmier: "Thus, McGregor suggests that [(1)] management by direction and control —regardless of whether it is hard or soft—relies on motivational methods that are relatively ineffective [and (2)] since the higher order needs are now the relevant ones for most people, commitment to jobs is increased when the methods and procedures allow individual judgment and choice"

(KAZMI69,224–225). According to Theory X, workers had to be managed and controlled actively because otherwise, they could be expected to work against organizational needs. Theory Y states that this is not only ineffective but counterproductive.

The rationalization for the required paradigm change was based on the hierarchy of needs that we discussed in Chapter 3. Although we cited several counterexamples to this structure, its overall applicability is generally valid. Thus since primary needs were basically satisfied in the workforce after World War II, they no longer provided the degree of motivation that was present in prior generations. However, continuing to be treated as those "lower on the food chain," workers responded defensively with behavior that maximized their own perceived interests as opposed to those of the organization. This clearly ran counter to Theory Y, which taught that if workers would be given the opportunity to develop their abilities of self-expression and self-development, the net effect would be a reorientation of worker motivation, which would lead to major improvements in morale and productivity.

Before showing that Theory Y is an essential foundation of technimanagement, it is of interest to wonder why it has largely disappeared.

4.4 What Happened To Theory Y?

Although technimanagement is totally consistent with Theory Y, it would appear from a review of contemporary management literature that this is not inconsistent with most popular management philosophies. But reality must be faced: Although Theory Y became a fad, it never became the dominant paradigm according to McGregor's definition. Theory Y has been given a tremendous amount of lip service over the past 35 years, and there is no question that this indoctrination has had the effect of paving some major highways away from Theory X. But while many have embarked on these highways, few have arrived at the Theory Y destination. We need to evaluate the reasons for this to appreciate its applicability to technimanagement.

The natural question is: If Theory Y is so good, why is it not being generally applied? The response from management: "It is!" And therein lies the first problem: the inability to understand the depth of the changes required under this new and totally different paradigm. An underestimation of the mammoth effort and extensive ramifications of Theory Y implementation could easily result in an entire company thinking that they are practicing Theory Y when in fact they are much closer to Theory X. However, the lowest-level subordinates have no problem in seeing that the emperors are wearing no clothes. Further evidence that Theory X is alive and well is presented in the adjacent display.

> **Theory X Is Alive and Well**
>
> Some examples of symptoms are:
>
> - Arbitrary work assignments
> - Management statement: "You can only take Theory Y so far"
> - Management attitude: You have to get your subordinates to "buy in"
> - Direct work measurements on subordinates
> - Procedures to control the actions of individuals
>
> Quotation from one of the most popular current references on the management of software design and development:
>
> > There may be a sleeping giant inside your own organization, ready to awaken when it is in danger. The giant is the body of your co-workers and subordinates, rational men and women whose patience is nearly exhausted. Whether they are great organizational thinkers or not, they know Silly when they see it. And some of the things that do most harm to the environment and sociology of the workplace are downright silly. (DEMAR87,173)

We do not dispute that there are some very effective Theory Y managers as well as some organizations that operate predominantly under a Theory Y philosophy. However, it should be quite clear that Theory Y has not become the predominant management paradigm. One difficulty in this process of identification is that all organizations are somewhere between total X and total Y. It should be quite clear that what we call a Theory X organization might have some Theory Y attributes, and vice versa.

The failure to assess an organization's relative X–Y position properly has little to do with upper-level management (i.e., organizational) policy, which often espouses Theory Y and implements such things as open-door policies to circumvent the fact that their lower levels are not practicing it. These, along with Theory Y training and motivational programs, do little more than put a veneer on Theory X and disguise it to look quite like Theory Y. This misperception is extremely counterproductive, since under that veneer it is the same old paradigm. If it quacks, flies, has webbed feet, and walks like a duck, calling it a Doberman does not make it a watchduck.

The primary deterrents to moving to Theory Y are the egos of most managers. Think about your first promotion (or appointment) to management. Why did you get the promotion? Isn't it because you knew more about what is going on than those who competed with you for the promotion? (We recognize that this is not always reality, but which managers among us are willing to face this reality personally?) Shouldn't you be proud of your new position? After all, you earned it! If you weren't good, you wouldn't be where you are. So don't let your subordinates tell you anything!

If you have never felt this way, chances are you have never gotten a promotion to a major position of power over others. All of us are Theory Y people before we are in the driver's seat. However, once promoted, all advice becomes implied criticism (backseat driving), and the acceptance of such advice is often perceived by managers as a sign of their own weakness.

[This regression from Theory Y to Theory X caused by promotion (generalized: success) explains much about the evolution of both individuals and organizations. Coupled with the ego problems discussed above in conjunction with the Hawthorne effect, it probably explains the reason that most badly run companies are in the shape that they are in today and the reason that most well-run companies may not enjoy that status tomorrow. More on this in Chapter 13.]

A second deterrent to successful Theory Y implementation was discussed by McGregor and quickly discarded: the *problem individual*. He realized that if there were people within the organization who were intentionally engaged in sabotage, they would obviously not respond to Theory Y. The solution seemed simple enough: Fire the person. (But like most simple solutions to complicated problems, it isn't.) Questions: What if there are several of these "saboteurs," some of which are undercover? (In some extreme cases, management might perceive this to be the entire workforce.) What if they are not overt saboteurs but only marginal troublemakers? How do you quell these troublemakers without stamping out creativity? What if their particular technical services are scarce and really essential to the organization? What if tenure or other needs of the organization make dismissal infeasible? Is the method for building a Theory Y organization the systematic elimination of everyone who does not fit within this management paradigm? This is akin to the (hopefully past) practice within communist Russia of declaring political dissidents to be insane. In fairness, we know that this implementation technique is absolutely not what McGregor had in mind. But there is no doubt that problem individuals can cause precipitous regression to Theory X.

This leads to a third major cause which is not theoretical at all—it is operational. Managers were stuck in their preconceived paradigm, which included a very large dose of blaming all problems on the workers (or possibly the labor movement in general). As if this were not enough, workers were identically entrenched in their determination to blame everything on management. Theory Y sounded great to them: Why wasn't management practicing it? So in many large companies they organized themselves into unions and demanded their share of the "pie." Of necessity, and often due to past injustices, this type of behavior was rewarded! Thus we see why it was repeated. The general control process for this behavior was far too slow, and it was not until the late 1970s and early 1980s (20 to 30 years later) that unreasonable worker demands led to the demise of many American production facilities. The overwhelming tendency still is to blame this on the union, the government, or a foreign country. Not, we hastily add, that there isn't a surplus of blame to go around. But ultimately it is management who must accept the blame for failure to provide the mechanisms to avert the common disaster for both management and labor.

Let us attempt to analyze one step further the cause of this phenomenon. In defense of McGregor, we might say that Theory Y was never truly implemented. In defense of management, we could argue that the mindsets of both management and labor prevented this from occurring. But what good is something that is theoretically correct if it is not practically useful? This brings us to a major theoretical finding with regard to management control over Theory Y implementation, namely:

1. Theory Y motivations are derived from the realization of rewards based on higher-order needs.
2. Managers cannot satisfy higher-order needs directly as they could with primary needs.
3. Therefore, managers cannot directly establish and operate the control systems (mechanisms) essential to the implementation of Theory Y.

This is a basic cause of the demise of Theory Y and subsequent attempts to water it down, compromise it, or make it "practical."

Attempts were made by the most progressive managers to satisfy the higher-order needs directly. That is, an attempt was made to incorporate them into the primary control system of each worker. This failed, however, because management continued to view their efforts in terms of their former reward paradigm, not in terms of meta-control. Some who did understand higher-order needs attempted to administer them in the same way that traditional organizations had always administered rewards, and that just did not work. These efforts were (and still are for the most part) nothing short of attempts at manipulation. When they do seem to work, it is only over the short term. As soon as the workers perceive what is being done to them, they turn the tables on management and become the manipulators rather than the manipulated. The initial premise of Theory X (i.e., that the workers are stupid) is amply proven to be totally false, as the workers prove that it is management who is being manipulated.

Now let's try to keep from playing that role ourselves, for it should appear to the reader at this point that we have flushed Theory Y down the tubes at the same time that we have subscribed to it. The final flaw in McGregor's reasoning is that he viewed unskilled laborers and their managers in terms of himself (an error that all of us make much too often). In retrospect we should recognize a difference. I am highly motivated to write this book but not because someone is standing over me forcing me to do it. In fact, I am practically assured of not getting any credit for it on my current job, and (at first writing) I had no assurance that it would be published. Why, then, am I doing it? (The answer is irrelevant, and I am not sure I could articulate it fully—and further, who cares?) The point is that I cannot expect auto workers in Detroit who turn bolts hour after hour to have the same motivation for their work as I have for mine.

We are not arguing that quality circles, job enrichment, job enlargement, and so on, will not assist in providing greater motivation and morale. However, the excitement for those types of jobs may never be the same as the excitement that managers feel for their jobs. We might wish that this were not the case, but we all know that it is. The recognition of this reality is the first step toward understanding the inapplicability of Theory Y to the workforce in general. We cannot judge the motivations of others based on our own motivational systems.

Now that we have reviewed some of the reasons that Theory Y has not been implemented successfully, we must explain our optimism for its relevance to technimanagement.

4.5 Technimanagement and Theory Y

We demonstrated in Section 4.4 the multitude of reasons that McGregor's excellent theory has not created nirvana in all aspects of our economy. The continuing unemployment, labor unrest, and lack of competitiveness in our society is ample evidence of the following: (1) no attempt was made to apply it, or (2) the attempts that were made failed. Reality is an indeterminable point somewhere between these two extremes. However, we do not believe that this will alter the major conclusion of this chapter, which is summarized by the following:

> McGregor's Theory Y provides the major premise of technimanagement.

McGregor's assumption that the worker is motivated like himself is obviously correct when restricted to people like McGregor, and this is exactly how we have defined the members of technical organizations. Thus by our definition (see Chapter 1) of the professional aspects of the people who report to the technical manager, we have essentially defined this to be true.

You might also wish to go back through the section above which critiques the application of Theory Y and note that none of the criticisms that have made Theory Y inapplicable to the general workforce apply to technimanagement—except one. The one exception has to do with that introduced in Chapter 3 and elaborated on further at the end of Section 1.3 (i.e., the ego of the manager). In this regard, technical managers are no different from all other managers. However, if this can be *controlled*, Theory Y will deliver the goods. But this is a very big "if." Much of the rest of this book has the underlying theme of making this "if" become reality.

5

Decision Making and the Principle of Authority

5.1 Definitions: Authority and Stability

The statement that "McGregor's Theory Y provides the major premise of technimanagement" implies our great confidence in McGregor's work, and we urge everyone to become totally familiar with his writings. It also implies the presence of a number of minor premises, or principles, which are necessary to formulate the complete technimanagement philosophy. We begin by introducing one of the most counterintuitive principles of management, which we call the *principle of authority*. Consider the following statement excerpted from a previous work by this author: "Authority only exists to the point where subordinates will allow it to exist" (BROWN76,31). This necessarily implies that authentic (as opposed to perceived) authority in any organization ultimately proceeds from the bottom up. We will show that this is universally true despite the reality that *formal authority* (i.e., that officially bestowed by position within the organization) generally proceeds top down. It is this apparent contradiction that makes the principle of authority so counterintuitive. However, since it is the understanding of reality that empowers the manager, the assimilation of this principle is essential to managerial effectiveness.

To clarify this principle further, we might start by requesting the reader to submit a counterexample. The most extremely repressive situation that we can imagine is the military of a totalitarian government. Certainly, officers have the authority to punish subordinates who do not follow orders. So, in an extreme example, a soldier who disobeys an order might be shot. However, this is far from authority to make that person carry out the order. Instead, the commanding officer merely presents an alternative that is less desirable than the order. This is different only in degree to the many relatively undesirable things that we all do every day.

Indeed, we might prefer to go fishing every day of our lives, but the consequences of that action make us *want* to go to work. Thus even this extreme example of constraint fits our principle. The fact is, you cannot make people do anything that they do not want to do; and when they take actions that are undesirable to them, it is only because they perceive the alternatives to those actions to be even less desirable.

If this principle holds for these very extreme counterexamples, it should be an accepted fact in technical management. We are not using the threat of force to coerce action; the maximum (management-imposed) punishment is the loss of the job, which should not be too much of a threat to most professional-level personnel (unless they have been fired several times in the recent past). In short, if managers are going to get cooperation from subordinates to achieve organizational goals, it will be because there is a willingness on their part to yield to management authority over them. Essentially, we are affirming that ultimately, *they* are in control. (There are other motivators that are potentially stronger than job loss, such as peer pressure, which we will address in subsequent chapters.)

Intuitively, there is a continuum between the extremes of coercion and free will. Managers can motivate behavior by applying rewards and punishments along this continuum (recall that a punishment might be the mere absence of a reward). The extreme example given above was clearly one of coercion: making the alternative to a given behavior (in this case death) less desirable than the behavior itself. Although the authority is still given (the subordinated could elect death, and many have!), it is given under duress. Contrasted with coercion is the subordinates' belief that management is acting in the subordinates best long-term interests. This is the essence of leadership.

Since pure free-will authority is derived solely from the assurance of the subordinates that the manager is acting in the subordinates best long-term interests, any deviation from free-will authority affects this confidence. To the degree that this confidence is deteriorated, authority becomes *unstable*. *Stable authority* is essential to the effective use and accrual of power. The adjacent display defines the concept of *stability of authority* in more detail.

Note that the difference between coercive authority and free-will authority is solely in the mind of the subordinate, not necessarily in the nature of the alternatives or the resulting behavior. In the coercive case there is a choice between two undesirable actions, and the one that is least desirable is chosen by the subordinate. In the case of the exercise of free-will authority, there are no "undesirable" choices. The subordinates accept the authority of the managers because they feel that it is in their long-term interests, not because there is another less-desirable alternative which is going to be imposed if they do not accept their managers' edict.

> **Stability of Authority**
>
> **Definition**
>
> > The stability of authority is a function of the degree to which subordinates see decisions made to be consistent with their long-term interests.
>
> **Implications**
>
> - The authority of a manager can build or deteriorate based on the decisions made.
> - Stability of authority determines its power-building capacity.
> - Stable authority builds power.
> - Unstable authority deteriorates power.
>
> **Power determinants**
>
> The ultimate power of managers is determined by the degree to which their subordinates will continue to recognize their authority.
>
> **Types of authority**
>
> - Coercive authority
> - Free-will authority
>
> **Relationship between Authority Type and Stability**
>
> > The stability of authority is directly proportional to its free-will characteristics

With these basic definitions in place we can proceed to a deeper understanding of the variable nature of authority.

5.2 Authority and Organizational Evolution

Although the principle of authority and its ramifications should be readily accepted, generally they are not. Consider the frequent references supporting top-down bestowal of authority which pervades the popular contemporary management literature (e.g., that authority is given from above). In this chapter we explore the reasons for this misperception and the problems that arise from it. Before getting into these, however, we want to establish the volatile nature of authority and thereby establish its relationship with the abstract entity that we call leadership. To do this we introduce some terminology that will help to delineate between the concepts of authority and leadership. According to Kast, there are three types of authority (KAST70,318): *charismatic, traditional,* and *rational-legal.* These tend to evolve, respectively, over time.

New organizations are usually formed by *charismatic* leaders who assume authority by virtue of their inherent (not necessarily natural) leadership abilities. These individuals are usually closely associated with a cause, good or bad, valid or not. They have the capacity to enlist others, not necessarily to themselves but to their cause (although sometimes it is difficult to differentiate between the two). Those initial followers yield to this authority because this person best represents (i.e., communicates) the cause to which they are all committed (e.g., making a profit, the production of a superior product, a social cause, a political or religious idealogy, etc.).

As the organization ages it will either be successful or die. If it is successful it will grow, and many of the lieutenants of the charismatic leader will assume positions of formal authority. It is at this point that the *traditional* forms of authority begin to take over. The lower levels recognize the necessity for this structure, and although these are not charismatic leaders (they might not be leaders at all), they are still allowed to have authority by the lower levels in deference to the formal positions that have been bestowed on them from "above." In other words, there is a type of *organizational momentum* that carries over from the charismatic leader to the succeeding generations who are appointed by the organization. However, generally these managers are much less charismatic. At the same time, the followers now tend to be attracted more by the success of the organization itself than by the organization's cause, which was the original source of its success. Thus it is the mediocre followers as much the noncharismatic managers who evolve the traditional form of authority.

Finally, when this organization ages a bit more (and this occurs at radically different rates for different organizations), the principles that led to the formation of the original organization are all but abandoned. The organization becomes an end in itself, and the authority type evolves to the *rational-legal* authority, which is necessary to assure survival. The rational part of it stands almost completely opposed to the emotional, which is the source in the charismatic type. Indeed, it is largely based on rationalization (and this is meant in a most negative way). The legal aspects are imposed by the bureaucracy that grows up, which is also rationalized to be necessary to support the larger organization.

We take up the theoretical basis for this implied principle of the evolution of organizations in Chapter 13. At this point our goal is to demonstrate that authority occurs with a variety of degrees, and it is recognized from the bottom up with a variety of motivations. These have two potential ranges: (1) from a purely emotional base to one that is almost completely objective, and (2) from those that are purely altruistic to those that are completely self-serving. (There is no implication that these two ranges are in any way correlated.)

With this introduction we can compare some concepts of leadership with those of authority, consider the relationship between authority and responsibility, and determine how this affects the decision-making process.

5.3 Authority versus Leadership

The subject of leadership is extremely important, and it deserves its own detailed consideration (Chapter 21). Many of these details cannot be communicated effectively prior to establishing considerable additional groundwork. However, in attempting to clarify the principle of authority, the relationship between authority and leadership cannot be avoided. Indeed, these two are often used synonymously, but the results of confusing these two concepts speak for themselves. The purpose of this section is to clarify the basic differences and to introduce the role of leadership within technimanagement.

Let us give some rudimentary sense of the intuitive meaning of leadership to establish this contrast. It is probably much more intuitive that leadership proceeds from the bottom up than that authority does. Indeed, this leads to the following principle:

> The very greatest prerequisite of an effective leader is the keen ability to determine who is willing to be led, and under what circumstances.

At the point when their leadership is no longer required, effective leaders withdraw to a support role (in strength, not weakness) and there wait patiently for their next opportunity to contribute by leading. This is contrary to the current concepts taught within many proactive assertiveness training courses, which can seriously cripple their adherents by creating a constancy of aggressive conduct. There is a time to be assertive, but improper timing can make insistence extremely counterproductive.

We fear that the principle discussed above can easily be misinterpreted to support short-term opportunistic leadership. This is grasped by those types of people who wait for an issue to surface, determine that there is general support for it, and then ride it for all that it is worth.

Because they are issue- rather than principle-driven, their populist position is rarely optimal. Instead, it strives to convince the maximum number of people over the short term and thereby takes advantage of natural bias and discontent (in the extreme, mob rule). At the first sign of potential failure, however, a scapegoat or excuse is sought, and our leader is quickly off to the next issue. We cannot escape classifying this short-term, self-serving hypocrisy as leadership, because there is clearly a leader, and others are certainly following. All we can do is to define this and all other such perversions as being *ineffective leadership*. Their popularity lasts only as long as they are not recognized for what they are. Unfortunately, all too often this does not occur until they cause the organization to self-destruct (and even then, some go on to feed off another group).

The difference between this and *effective leadership* is one of orientation. Ineffective leaders use the organization for their own personal benefit; effective leaders serve and support the organization. They are guided by principle, not self-serving reactions to issues. As such, they assume the long-term view toward attaining the goals of the organization and thereby serving its members. Consider the following quotation from Tom DeMarco: "Think of a leader as one who helps you to take steps that are difficult in the short term but helpful in the long term" (personal communication, 1/18/92). This is a very insightful, yet counterintuitive description. People will naturally, almost by reflex, do what they perceive to be in their immediate short-term interests. This behavior is almost invariably counterproductive, and leadership is required to educate and then motivate them to sacrifice in the short term in order to achieve greater benefits over the longer term. *Effective leadership* is the capacity to get followers to take actions that are in their long-term rather than their short-term interests.

Inevitably, the exertion of such leadership requires the acceptance of a change from the well-established paradigm oriented toward immediate rewards. This leadership cannot just "go with the flow" (as the opportunistic hypocrites do); but neither can the change be coerced! Thus effective leadership strives for an optimal balance between coercion and pure democracy that will maximize the total benefit to all members of the organization and thus to the organization itself. In discussing the relationship between change and leadership, Whitehead stated: "Change, to be acceptable to a group, must come from within, and must appear as the visible need of its present activities. . . . So management in industry can lead its groups to just that extent to which it is itself accepted by those groups, and it can lead no further; anything beyond that will be resisted as compulsive interruption to social living" (WHITE36,110). Again, this reinforces our concept that leadership, like authority, proceeds from the bottom up. This brings us to a point where we can effectively distinguish between these two abstract entities.

To this point the similarities between leadership and authority have been cited. Are they then the same, as some suspect? The answer is clearly no, for:

> Although authority proceeds directly from leadership, leadership in no way results from the possession of formal authority.

We might view it as being analogous to a rachet wrench: Leadership can force authority, but the wrench is worthless in the other direction. The most grievous upper-level management error is to assume that because they have "given" someone authority over others, that person then magically becomes the leader of these employees. First, although upper management can place lower-level managers in a position where they can determine merit pay raises, work assignments, and so on, they hardly give them authority over their subordinates; only the subordinates can give that. In fact, one of the surest ways to kill a person's leadership capabilities is for that person to be perceived as being too closely aligned with upper management. This is because true leadership emanates from the informal rather than the formal organization. Even in cases where authority is "given" by the lowest levels (as in the case of the election of the U.S. President), this endows neither the capacity nor the acceptance to qualify this person to be the leader at any given future time. If she or he is a leader, it is only because the followers continue to allow it to be so; this has been proven all too often in recent history (and it is quite evident at the time of this writing).

We will revisit these first principles of leadership and the informal organization in subsequent chapters. At this point we turn our attention to one of the most vexing perceptions of lower managers.

5.4 Responsibility without Authority

Probably no other complaint is heard from lower or middle managers as much as having been charged with the responsibility for some function without being given the authority to accomplish it. Consider the following quotation from Keith, which reflects this commonly accepted theory: "The failure of an organizational pattern may usually be traced directly to the fact that responsibility for some activity has been placed in the hands of an individual, but at the same time this individual has not been given sufficient authority to allow him to exercise command or control over men and machines so that the task can be accomplished as management desires. It seems foolish that such a situation can exist" (KEITH71,174). Indeed it does! Perhaps under Theory X, where authority was equated with coercion, there was some legitimacy to this complaint. We cannot argue with this. All that we can state is that such a statement has no place in the realm of contemporary technical management. Let us explore the reasons for this.

The major premise in the statement above is that upper management has the power to "give" greater and greater degrees of authority to lower-level managers. This premise was shown above to be false. The authoritative power to exercise control over the behavior of other people can only come from those people themselves. Further, the extent to which reward and punitive discretion can be bestowed by upper management is quite limited. The ultimate punitive power is that of firing, which even within itself might not be that undesirable to a professional who has a number of employment options. (Rarely in any humane organization is someone not given the option to resign when the "handwriting is on the wall.") However, even this is restricted by legal (e.g., discrimination) and morale considerations, and thus it is generally closely controlled with systems of warnings, probationary periods, and so on, by upper management. The only reward that can be given from above is that of promotion and financial remuneration, which both have their inherent limitations within the technical organization.

Although clearly the ability to control rewards and punishments can be given to lower levels of management, the perception that this automatically produces authority shows another erroneous management perspective. Quite often controls (i.e., limits of authority) are placed on lower-level managers to prevent them from abusing their punitive and rewarding powers for this very reason. Upper management might recognize through experience that when lower-level managers are given full discretionary powers over their subordinates, the use of this power typically results in counterproductive actions. Rather than dealing with this abuse of power directly (which would require definitive action to discipline lower-level managers), the bureaucracy usually imposes rules that apply to all managers, thus producing a multiplicative negative impact throughout the organization.

The solution to this apparent dilemma of the lower-level managers lies in their recognition that a number of rewards can be administered that are not "given" from above. As an example, consider the flexibility that a manager has to allow employees collectively to distribute and design their own work tasks. This is a major reward to most professional personnel, since their specific task assignments largely determine the enjoyment of their work. This might be questioned as an erosion of the authority of the manager, since the "authority" for this is distributed to the subordinates. As counterintuitive as this might seem, it actually results in an increase of authoritative power! First, it prevents the disempowerment that inevitably results when the manager dictates job assignments (even if these should

Sec. 5.4   Responsibility without Authority

match subordinate desires). Second, it assures the maximum motivation and satisfaction of the subordinates, thus increasing their inclination to support the manager in future decisions. Recall that in the technical organization the only type of authority that is of use to the manager is free-will authority, as defined above. Whereas coercive authority might bring about a favorable change in behavior in the short term, its continual use will so deteriorate the power of the managers as ultimately to render them impotent.

(We recognize that the bureaucracy will often establish rules to counteract the flexibility of lower-level managers to exercise these types of rewards, which is indeed a limitation of authority. However, the options of the manager cannot be totally overwhelmed by the bureaucracy. Also, good lower-level managers will intentionally remain ignorant of such rules for as long as possible. Once their power is built by the success of this ignorance, the rules will collapse of their own weight.)

The example given above used the determination of work tasks as a reward. We wish to generalize this to produce a central principle that not only provides maximum motivation to subordinates but also enables the manager to create additional authoritative power. This is summarized by the following concluding statement, which we call the *principle of free-will authority:*

> Technimanagement provides rewards by means of distributed decision making; in the meantime, the free-will authority of managers increases to the extent that organizational goals are fulfilled through decisions made by their subordinates.

This profound principle is totally counterintuitive to all Theory X managers. The extent to which it is believed (and hence practiced) is an excellent metric of the degree to which a manager has embraced technimanagement.

We recognize the weakness of obtaining concurrence by intimidation, and that was not the intent of the last sentence. The burden of proof of the statement made above is on us, and we will not only prove it but show that it is a truism. The obvious question is: How can you be assured that subordinates will fulfill organizational goals when they are allowed to make a large proportion of the decisions? This is simple: *Enable their full participation in setting organizational goals!* This is the essence of goal integration since it not only brings the subordinates' goals in line with those of the organization, it simultaneously brings the organization's goals in line with the subordinates' goals. (For those operating within the larger context of a Theory X organization, the "organization" here might be restricted to the organizational component. The principle is the same.)

To continue our proof, we must establish the relationship between free-will authority and distributed decision making. This has to do with a major principle of conservation of power that is elaborated on in Chapter 19. Briefly stated, power once discharged takes concerted effort over time to rebuild. When managers take the prerogative to make decisions independent of their subordinates, it costs them some of their authoritative power. Dictatorial decisions that tend to meet with all subordinates' satisfaction are essentially break-even, since this is what is expected. However, if only one person is dissatisfied, this can lead to a severe loss in the stability of authority (i.e., the potential to exercise influence in the future). Thus the more decisions that are made independently, the greater the potential for the manager to be seen as being arbitrary and possibly even capricious. The more difficult and controversial the decision, the more power that is lost and the less stable the authority becomes. Clearly, the more the decisions are effectively distributed to subordinates, the more power (stable authority) that is retained by the manager.

It might seem that we are advocating that managers abdicate their decision-making responsibility. Just the opposite is true. For the essence of effective decision making is the establishment of the meta-control system, not the micromanagement of subordinates. In the example given above, this would involve the creation of a mechanism by which the subordinates could perform their own job assignment without creating friction among themselves. Further, there will be times when the manager's decision on an issue will be essential to the process. *It is to enable these key decisions to have force of implementation that the building of authoritative power is required.*

Relating these concepts to the evolving organization (Section 5.2), charismatic authority is almost purely free-will. The relationship between leadership and authority characteristics is not coincidental. In the classical evolution of organizations, charismatic gives way to traditional authority, which is a combination of free-will and coercive authority. Although the bureaucracy might carry with it many programs to prevent coercive authority from existing, most mature organizations still rely heavily on this evolved variation of Theory X. Unfortunately, this is the training environment for most technical managers today. However, technimanagement creates a transformation of this type of work environment toward the ideal of the pure free-will authority situation. The fruit of initiating this transition can be obtained in a relatively short time.

Authority implies the power to make and implement decisions. It is appropriate, therefore, that we explore the decision-making process before we conclude this chapter.

5.5 Decision-Making Process

Decision-making techniques are important and we consider them in Section 5.5.2, but first, we need to consider the meta-decision necessary to resolve an important question.

5.5.1 Who Makes the Decision?

The discussion above implied that technimanagement involves subordinates in most decisions, while others are retained by the manager. This may also be the perception in many organizations whose leaders truly believe that they are practicing Theory Y, but who are in reality dominated more by Theory X. The key distinction here is contained in the answer to the following, which we call the *meta-decision question: Who determines which decisions are made by the manager and which are made by the subordinates?*

Most organizations have some type of committee structure that attempts to get a wide range of people, often at different levels, to "buy in." However, ultimate authority for the decision rests with the manager who appointed the committee. In other words, if the committee comes back with a decision to which the manager does not agree, she or he has the power to overturn it. This overriding might proceed in a subtle way, since the decision maker recognizes that the committee perceives that they have put much work into their recommendations, and they are not about to be persuaded to alter their recommendations. Often, the issue is just deferred ad infinitum.

All but the naive have seen the effects that this has on the future functioning of committees. Within this structure, the smarter managers will only appoint committees that they know in advance will bring back the "right" answer (i.e., they load the committee to their advantage). Within the committee itself, the experienced people adopt a "why bother?" attitude and give lip service to the committee by anticipating and then adopting the management line. After all, why bother? The demotivating effects are obvious: Creativity is stifled, and the mediocre results that follow are quite expected.

This is the result of the Theory X mentality coupled with all of the new-age management methods of manipulation. Clearly, the answer to the meta-decision question in the traditional organization is that this is a decision of management. However, we contend that as long as the manager dictates who makes the decisions, the organization is still de facto Theory X. Although management gimmicks might be attempted to make people feel good about the situation, this management style is destined to fail in the technical organization. This leads to the following principle of meta-decision making:

> Although it is essential that technical managers retain the authority to make a selected subset of the decisions in behalf of their subordinates, the ultimate responsibility for determining this subset lies with the subordinates.

This is not a dynamic decision made on a day-to-day basis. Rather, it must be worked out beforehand (an excellent theme for annual or biannual retreats), and then it should not be modified for a given period (until the group can reassess the wisdom of its prior decision).

In other words, we have created a control (actually meta-control) structure for determining *who* makes the decisions. This is a vital element of the meta-control system. As an example, the group might decide that at least for the foreseeable future, the manager will be responsible for annual reviews and raises. They may not wish to spend the time in establishing a peer-review system, or they might resolve that the manager has a better global view in this regard. However, after this runs for one or two years, they become dissatisfied and decide that they want to tackle this one themselves. So, together, they work out a peer-review system which they then implement with the degree of management involvement which they determine. (Further qualifier: When we say "they" in reference to distributed decision making, this requires the creation of teams, the membership of which seeks both vertical and horizontal integration, under principles discussed further in Chapters 14, 15, and 24.)

The example given above is at the extreme to demonstrate that nothing is sacred. In fact, the withholding of authority for certain decisions, especially with regard to the reward system, simply retains the Theory X orientation of the organization, regardless of what else is attempted for show. This does not mean that total sharing of authority will occur overnight, however. It might be years before an organizational component is ready for the advanced transition to technimanagement, which implies that the subordinates are integrated into the meta-control aspect of determining who should make the decisions. However, for organizations of professionals, this transition process is not only realistic, it is essential. As we shall see, the primary means of control will be peer pressure (using the term in a positive sense, where we assume that the vast majority have integrated the interests of the organization with their own). The greatest challenge for technical managers today is to engineer a transition to this ideal. As stated in an earlier chapter, this should be done without much concern for the outside world. When it works so well in your component of the organization, others will have to emulate it, in self-defense.

5.5.2 Decision-Making Mechanism

Although totally distributed decision making might be the ultimate goal, this does not relieve the manager from the problems that decision making entails. Certain things become easier, but the efforts required in coordination greatly increase. Because of the increased free-will authority involved, the leadership aspects of the manager take on a much higher importance than the raw exercise of power so predominant under Theory X. In short, the technical manager will be involved in all decisions even though not making nearly as many as in the traditional case. The purpose of this section is to give guidance to this process. This will apply for those decisions that remain with the manager, either through assignment of the organization itself or with regard to the establishment and functioning of the meta-control system. The process of bringing about consensus in group decision making is discussed in Chapter 14.

First recognize a major principle of decision making which is so important that we state it up front:

> The timing of the decision is as important as the decision itself.

This has a strong impact on the determination of "who" is going to make the decision. What this implies is that if the decision is made at the right time, its probability of being correct will go up dramatically. Conversely, if its timing is considerably off the mark, its chance of being correct will be very low.

The reasons that timing is so important are fairly straightforward. However, bad timing is probably the cause of far more bad decisions than is poor judgment (although bad timing shows poor judgment). There are two types of timing errors: too early and too late. Decisions made too early do not benefit from information that could be accumulated. They also create a "tipping of the hand" to potential adversaries. Finally, hasty decisions tend to be made by the wrong people. On the other hand, if decisions are put off too long, they get made by default. That is, the organization will create a situation quite different from that obtained by taking a proactive approach. Ultimately, reactive damage control (firefighting) will be required.

A second major point that is often ignored in decision making is that decisions are always a selection from a set of imperfect alternatives. Rarely, if ever, does the decision maker have total discretion to implement from a continuous range of alternatives (i.e., set things up exactly as you would want them). Instead, there will be a large number of constraints on the solution space, perhaps due to past mistakes or current resource limitations. When presented with a problem, do not search for an unrealistic ultimate solution and then fret because it is practically infeasible. (This is one of the major sources of time dissipation by committees.) Be proactive: Ask your advisors to enumerate the alternatives. Then proceed to make sure that because of their perspectives, they have not overlooked some alternatives, as so often, the thoughtful definition of alternatives will produce only one reasonable course of action. In this case it is really your subordinates (i.e., those who formulated the alternatives) who should be credited with the solution. The adjacent display suggests a procedure for decision making.

Suggested Decision-Making Procedure

- *Define the problem thoroughly.*
  - *Be sure that there really is a decision to be made.* Many problems that seem to demand attention might better be left to other mechanisms for their resolution. If the timing is such that it is preferable to defer action, a time should be designated at which the process will be resumed.
  - *Be able to state and document the problem.* A precise definition of a problem is 95 percent of its solution.
- *Assemble alternatives from as broad a base as possible.* Do not jump to conclusions or allow a skewed viewpoint to be overly influential.
- *Weigh the costs and benefits of each alternative.* There are very few win-win situations (if you identify one, the decision-making process largely vanishes). Most of the time any action that you take will have a downside. Examples: (1) in any resource allocation, cost is always a downside, and (2) any political victory tends to empower your adversaries.
- *Try out alternatives on advisors, and evaluate them as scientifically as possible.* Studies might be required. Consult with as wide a range of advisors as possible, especially among your subordinates and critics.
- *Select and implement the alternative.*

The decision-making process is a control system in which the measurement consists of an assessment of the current state of the organization and the correction is related to the impact of each alternative on the future. All decisions are based on an imperfect assumption of the effects that some action will have on the future (usually based on past experience). Perfect prediction of the future would make decision making unnecessary (that's the fun of it). This past-to-future extrapolation is the key reason that timing is of utmost importance in decision making. Astute managers will never make a decision before it is totally necessary, since they recognize that additional circumstances will inevitably arise that will make the premature decision obsolete (maybe even seem stupid). On the other hand, it is imperative that a decision not be delayed to the point that its failure to be implemented is the cause of another, even bigger problem. The balance between these two timing errors usually determines far more about the success of the decision than the decision itself.

In the distributed decision-making paradigm, the manager should take the leadership in either guiding the process or establishing mechanisms for continually addressing problems of various types. Thus you will be involved even though you might not actually be "making" the decision. Recognize in this regard that the responsibility for the decision is still yours. It is your confidence in your subordinates that will convince you that their decision is better than yours could ever be. However, when things go wrong (and they inevitably will) it is essential that you take the blame, not as a means of manipulation, but in honest recognition that had you provided better leadership and influence, a better decision would have been made.

In the following two chapters we apply the principles established thus far to the exposition and evaluation of two popular management techniques: management by exception and management by objectives. Both of these are attempts to translate delegated authority and responsibility into actions that satisfy organizational goals. We shall see how they can be modified to fit within the principles of technimanagement.

6

Management by Exception

6.1 Definitions

At this point in our discussion of technimanagement we wish to consider two tenets of traditional management: management by exception (here) and management by objectives (in Chapter 7). The purpose of these two short chapters is to compare these techniques of traditional management with the principles of technimanagement. As this is done, we hope to demonstrate that most of the management techniques applied in traditional management have application in technimanagement. However, the orientation and interpretation of these techniques must be questioned as opposed to the technique itself. Thus it is not a matter of throwing away all traditional techniques to initiate the transition to technimanagement. Many of them will serve not only in the transition but also once the organization fully matures. The objective is to take advantage of the wisdom derived from past experience to make possible continuous improvement in the management structure itself.

To begin, consider an operational definition of *management by exception* (MBE):

> MBE consists of a clear definition of those areas over that the subordinate has flexibility to exercise judgment and make direct decisions; the subordinate must seek upper-level review and approval for decisions that must be made outside these areas.

This approach to management goes back to Old Testament times (Exodus 18:13-27), and it has generally been accepted in all traditional management approaches. In fact, the complaint that an individual manager has responsibility without authority is usually due to a feeling that the MBE definition of areas is too restricted. However, this definition is rarely documented. In most organizations this would be practically impossible; however, certain rules emanate from the bureaucracy, especially in response to perceived abuses. Even in organizations that might attempt a complete MBE documentation of authority/responsibility, those areas outside the document are handled implicitly.

Consider how an organization might function in the absence of MBE. One of the primary problems in the traditional management organization is that the MBE areas are not totally defined. However, the manageability of the traditional organization is clearly related to the clarity of the MBE guidelines. Without clear MBE definitions, two types of errors will be made: (1) decisions are made at inappropriate levels of the organization, or (2) decisions fail to be made altogether, bringing about whatever default circumstance happens to result. Usually, this default situation is the delay of the decision until a critical situation arises, which distracts attention from long-term planning to short-term problem solving (firefighting).

It should be clear that the total elimination of MBE (either formalized or implicit) would be fatal to the traditional organization. The question is: Should it be modified for technimanagement? The answer is yes—MBE must be adapted to the new paradigm. This adapation is quite simple when the principles presented in the earlier chapters are applied.

6.2 MBE Under Technimanagement: MBNE

The primary principle is that of negotiated authority. Under the traditional management paradigm each manager determines what to delegate and to whom. Managers determine which of the decisions they will make and which will be made by their subordinates. MBE generally holds that this is the prerogative of the manager. Although there might be some interaction to obtain information about subordinate desires and abilities, the ultimate decision always rests with the manager.

In most technical environments there is a serious question as to whether any manager has the ability to assess the intricate activities of each subordinate accurately to delineate the optimal areas of decision making. The fact that a precise definition of the exception rules rarely exists is adequate testimony that this question is warranted. But if the manager does not have this capability, who does? Again, we appeal to the principle of negotiated authority introduced in Chapter 5 to formulate the following principle of management by negotiated exception (MBNE):

> The guidelines for subordinate decision-making limits should never be viewed as fixed. Instead, they should be determined by recurring negotiations, which should have as their objective establishment of the optimal distribution of authority and responsibility for that point in the organization's transition.

Under this principle the manager and the subordinate(s) collaborate as equals in determining who is best able to make individual decisions as well as those that require additional collective communication. After all, if the manager and the subordinate both understand the overall mission of the organization and the particular project, why should subordinates not know when they are "over their heads?" General guidelines should certainly be negotiated and thoroughly understood, especially at the extreme points where the subordinate should always and never make decisions. However, there should be a fairly broad "gray" area in which subordinates can both grow and adapt to the needs of the dynamic organization. This is quite different from the rigid, management-defined MBE of the traditional management paradigm, which is largely based on Theory X.

The principles in this chapter will continue to view the manager–subordinate relationship in the traditional one-on-one paradigm. However, it is informative to look briefly toward future chapters in which this model will be modified considerably by group decision making and distributed management. Thus the negotiation (N) within MBNE will be performed largely with groups as opposed to people. As the transition progresses, groups, consisting of people from several levels, will be charged with the responsibility and given the corresponding authority to make and implement decisions. (This process is commonly called *empowerment*.) In addition, there are some decisions that the entire organizational component will make collectively (or conversely, it could be said that no one makes them; yet they do get made). We concur with traditional management doctrine that if everyone is responsible for something, effectively no one is responsible for it. Thus in these global decision-making situations the manager retains responsibility (which goes without saying); however, the manager will seek and obtain a consensus before implementing the decision. To keep the principles of this chapter from becoming overly complicated, we will not consider group decision making any further at this time.

The alternative to MBNE is a stifled organization in which everyone knows their limitations and no one ever violates the bounds of the authority that they are given. While this is the Theory X manager's dream, it is a worst-case scenario in technical management, since some of the most dramatic breakthroughs result from what is clearly "insubordination" by the Theory X definition. We put that last word in quotation marks since the engineer or scientist who makes such a breakthrough, while seemingly rebellious to his or her immediate supervisor, might nevertheless be in complete accord with both organizational and project goals.

Nothing will delight personally secure managers more than when one of their subordinates assumes authority over a given situation and makes the right decision. Insecure managers will view such behavior as a threat to their positions. Just the opposite is true! This is what is generally called *taking the initiative*, and it is a clear sign of growth. Most managers will agree enthusiastically with this philosophy. However, the true test of their agreement (or hypocrisy) occurs when the subordinate fails to make the "right" decision and the resulting consequences must be faced. In the traditional organization, this calls for disciplinary action, blame and punishment; and you can be assured that no further subordinate initiatives will be taken. In the future, if the manager does not think of it, it will not get "thought of," and the area of delegation will shrink from the cold water thrown on the flames of initiative. Subordinates will retreat to the routine tasks which they have been disciplined to believe to be "their jobs" and will leave all unnecessary decision making to others.

As opposed to blame assignment, MBNE uses mistakes for organizational improvement. They are factored into future negotiations to determine modifications that might be required. However, short of malicious intent, the assumption is that the decision to make the decision (i.e., the meta-decision) was right. That is, aggressive assumption of authority is encouraged. The error could indicate that the decision was made at the wrong level. However, it could also indicate several other things: (1) the failure of the manager to anticipate the decision—thus indecision on the part of the manager provoked the subordinate's action, (2) the decision made was consistent with what the manager would have made in any event, or (3) it was a normal error which is expected to lead to growth— and thus it is not expected to be repeated. As opposed to retreat, these reasons support the subordinate's continued assumption of authority. Thus, using the error as a cause for discipline would be quite counterproductive in such cases.

In technimanagement there is no search for blame; the responsible manager assumes it all. If for no other reason, it can be attributed to lack of effective leadership. The faulty exercise of authority is seen as a learning situation. The subordinate is not blamed but asked: What did we learn from this episode? This might result in a redefinition of the MBNE guidelines, or it might be the training ground for their further and more perfect ongoing negotiation.

The ultimate question is: How do you control an organization in which everyone is encouraged to negotiate the boundaries of their authority? Would this not lead to chaos or anarchy? This question implies that chaos and anarchy are not symptoms of current technical organizations (so we might insert the word *more* before chaos). To answer, we must unask the question, since it demonstrates a misunderstanding of control within technical organizations. We saw in Chapter 2 that control of the technical organization is not something that can be imposed. Recall that the objective of management is to establish the meta-control system. It is to initiate the transition and then move the organization to a point at which decisions are made by the most effective mechanisms possible. (Subordinate blame should be recognized as a major deterrent to progress in this direction.) Once this system is in effect, the individuals within the organization will not seek anarchy, they will seek (and find) leadership.

6.3 MBNE Deterrent: Resentment

The principal deterrent to the implementation of MBNE is fear within upper management caused by lack of trust of subordinates. This fear is justified when there is evidence in the ranks of resentment toward (suspicion of) management. Elimination of this resentment is critical to the transition to technimanagement. Often, management does not fully perceive the resentment that exists until it precipitates dramatic action (e.g., resignations or moves toward unionization). Once it is perceived, managers generally mistake its root cause. Let us consider these two problems in turn.

Management ignorance. There are several reasons that managers do not perceive resentment on the part of subordinates. Probably the most significant of these is that the information feedback they get from subordinates is thoroughly filtered of any hint that this is the case. This itself is a symptom of Theory X, and it is the result of conditioning since about the fourth grade (prior to that, work is largely in groups). Most subordinates have been conditioned to please; the last thing that they want to display is a "bad attitude." So for the most part, the only attitude that is displayed is a "good" attitude. This is reinforcing to the manager's ego, which is a second cause of this misperception. After all, how could anyone resent what he or she is trying to accomplish? Is it not for the good of everyone, especially the subordinates? If they had a problem, why wouldn't they just say something?

We have surfaced a paradox of management naivety. The first is that those who are attempting to practice open and participative management are most apt to have the attitude discussed above. Or, put in more succinct negative terms: *Manipulators tend to be paranoid.* So when resentment rears its ugly head it just confirms the beliefs which they (the manipulators) had all along, which leads to further regression to Theory X. Those who are doing everything they can to support their subordinates, however, are likely to be blindsided by resentment, regardless of its cause. The solution is for managers to be neither naive nor paranoid but to face reality. Reality: Negative subordinate attitudes from past mismanagement have crossbred with the misperceptions of current management motives (assuming that they have improved) to produce a hidden harvest of resentment that can surface at any time. This might be directed toward management in general, but do not be surprised if it is toward you in particular. This should not make managers paranoid (or manipulative); it should prepare them for dealing with this serious problem. Resentment of past practices can provide a strong motivation for change, and thus it can be utilized for its positive attributes.

Much of the problem has arisen by a failure to deal effectively with exceptions. That is, in an effort to prevent direct confrontation with a person who is perceived to be misbehaving, general rules are established that must apply to all. The traditional management attitude that has evolved is: If one person can ruin the soup, prevent everyone (or all but one) from fiddling with the ingredients! Most bureaucratic rules are created as an outgrowth of this attitude, which places undue restraints on everyone in the organization. Although intended to prevent confrontation, this approach cannot help but build resentment.

If managers are ignorant of resentment, why would they condition their management strategy on the fact that such resentment might exist? This is an area of potential countercontrol, for the worst effect is to assume an attitude that does not exist—whether resentment or contentment. However, the solution to the puzzle is in the definition of the *cause* of the resentment in the first place.

Cause of resentment. We will now formulate a counterintuitive truism, which we will call the *resentment truism*, then demonstrate its obvious validity:

> Unless managers honestly believe that they are the cause of resentment, there is absolutely no way that they will be able to devise measures to eliminate it.

By contradiction, if the managers are not the cause of the resentment, then there is absolutely no way that they can eliminate it. On the other hand, if they are attempting to eliminate it, they must admit that had these same countermeasures been taken earlier, they would probably have worked. Otherwise, why try? Thus the absence of these countermeasures must be the cause. The more Theory X an organization is, the more management is perceived to be "in control," and thus the more they have no one to blame but themselves.

(We recognize that some resentment is totally unfounded. Management tendency is to conclude that *all* resentment is unfounded. The resentment truism does not argue that managers cause all resentment, only that if they do not recognize themselves as the *cause*, there will be no incentive or efforts to improve in this regard.)

The resentment truism applies doubly to the technical organization, since consistency of purpose is the rule of professionals rather than the exception. However, have you ever seen a situation in which management takes the blame for such resentment? Again, the ego of most managers just will not allow it. Thus we have the following most common scenario: (1) unknowingly management initiates actions, procedures, and a general style and approach toward decision making that suppresses rather than liberates the creative initiative of the subordinates; (2) a general resentment to this builds up on the part of the subordinates, but feedback is filtered to prevent this from being manifested to management except in some exceptional cases; (3) management interprets these cases as the underlying causes rather than symptoms of a flawed management system (e.g., Theory X orientation); (4) thus these exceptions are viewed by management as challenges to the system and insurrection against the general welfare of the organization, and they take action to eliminate the problem, usually by invoking rules; (5) the resentment builds to a point where some major precipitous consequence results, at which time it is too late to reconstruct the ideal situation; and (6) all of the above further confirms management's (false) belief that the subordinates really cannot be trusted. Hence traditional MBE unintentionally produces a default meta-control system that results in countercontrol. The solution requires that we break the link between mistrust and fear.

6.4 Toward a Solution

We stated above that in such an organization it is understandable that management will not want to negotiate the type of flexible MBNE guidelines that will lead to effective technical management. Are we then locked in a box? Unfortunately, in many traditional organizations the answer is probably yes. However, in technical organization a transformation from traditional MBE to MBNE is not just possible, it is essential. How can such a transformation be accomplished?

The answer is leadership, which we take up in subsequent chapters. To whet your appetite at this point, we can link effective leadership to the recognition of the principle of authority discussed in Chapter 5, and at the same time deal with fear. Just as sheep cannot be driven, neither can intelligent professionals be motivated by fear. While the sheep analogy is limited in application, the notion of "driving" based on fear is the primary reason that MBE will not work in its traditional top-down imposition. Let us summarize it with a simple statement which we will call the *principle of fear*:

> Fear within management is the sole cause of management by fear.

A fearful leader cannot help but manage by fear. The most effective leaders know no fear because *they are driven by principle, not by the assessment of personal consequences.*

If technical organizations are to be transformed to liberate the human element, it will only be by bold and fearless leadership which is not afraid to allow the people in the organizational components to function to the full extent that their decision-making abilities will allow. This means that they are not afraid of "losing control" or having to accept the consequences of their subordinates' actions. This means that they are not afraid of taking the blame when this happens; thus they are not afraid of their subordinates nor of their superiors.

Before leaving this subject, the responsibilities of the subordinates under technimanagement must be understood. There must be a subordinate perception that the organization is moving toward the distributed management paradigm which is inherent within technimanagement. They must be willing to be led. They must be willing to recognize those points (in time and function) at which decisions are better left to upper levels of management. They must be willing to negotiate these decision differentials, and they must be willing to grow and take the initiative. When attaining positions of management, subordinates must be willing to empower as they have expected to be empowered. Most important, they must be willing to take the responsibility for their decision making and not constantly blame their management for everything that goes wrong. This can be accomplished by instilling the same no-blame philosophy in subordinates as they assume more and more decision-making authority. This, in turn, is conditioned by acceptance of the idea of a growing and learning organization on the part of the informal leadership of the organization (which we address in subsequent chapters).

The process of moving an organization from standard MBE to MBNE is one that should take at least five years for most organizations and possibly ten for those that have been heavily characterized by top-down control. It is not something that is done in isolation but is, rather, an integral part of the establishment of the meta-control process. This transformation must begin with the acceptance of the principles of technimanagement, which includes the proper application of exception principles.

In conclusion, management by exception can be an excellent and powerful tool of technimanagement. However, it cannot be implemented top-down as in the traditional Theory X organization. The decision-point differentials must be negotiated, and they must be flexible, allowing for continuous growth and experimentation as new, uncharted water is encountered.

7

Management by Objectives

7.1 Definitions

Management by objectives (MBO) is a technique that was extremely popular in the 1980s but has now been overshadowed by other management fads. This is unfortunate because, like MBE, MBO has many positive features that should be incorporated into technical management. It also has a potential for abuse. However, before we get into the pros and cons, let us get a clear understanding of what we mean by MBO. Consider the following definition:

> Management by objectives (MBO) is a management technique that requires definitive written objectives for specific periods of time and assessments of their completion at the end of that time span.

Note that this definition can apply to organizational components as well as to individuals. The astute reader will recognize that this is nothing more than an application of our definition of control. The essential *goals* element is translated into definitive written objectives whose accomplishment can be clearly assessed. Thus it is important to recognize that the imposition of MBO is clearly an attempt to bring control to organizational activities.

The difference between MBO and "business as usual" is in the three detailed formal requirements: (1) written, clearly set objectives; (2) a time element; and (3) the necessity for assessment. Let us consider each of these manifestations of control in more detail.

Two terms will help us understand the written objectives requirement: milestones and deliverables. A *milestone* is a particular recognizable point (instant in time) in the life of a project. It can be identified clearly by the occurrence of an event, possibly the approval of a specification or the generation of a deliverable. A *deliverable*, on the other hand, is a product. It is tangible: usually a document, a functioning piece of software, or a prototype. This brings us to the first axiom of MBO:

> All projects should have their goals translated into objectives that can be allocated to the teams (and in turn to individuals) in such a way that each team's and person's milestones and deliverables are clearly understood.

This common "understanding" must exist on the part of both the managers and the subordinates. In technical organizations this is virtually impossible without some form of documentation.

The second requirement is a specification of time. Without such there is no way that goal accomplishment can be assessed. Even in the less formal (i.e., undocumented) approach, some time specification is understood. Without MBO a manager might get impatient, whereas a subordinate might feel that things are progressing quite well. After patience runs out we get the classical question: Don't you have that done yet? By specifying a definitive time, MBO creates an understanding between all parties as to the expectations.

Finally, there is the formal assessment. Under MBO everyone understands that such is in order, and the assessment itself is not seen as a special inquisition. Those being assessed generally are not as defensive as in the situation when these formal steps are not defined. This formal step may greatly increase the communication within an organization, since under other systems there may not be communication until there is confrontation. This formal requirement establishes a particular time for the assessment, and for this reason communication will take place one way or another at this time. Finally, the assessment can be used for the administration of rewards, and thus MBO tends to bring about a synthesis of control and reward theory.

7.2   MBO and Technimanagement

If all of this sounds great, it could be for one of two reasons: (1) the reader is still locked in a Theory X paradigm, or (2) the reader is viewing *control* and *rewards* consistently with the principles described in earlier chapters.

Sec. 7.2   MBO and Technimanagement   85

(Obviously, we hope that the second is true.) This dichotomy demonstrates that MBO, in and of itself, is neither good nor bad. If it is employed with the traditional manipulative meaning of control and rewards (as is predominant in most organizations today), it will bring about further discord. However, if the terms *control* and *rewards* are applied as defined for technimanagement, MBO can return considerable benefits. Let us consider the most common (i.e., the former) type of application first. Deming, the father of total quality management, has approached this in his typical "take no prisoners" style, as we see from the following quote:

> Management by fear would be a better name, as someone in Germany suggested. The effect is devastating.
>
> - It nourishes short-term performance, annihilates long-term planning, builds fear, demolishes teamwork; nourishes rivalry and politics.
>
> - It leaves people bitter, others despondent and dejected, some even depressed, unfit for work for weeks after receipt of rating, unable to comprehend why they are inferior. It is unfair, as it ascribes to the people in a group differences that may be caused totally by the system that they work in. (DEMIN85,8)

We can certainly see the validity of this comment when MBO is not applied properly. The problems inherent in sacrificing long-term goals for instant gratification provide a recurring subtheme of this book, as does the necessity to eliminate fear. Also, we can recognize the downside of assigning blame as another potential shortcoming of MBO. However, we have only touched on some of the sociological aspects of teamwork as opposed to rivalry and politics (more in later chapters).

Clearly, Deming is addressing the typical way that MBO has been applied in industry (not limited to technical organizations); however, all of his concerns are quite valid. But Deming's approaches to control (continuous progress toward improving quality) cannot avoid the setting of definitive objectives and an assessment of their attainments. So what's the difference?

Plenty! The key differences are much the same as those put forward in Chapter 6 between MBE and MBNE (management by negotiated exception). The objective is to enable MBO to be consistent with the principles of technimanagement. Figure 7.1 summarizes the differences between the traditional and technimanagement approaches toward MBO. These substitutions transform MBO from a dictatorial system to one of distributed decision making. The primary objective of this transformation is to free all employees to use all of their creative abilities.

APPROACHES TOWARD MBO IMPLEMENTATION	
Traditional	**Technimanagement**
Management defines the objectives	Objectives are negotiated, and they are determined mostly by the individuals who will use them to guide their activities
Assessment for discipline and reward purposes	Assessment is performed by the team and only used for team and individual improvement
Concentration is upon short-term objectives	Short and long-term objectives are defined in terms of the short-term (milestone) components in the attainment of a strategic plan
MBO objectives and metrics are generally restricted to individuals	The objectives are developed for and by the team, as are the metrics to gauge their accomplishment

Figure 7.1 Traditional versus Technimanagement MBO

The modifications given in Figure 7.1 counter all of Deming's objections to MBO, since they are lodged against the traditional management applications, not those guided by the principles of technimanagement. This can be seen by going through Deming's criticism systematically to see if there is a countermeasure qualifier to each item in Figure 7.1. However, even in its modified version, MBO requires justification. This is due to the time-consuming thought process and documentation that must accompany it. In the form in which we have proposed that it be applied, we believe that even Deming would agree that this is nothing but good planning, and it is an integral part of establishing the team and personal control processes. But before accepting it outright, let us explore further the complications anticipated when the transition is initiated from traditional MBO to its improved form.

7.3 MBO Complications

The immediate knee-jerk reaction that is expected when the new MBO approach is presented to the traditional manager is that it will result in a chaotic situation that cannot be controlled. We have already answered this contention in two ways: (1) top-down definition of objectives, even if workable, would not be wise in technical organizations since management's knowledge of systems design and development cannot hope to be a fraction of the collective knowledge of their subordinates; and (2) in reality, the manager does not control even in the traditional MBO approach. Although Theory X gives the perception of control, its long-term effects speak for themselves.

It is recognized that the subordinate's setting of objectives is not without problems. The preponderance of studies and the author's own experience have shown that the weakness which professionals have is in setting their goals too high rather than too low. Computer programmers, for example, are notorious for underestimating their times to delivery. They never seem to learn that just because something is possible does not mean that it is accomplishable in a short amount of time. Historical standards, if they apply to the proposed design or development efforts, should certainly be used for estimation to assist in this regard. However, they rarely, if ever, meet the needs exactly. The very nature of the technological organization necessitates that every new project break new organizational ground. Thus there is an unknown component that can only be handled by experience. So the background of the employees and managers involved must be applied to establish realistic objectives. However, for the most part this will be in the direction of keeping them from being too optimistic.

The fact that even the manager might not be able to keep the team from setting unrealistic objectives is an additional reason for not holding them strictly to their original estimates. A second one is that additional unanticipated time might be required to make a quality improvement in their product. Forcing time constraints cannot help but compromise quality, which has the potential of costing several times the investment in the future.

When an objective is not met, it should serve not only to guide the team in their work activities but also to guide them in the future setting of more realistic time objectives. However, any managerially imposed negative reward of not meeting a goal will only motivate the team to push for longer and longer time goals so that they will be rewarded for their attainment in the future.

Another objection to MBO under technimanagement, which we cannot avoid, is the danger that it will be misused by management toward regression to a carrot-and-stick approach toward motivation. This can be said for any metric, but clearly metrics are essential to control. Although quite valid, this is not an adequate response to the objection. MBO, as well as all other techniques that require measurements against objectives, must be accompanied by an understanding on the part of both the employees and their management that *the intent is continuous improvement, not blame assignment*. There is no substitute for trust, and if it does not exist from the outset, little can be accomplished by either the initiation or the abolishment of any technique.

As a final caveat, we should assume that this revised implementation of MBO will not please everyone. The natural reaction is to think that the group most uncomfortable with this will be management. Initially, this will probably be the case, since those who are not secure in either themselves or the principles of technimanagement will probably view this approach as a loss of control. Nevertheless, if they will give it a try, amazing things will start to happen. Suddenly they will find that their jobs are much easier and that a great burden of weight has been removed from their backs since productivity increases. On the other hand, the planning and estimating functions have been transferred to the teams, *increasing individual work requirements*. In the intermediate term the team members will feel somewhat burdened by this new responsibility. After all, isn't this what you are getting paid the big bucks to do?

The answer to that question is no. You are getting paid to assure that the organization runs in its most efficient way. This is not an easy thing to communicate, however. This is a function of the current organization structure, its culture, your management, the skills of your subordinates, and most important, their maturity in self-management. If this sounds complicated, it is because it is complicated. If a computer could resolve all of these variables and determine the optimal extent to which these decisions should be shared, it, not you, would be managing your operation.

In summary, in Chapters 6 and 7 we have alluded to a notion of transition which will not be fully detailed until our concluding chapter. However, it is clear that the major changes required from the traditional applications of MBE and MBO will not occur optimally overnight. We can further generalize this to apply to most traditional management techniques, which we will call the *principle of transition*:

> Most traditional management techniques apply to technical organizations when modified to conform to the principles of technimanagement. A period of transition and maturity is required to bring about this modification. During this transition, the favorable characteristics of the technique should be retained and emphasized, and the unfavorable aspects should gradually be deemphasized and ultimately eliminated.

Important to this principle is the idea that while the organization is going through a complete reorientation toward distributed management, it does not have to eliminate all techniques that have served it in the past. Rather, most of these techniques can be adapted to the philosophy of technimanagement in a straightforward evolutionary manner.

8

Optimization, Equifinality, and Systems View

8.1 Definition of The Problem

We need to pause at this point in our search for wisdom and interrupt our discussion of management to consider some more general theoretical concepts. Recall that when we introduced some of the very basic concepts on which technimanagement is based, we found that they did not originate within management science. As examples, the elements of control apply to any system that might be subject to control, and the theory of rewards is a basic tenet of psychology. The fact that we utilize principles derived outside the study of management is not important. What is essential is that we base technimanagement on established principles that are universally true, as opposed to fads, cults, anecdotes, or personalities.

In this chapter we introduce the three closely related concepts given in the chapter title. This is necessary because we live in a time when very simplistic thinking is propagated by one of the greatest influences on our reasoning habits: the conditioning produced by our mass media, television (TV) in particular. Depending on the saturation (time consumed by TV/movies), the extent of conditioning can be almost to the point of brainwashing. The concepts in this chapter are not at all difficult to understand. What is difficult to understand is that this influence is grossly underestimated by most of the general public, *which is the population pool from which all of our managers are taken*.

The symptoms of simplistic thinking include, but are not limited to, the following: (1) hasty generalization, what we will call *proof by anecdote*; (2) binary thinking; (3) acceptance of propositions without proof; and (4) tunnel vision. It would serve us well to define these errors in logic before we present principles for their extinction. Instead of a textbook definition, let us use the medium that propagates these logical flaws to provide examples so that we can gain an appreciation for a major component of the cause itself.

Consider the following scenario, variations of which play themselves out dozens of times a day before millions. A talk-show hostess/host interviews a social scientist who has just conducted a very detailed study demonstrating the increased sense of security attained by children in two-parent homes as opposed to single-parent homes. After this somewhat scientific discourse, a member of the audience states that her uncle Joe raised 10 children and they turned out a whole lot better than most two-parent children that she has known. In fact, one of them became president of a major corporation. The audience applauds vigorously, obviously accepting the conclusion: Single-parent-raised children are just as secure as two-parent children. Consider the following logical flaws made:

1. *Hasty generalization (proof by anecdote).* The member of the audience is given equal credibility with the expert guest. Of course, being an "expert" does not make one incapable of error, but neither should one anecdote even be compared to the results of several scientifically controlled studies.

2. *Binary thinking.* The correct conclusion was not drawn from the anecdote (i.e., that there is at least one special case in which a single parent did as well, if not better, than many two-parent families). Instead, the social scientist's conclusions are either accepted or rejected wholesale. In fact, nothing that the social scientist said was contradicted by the anecdote, and he even stated that there were exceptions to his general conclusions. However, the audience reaction from this point forward was more than negative—it was generally hostile. This shows a laziness of thinking that is the first step toward mob psychology (very few people failed to applaud the anecdote).

3. *Acceptance of premises without proof.* No one even questions whether there is an uncle Joe, if he was an "average" guy, or if the entire story was not just made up. Of course, we all probably have an "uncle Joe" for whom things turned out just fine. But how many anecdotes are there on the other side of the picture, and why are they not presented? Even when they are, where does this leave us? The battle of anecdotes only demonstrates the inability to formulate realistic opinions based on a unified set of underlying principles.

4. *Tunnel vision.* This is the flaw of too much focus; we can think of it as locking in on a narrow cause—effect mechanism (also called "not being able to see the forest for the trees"). In this case the rest of the picture includes all of the other factors that influence the emotional and mental health of children: (a) the child's peer group: friends, cousins, brothers, and sisters; (b) the influence of other adult relatives: grandparents, aunts, uncles, etc.; (c) the general home environment: the presence/absence of alcohol, tobacco, other drugs, TV, movies, spouse/child abuse, etc.; (d) the time and activities that are shared by parent(s)—we could go on, but hopefully you get the picture. A related problem is also noted when two or three alternatives are proposed for a decision. Somehow we seem to "lock in" on these alternatives and decide between them rather than enlarging our purview toward formulating others.

The anecdote given above was used to exemplify our impression of the common nature of these logical flaws. This anecdote does not illustrate with what frequency these types of demonstrations occur, nor do we believe that this is at all an extreme example. (If you have not already, we invite you to turn on your TV to find this out for yourself. It is our experience that such abuses of logic are the rule rather than the exception in our pop culture, which is so highly influential on many, especially the young. The examples in the afternoon talk shows are particularly bizarre.)

In addition, there are the half-hour sit-coms, which generally formulate a problem in 15 minutes and then concoct a simplistic solution in the other 10. Spliced into these interest grabbers are commercials which essentially do the same thing but in 15 or 30 seconds. They present a need (problem) followed by an airtight solution. Sporting events and movies can also contribute to this fantasy world if not kept in their proper perspective (see the display below).

> **What About Sports?**
>
> - *Our sports culture.* The number of times that management principles are illustrated with examples from sports establishes our case. Even the key land battle of the Gulf War was explained to the press and public using the analogy of a "Hail Mary" football play. We agree that there are any number of excellent sports analogies that can and should be made, especially in the areas of leadership. But it is important to realize where these break down—life is just not as simple as any game!

> **What About Sports? (Continued)**
>
> - *Reinforcement of simplistic solutions.* The conditioned belief that everything can be resolved in a couple of hours of visible effort does not reflect reality. The players realize this, but does the audience?
>
> - *The win-lose mentality.* In all sports, winning requires that someone else lose. This can create a distorted view of how life and business ought to work. When we begin to derive more pleasure from others' losses than from our own successes, something is really sick!
>
> - *The importance of being "number one."* The greatest problem is a failure (unwillingness) to separate perception from reality. Reality: The number one team is rarely the definitive number one. To quote a much wiser source: "I returned, and saw under the sun, that the race is not to the swift, nor the battle to the strong, neither yet bread to the wise, nor yet riches to men of understanding, nor yet favor to men of skill, but time and chance happeneth to them all"—Solomon. Yet who cares?—being perceived to be number one has become much more important than doing your best.
>
> - *Qualifier.* There is an optimal level of emphasis on sports that is healthy, entertaining, economically stimulating, and building of character to the athletes and spectators alike (see Section 8.2). Clearly, we believe that a subculture of the American public has gone far beyond this optimum to the point where sports has become their predominant religion. Based on some recent incidents at international soccer matches, Americans are not unique in this regard.

Any moderate exposure to these seemingly innocent forms of entertainment will certainly not cause a distorted view of reality. However, social scientists tell us that the vast majority of our population has been bombarded with literally hundreds of thousands of repetitions of these treatments during their formative years. Is it any wonder that most will not sit still for a meaningful, orderly prolonged debate to resolve the many complicated problems that our government and industry face? Instead, we base our most important political decisions on sound-bites and 30-second commercials. Can this influence be ignored in shaping the managers of tomorrow? To neglect its effect on our ability to resolve important decisions is to turn our backs on reality.

This problem manifests itself further by the virtual absence in academic and business settings of discussions of the decision-making process itself. It is virtually impossible to engage in such philosophical discussions, mainly because of the anecdotal-based reasoning that seems to arise immediately. Discussions regarding meta-decision making (i.e., the discussion of the appropriate thought processes to support effective decision making), are usually short-circuited by the tendency to focus quickly on (and criticize) the *subject* of examples presented rather than the *principles* they illustrate. This is an example of tunnel vision caused by the tendency to jump to a proof by anecdote.

This author could be accused of the same simplification in the example given above. However, recognize that it was not our intention to make any statement at all with regard to the single-parent issue. Rather, it was used to illustrate a principle with regard to decision making and the formulation of opinions. Focus on the validity of the opinion expressed sidetracks the purpose of our present discussion. We could equally well have had the expert on the other side; or we could have used any other of the multitude of subjects under discussion in our society today. We will allow the reader to judge whether this is a hasty generalization or an attempt to exemplify a general principle.

We recognize that many of our readers (especially those with the patience to bear with us to this point) are immune to the subtle influences of our popular entertainment. Perhaps you never watch TV, or perhaps you restrict your watching to particularly edifying segments only. Perhaps you are of such a turn of mind that these things would not have an influence on you in any event. If so, you will not be nearly as subject to these influences. (On the downside, however, neither will you be as inclined to understand those who are.) Regardless of your inclinations, however, in establishing the underlying causes for our management problems, the large influence that such repetitive reinforcement makes on the vast majority of the minds in our society cannot be ignored.

It would certainly be a hasty generalization on our part to conclude that *all* of these types of errors are brought about by the mass media, particularly TV. We are not claiming this. On the contrary, *the cause of the mass media is the culture of the civilization that brought it into existence and continues*

Sec. 8.2   Optimization   95

to support it. (This validates our point!) Certainly, a major share of the blame must go to the failure of our families, churches, and school systems to shape our culture and discipline our people toward more intelligent approaches to decision making. In the spirit of technimanagement, we all share the blame for not taking advantage of direct opportunities that we have to stamp out this ignorance or to provide adequate leadership and influence on those who do have such opportunities. Hopefully, the concepts presented in this chapter will not only be helpful in this regard but will provide terminology and principles that we can use to define additional principles of technimanagement.

8.2   Optimization

We have used the word *optimization* (and derivations of it) above, assuming an intuitive feeling for its meaning; at this point we wish to give it a more formal definition. The term *optimization* packages an extremely useful set of principles, most of which are neither properly understood nor applied by managers in general. As a class, technical managers probably have fewer problems than others have with these concepts, mainly because they are covered in many technical areas which include traditional operations research techniques. However, the unfortunate abuse of this word in the computer industry (e.g., "optimizing" compilers) and possibly elsewhere has tended to confuse the true meaning of the word with that of mere improvement.

With this introduction, allow us to propose the following working definition:

> Optimization is the process by which trade-offs are made between various decision parameters with a view toward obtaining the overall best possible solution.

Mathematically, the best possible solution is usually expressed in terms of an *objective function* (e.g., cost, benefit, profit, etc.) which we try to either minimize or maximize. We will not refer to the objective function as being optimal—it is either minimized or maximized. It is the solution (i.e., the decision parameters that are to be implemented) that must be optimized. Thus the set of these parameters that yields the best value of the objective function is, by definition, the *optimum solution*.

Stampede To The Extremes

> Extreme positions usually lead to extremely bad decisions.

Examples
- Freeze on all hiring
- Pressure to "fix" the site of a recent fatal accident which does not recognize the possibility that the funds consumed might disable several other more critical, life-saving projects
- Mandatory jail sentences for all drunk driving fills all jail space, forcing judges not to convict and consuming space that should be allocated to major felons

These approaches seek a *silver bullet* solution, not recognizing the damage that firing the bullet does as it ricochets off the internal organs of the organization.

Deceitfully Subtle Extreme Positions

Example: choice of whether to use a structured design technique or rapid prototyping in software development project

The *binary approach* would be to evaluate the costs and benefits of each and choose the one that would seem to return the best results for the organization and the project under consideration. The *optimization approach* is to look toward obtaining the best aspects of both approaches, possibly by rigidly designing those components that are based on established technology while applying rapid prototyping to those that require considerable innovation. In other words, it involves first a recognition that there is a continuum, and then an attempt must be made to determine the point along this continuum that will produce the maximum benefit to the organization.

There is an old self-proving truism: *Too much of anything is not good.* We cannot argue with this statement, for if the excess of anything were good, it would not be "too much." Most management decisions are not resolved by extreme solutions, or if so, the solution itself is probably quite trivial. However, there seems to be a natural human tendency to be *stampeded to the extremes*, as discussed in the adjacent display. This is because the acceptance of an extreme tends to require much less intensive thought than does the determination of an optimal solution. This brings attention to the following trap:

> To every extremely difficult, complex problem there is a quick, simple, easy-to-implement, inexpensive, wrong solution.

Sec. 8.2   Optimization

This is our paraphrase of the original statement by H. L. Mencken: "For every problem there is one solution which is simple, neat and easy to understand. The solution is wrong." The fact that this is so often quoted, and in so many variations, is evidence of recognition that we tend toward the simplistic—and the erroneous. Most optimization problems are extremely complex. In fact, without making the assumptions required to establish a mathematical model, these problems are not solvable. Even with these assumptions, complex algorithms are usually required to determine optimal solutions.

The difficulty in obtaining the exactly optimal solution is eclipsed as a problem, however, if managers do not even recognize that certain problems cannot be solved without optimization. For example, consider the determination of the point in time to stop designing and start developing software. In the extremes, we would either not design at all or would spend forever designing and never begin coding. This is obviously an optimization problem, where the objective is to deliver a product of definitive quality in the most efficient manner.

Theoretically, this problem can be modeled by plotting the various costs (Y-axis) against the time spent in design (X-axis), as shown in Figure 8.1. Assuming a constant design team size, the cost of the design component rises linearly over time. On the other hand, the cost of development is quite complex. With no design, the development cost will be extremely high. This cost comes down quite steeply over time, but at some point the marginal benefit of continuing to refine the design becomes negligible. The total cost is the sum of costs represented by these two curves, which can be plotted as *total cost*. The theoretical optimal point to stop designing and start development is the time point corresponding to the minimum point on the total cost curve.

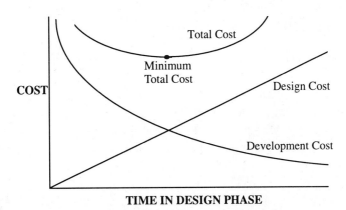

Figure 8.1   Optimal Point to Initiate Development

The data to apply this model are rarely available to an organization. Even if past data are available to measure costs of design, the effect on development cost of varying times in these design efforts would be almost impossible to obtain; and even if it were obtained in the past, it might not apply to new efforts in the future, given changes in tools, languages, compilers, experience, and so on. In addition, those experienced with software design and development recognize that there is no single golden time when all design ceases and all development begins. Rather, the design for certain components is completed and personnel might be assigned to other design, development, or possibly rapid prototyping activities (which might be a cross between the two). In a healthy organization this series of decisions would be made collectively by people distributed throughout the organization.

So, should we abandon the concept of optimization? If the argument above leads someone to this conclusion, this is the very thing that we are complaining about in criticizing the example rather than learning from it. The example was not given to provide a method for determining the optimal point. It was given to illustrate the concept of optimization itself, for it is only by understanding the concept that we can hope to move toward improved (if not perfect) solutions. This leads to the basic principle that we are trying to communicate:

> Managers who understand the concepts of optimization are much better equipped to make judgments than are those who do not, even if it is impossible to obtain perfect optimality.

Taking this one step further, we might say: especially if the data are not available to find the exact optimal point. Just as the manager who will not admit faults cannot hope to overcome them, ignorance of the concepts of optimization stands in the way of improved decision making.

The degree of fuzzy thinking with regard to optimization is seen by many common statements made by managers. For example, "we want to deliver the best possible product at the lowest possible cost." Is this just semantics, or do managers believe that it is possible to deliver the best possible product (i.e., that which is superior in quality in all respects) at the lowest possible cost? (We recognize that some *quality* proponents believe that there is "no cost for increased quality," but clearly this must be within some reasonable range at which optimization can occur, for it is clearly possible to continue improving a product forever and never release it for sale.) Correct statements that show proper understanding of optimization concepts include: (1) "we want to deliver the best possible product within our budgetary constraints," and (2) "we want to deliver a product of superior quality at the lowest possible cost."

Advocates within society rarely demonstrate a knowledge of optimization concepts. We observe this behavior commonly in most fields that have their advocates, such as traffic safety, the environment, health care, and gun control. We often hear: "If project X saves just one life, it will all be worth it." Perhaps so in an absolute sense. But if we are talking about the issue of allocating fixed resources, the allocation of resources to project X might actually kill people if more lives would have been saved by allocating those resources to project Y. Notice that we are not arguing pro or anti anything. Proponents on all sides of issues generally fail to take an optimal view, which generally falls somewhere between the advocated extremes. Politicians holding this complex middle ground rarely last very long, which indicates the depth of the problem in our society. However, managers who fail to hold the optimal middle ground will create far more problems for their organizations than they will ever solve.

We anticipate confusion between the terms *middle ground* and *compromise*. There is only one optimal point, and if a manager can resolve those decisions that get closer to it, compromise is not in order. Generally, the optimal point will be found to lie somewhere between the extreme points, in what we have called the "middle ground." The acceptance of a solution in the middle ground is a compromise only if your original position was at one at the extremes. Managers who understand optimization will hold these extreme positions only when they feel that they are optimal (which is possible in some special cases). Usually, they will start with a position that attempts to reasonably balance the extremes and then improve this view as more information is forthcoming (primarily from their subordinates). This is not compromise; it is a unique and defensible position.

As a final summary point, it should be recognized that misperceptions of reality can never produce optimal decisions (except purely by chance). The entire point of this section is to emphasize the necessity for an accurate perception of reality. That perception must include a realization that optimal results are generally not attained by extreme or simplistic decisions. But it must go beyond that realization to include some rudimentary knowledge of system structure and the parameters by which a theoretical optimal solution might be attained. In most cases this comes by concerted and intensive observation—experience. We hope that those readers who were not familiar with optimization will view decisions in a new light. Then when difficult decisions arise, they will identify them as optimization problems and proceed to consider the balance of decision parameters that this process requires.

8.3 Equifinality

While we stated that there is only one optimal point, there may be many ways to arrive at it. According to Kast: "The concept of *equifinality* says that [the same] final results may be achieved with different initial conditions and in different ways" (KAST70,128). This principle should be accepted without proof, since it is intuitively obvious in many areas. There may be two different routes for getting to a location, but the time and mileage are not measurably different. The key word is *measurably*. In reality, no two ways for achieving an end are perfectly identical. One has to be better in terms of cost or time. However, since optimization is such an imprecise art, in large complex problems there are generally many alternatives that produce effectively the same results. The splitting of hairs between them might be quite counterproductive.

The importance of equifinality is the realization of its existence and the broadened viewpoint that it gives to the optimization perspective. This will be especially important when we broach the subject of self-directed work teams in Chapter 15. Frequently, the fact that one of these teams works out its own methods for accomplishing optimal results is much more important than the method itself. From a motivational standpoint, an internally developed less-than-optimal method could prove to be superior to one that is theoretically superior but externally developed. This is not stated to promote any type of psychological manipulation of work teams. It is stated in realization that the freedom to control one's own actions is a most valued asset. More important is the vested interest in making a method work when it originated directly from the individuals involved. Finally, the work team can tailor the method to fit needs that are fully understood only by the team members themselves. When these issues are addressed honestly and openly, manipulation is avoided, and the wise manager will allow these teams latitude even when making deviations from the manager's opinion (catastrophic deviations are the exception, of course).

To summarize, the concept of optimization is essential to intelligent decision making. However, it must be tempered by a recognition that rarely can we be perfectly sure that a given set of decisions is truly optimal. This being the case, there may be several alternatives that will produce effects which are practically indistinguishable from those that are theoretically optimal. In these cases, sociological considerations should become the predominant basis for managerial decisions.

8.4 Systems View

By definition, *systems* are made up of highly interactive and interdependent components. Within complex systems, many of these components might be treated as a system within itself. Since the boundaries of any system can be extended or contracted, it is essential to effective communication that we define the system under consideration by identifying the set of components that compose it. The term *systems view* has generally been applied to the notion that the application of optimization within each component will generally not produce optimization at the full system level. Hence the exhortation: "We must assume a systems view." Since this tends to be counterintuitive, especially at middle-management levels, these concepts bear further discussion.

The major premise is based on *Bellman's principle of optimality* (BELLM57), from which considerable advancements in mathematical optimization algorithms have been generated. The principle, which applies to all stagewise decision processes, can be paraphrased as follows:

> No system can be optimized when one or more of its components are operating at levels which are not optimized with respect to the system.

The last phrase is most significant. Effectively, this states that it is impossible to recover from a component within the system that is not being optimized from a systems view even though it might be optimized within. For example, if members of a sales component put all their efforts into customer contacts and none into assisting with product development, their sales objectives (short term) might be maximized, but optimal results for the organization would not be produced. The systems view strives for *harmony* throughout the system. Harmony implies considerable coordination to assure that all components are operating consistently with the goals of the system.

Although all of this might seem to be intuitively obvious, it comes apart when department heads (component) are rewarded for the performance of their respective departments in isolation from the performance of other departments (components). Similarly, if we view each person within a department as a component of the "system," the counterproductive nature of competition between them becomes quite clear. Theoretically, the problem is one of suboptimization—lack of a systems view.

Thus harmony implies sacrifice on the part of (most or all of) the components to attain full system optimization. Rarely, if ever, can a system operate at an optimal level if each of its components is operating at an optimum level with respect to itself. However, the way that organizational rewards are typically administered tends to bring about these local optimizations. The result is a series of functional conflicts that tend to bring about further system degradation.

It is essential that the manager understand that a systems view is required to achieve systems optimization. However, this only defines the problem, it does not solve it. Kast presents the ramifications of the systems view as follows: "The *systems view* suggests that management faces situations which are dynamic, inherently uncertain, and frequently ambiguous. Management is not in full control of all the factors of production as suggested by traditional theory. It is strongly restrained by many environmental and internal forces. The technical system, the psycho-social system, and the environmental system all constrain the management system" (KAST70,134). Is it any wonder that we strive so often for local component optimization? As complicated as it is to optimize a component (e.g., a department, or even a person), the problems involved in system optimization are an order of magnitude harder. Indeed, management is like a large, complex, unstructured software system. Correcting errors in one module is as likely to introduce additional (albeit different) problems as it is to solve the original problem.

The solution begins by recognizing the problem source. If these problems are so hard, perhaps the solution lies in distributing decision making to those who understand the problems best. This reduces the problem to bring about, by proper rewarding, a systems view on the part of every person in the organization. Convincing them that self-sacrifice is in their own best interests is the difficult part of this chore, and it requires a complete revamp of the traditional management system. The heart of this involves a recognition of the informal organizations (Chapter 12). However, before getting into this, three other subjects require our consideration in the next three chapters.

9

Parkinson's Law

9.1 General Applicability

Parkinson (PARKI57) observed one of the oldest bureaucracies in the world, that of Great Britain, and came to several interesting conclusions. The one that is quoted most often in management literature involves the utilization of time. We paraphrase it as follows:

> The amount of time that it takes to perform a task expands to fill the time available to do it.

The actual quote is: "Work expands so as to fill the time available for its completion" (PARKI57,2). (We will allow the reader to ponder over whether it is the "work" or merely the "amount of time to perform the work" that expands.) We dedicate this chapter to exploring the validity of this law, its application within technimanagement, and the ramifications of these potential applications.

Managers are concerned with two sets of time allocations: allocations of their own time and allocations made by their subordinates. Since it is practically impossible for technical managers to dictate exactly how their professional subordinates allocate their time, this can be influenced only indirectly. In our discussion we try to keep the personal and subordinate applications separate, since there are significant differences between them and their applications.

To a large extent Parkinson's law has become fairly accepted management lore, implicitly if not explicitly. To the degree that managers perceive it to be true, we should expect them to attempt to set fairly rigid time deadlines in their efforts to increase productivity. After all, if the work time is going to expand, why not keep the volume into which it can expand to a minimum? We are concerned with this attitude because of its potential demotivating effect when applied to implementing management by (negotiated) objectives. Recognize that time constraints placed on the objectives are as important as the definition of the objectives themselves. Thus the validity of Parkinson's law will strongly affect management attitudes toward negotiating objectives with their subordinates and work teams.

If Parkinson's law were completely fallacious, it would not have been given so much credibility. We surmise that like so many other management maxims, Parkinson's law is widely accepted because most managers have experienced its validity in their own work assignments. So we must begin with the assumption that there is at least a kernel of truth to it, even though it is difficult to prove (or disprove) experimentally.

Few rules are always true, and it only takes one counterexample to demonstrate that it has exceptions. We can all probably cite an example of a project that was unexpectedly, but successfully, completed early. This would clearly violate Parkinson's law. Thus it is clear that Parkinson's law is, if anything, only of general applicability. It would serve us well to define just when it is applicable. We submit the following qualifier for consideration:

> Parkinson's law generally holds for tasks generated by the bureaucracy; it very rarely holds for tasks that involve creative team efforts which occur independent of the bureaucracy.

Before attempting to validate this qualifying principle, let us be sure that we understand the terminology thoroughly. The words *generally* and *rarely* indicate that we recognize exceptions to our qualifying rule; however, we contend (admittedly without hard data) that the qualifier holds over 90 percent of the time, and therefore it is quite useful in practice. The phrase *tasks generated by the bureaucracy* refers to tasks that do not contribute directly to the accomplishment of the organization's goals, and possibly they do not even contribute indirectly to the organization's goals (they may even detract from their accomplishment). Examples of bureaucratic requirements include the completion of annual reports, verification of time worked, and many accounting requirements. *Tasks that occur independent of the bureaucracy* refers to activities to accomplish the goals of the organization which are performed independent of bureaucratic intervention. Finally, the *team* aspect indicates that the project is too large or too complicated for a single person to complete within the project time requirement, which is characteristic of most projects in technical organizations.

There are some tasks that we have not considered because they do not fall in one of these extreme classifications. For example, there are largely individual tasks that do not require a large amount of creativity and may also have some partial bureaucratic content. To these we have not given an opinion, and we might conclude that Parkinson's law might hold half of the time (in which case it is of no value in assisting our time allocations).

In technical organizations, the lower the level of management, the more creative the activity that is managed. Conversely, the higher up the organization that we go, the more activity that we have defined as bureaucratic. Thus, as compared to their subordinates, a much larger proportion of the tasks that technical managers are called on to do themselves fall into the category of "tasks generated by the bureaucracy." We must therefore conclude that while many of the tasks that the technical manager will perform fall within the purview of Parkinson's law, most tasks which they are managing do not.

The first bit of evidence to support our qualifying rule is Parkinson himself, who was studying a bureaucracy and not a technical organization. His observations have been and must be given credibility, but only in the context within which they were derived. The second argument comes from the nature of bureaucratic efforts: (1) they are driven largely by an attempt to control individuals within the organization as opposed to accomplishing the ultimate goals of the organization by stimulating creativity, (2) they are often open-ended tasks which can be completed to varying degrees of development, and (3) there is a good chance that the results will never be used. The second of these factors implies that the time spent on one of these tasks might range from as much as an entire career to as little as zero (skip it altogether, or possibly assign it to nonprofessional staff).

Collectively, these factors produce an attitude toward work which is completely different from that which professionals typically have toward their creative activities. There is a tendency to put these types of things off until they are pressing and then to work on them as long as time allows. (We commend such an attitude!) However, there are those who do not recognize the higher priority that should be given to nonbureaucratic functions, such as mentoring and leadership. They allow these relatively nonproductive tasks to expand in a variety of ways, such as (1) striving for perfection in communications with the bureaucracy, (2) not trusting nontechnical subordinates with these tasks, and (3) refusing to fight against ridiculous requirements (at key times to the point of just refusing to do them), and so on. Ultimately, these obsessions take over, and Parkinson's law prevails—they soon find that they have no time left over for the important things. What is worse, those who seem to have time for nothing other than bureaucratic tasks soon begin to view the endless flow of virtually meaningless paperwork with a "why bother?" attitude. After all, if you finish one job, there are just a dozen more in the stack. Ultimately, the endless flow of tasks become largely deadline driven, apparently reinforcing the validity of Parkinson's law.

We hope that these qualifiers have put Parkinson's law in a place where it can be useful in assisting time allocation. In the next section we consider one more aspect which helps understand the true nature of time requirement expansion and compression.

9.2   Compression Corollary

If your technical subordinates are spared from all but the totally essential administrative chores, they will be spending at least 95 percent of their time on productive design and development activities. These generally do not fall under Parkinson's law, for reasons that are virtually the opposite of the reasons that bureaucratic activities obey it. We say "generally" because this is largely a function of the maturity of the work environment, which is determined by the degree to which the transition has progressed to technimanagement. If, in fact, you notice that Parkinson's law does apply, your subordinates are treating their work like bureaucratic chores. Needless to say, this is a very unhealthy situation, and immediate steps need to be taken to apply the principles of technimanagement to set a healing process in motion.

The effective and enthusiastic use of time provides a potential measurement element for a control system that might be applied in moving the organization toward greater productivity. However, this is rarely, if ever, realized. In fact, the opposite is too often true. Managers will take a perfectly good team of technical people and impose time constraints on them, if not based on a misuse of Parkinson's law, then for reasons that appeal intuitively to them. The counterproductive nature of this type of control was discussed in some detail above when management by objectives was discussed in Chapter 7, and we will not cover that ground again. It is sufficient to summarize here that self-directed and self-motivated work teams do not need to have the imposition of external time standards any more than they need the imposition of objectives. However, this does not mean that they will have no concept of time or that they will not have a work schedule to which they will compare their activities on a continuous basis. But they will be doing it, and collecting data on their activities, for purposes of self-improvement as opposed to management control. (The former is an exciting experiment, the latter a bureaucratic requirement.) At the same time, the schedules are determined by the joint cooperative efforts of managers and their subordinates.

Sec. 9.2　Compression Corollary

The creative activity that is required in design and development work cannot be performed like administrative or bureaucratic work, no matter how much we proceduralize these processes with methodology. They consist of a series of mental processes, most often requiring what are commonly called *breakthroughs*. One of the worst deterrents to the accomplishment of breakthrough reasoning is time pressure. For once time becomes a central consideration, it distracts attention from the major project goals. The quickest and most dramatic breakthroughs will occur when creative minds are free to concentrate fully on the job at hand. No matter how tempting it is, time pressure can only disrupt this process. The management of individuals who are responsible for creative work is not the same as that of those who perform fairly routine activities, and the applicability of Parkinson's law is one of the major differences.

Most often creative activity is inadvertently interrupted, not by management-imposed time pressure, but by the demands of another project which might have been unanticipated. This can be misunderstood as a valid demonstration of Parkinson's law. Of course, it is not, since the work is not expanding to fill the time; rather, it is being compressed into the time available. The difference is in the cause–effect mechanism. To clarify this, let us offer what we will call the *compression corollary* to Parkinson's law:

> Work is compressed into the time allowed for it.

This idea holds for almost all creative activity, even when it is not interrupted. The reason for this is that we can never say that a project is definitively concluded. The decision as to the points in time to make the transitions from design to development and from development to release to the public should be based on complex optimization considerations (see Chapter 8). We would always like to have more time, and given more time, generally a better product could be designed or developed. But reality dictates that these processes be terminated at a certain point. At that target point we "package" the product and essentially compress our efforts to "get it out."

The compression of activities can certainly have a debilitating effect on an organization. Or, it can be used for further motivation, provided that it is due to legitimate circumstances and not due to some manipulation by management. We recall and paraphrase a favorite saying attributed to Vince Lombardi, the legendary coach of the Green Bay Packers in their world championship days: "We never lost a game, we just ran out of time." Time constraints are a reality in the real world as they are in the simplified world of football. The design and development organization (system) must exist in and interact with the broader environment that supports it. The optimization process through which the timing decisions are made for various design and development projects must consider data obtained from this environment as well as those from the organization itself. (And there is an optimal degree of the use of sports analogies as well.)

This is the essence of effective use of the compression corollary. It recognizes that we do not live in an idealistic world where time exists to generate a product of perfect quality and total concurrent satisfaction of all our customers. Rather, time decisions must be viewed as an optimization problem in which balances must be made simultaneously among time, cost, quality, customer satisfaction, alternative products, additional features, and a whole host of internal organizational factors.

9.3   Applications to Time Management

The technical manager's job is subdivided into two types of activities that compete for the manager's time: (1) those required by the bureaucracy, and (2) those that support the team being managed. In the short term the first of these gets the greatest visibility, and possibly rewards, by your management. As discussed above, some managers get so involved in it that it consumes virtually all of their time (they ultimately evolve into bureaucrats). This is unfortunate, for if priority is given to bureaucratic activities, the "real work" (i.e., that done by your subordinates) suffers. Giving priority to supporting your subordinates will result in the *intentional deferral* (not procrastination) and compression of bureaucratic chores. Sometimes they might even go away; at other times their deadlines will force them to take time away from your more productive activities. But this is largely unavoidable; indeed, some of these administrative activities are really necessary. Recognize and use Parkinson's law and the compression corollary to keep yourself from going beyond the amount of time that you allocate to your administrative chores.

The old adage "put first things first" certainly applies in time management. As an example, the author has written two textbooks. These were long-range projects taking at least two years each. If the 10- and 15-minute jobs were consistently placed in the way of these efforts, the books would never have been written. Instead, the first two to three hours of each day were dedicated to writing and producing a given page count per day—*whether I could afford it or not*. Some things never got done. But the two books got written. When you put your bureaucratic chores second to your most important job—providing support for your subordinates—some of the bureaucratic chores will slide off the edge of your desk into the trash. This will give you some short-term pain. We realize that it is a leap of faith, but the productivity returns over the longer term are immeasurable.

Provided that you attend to the job of management, your own time allocation is relatively unimportant compared to that of your subordinates. For the most part, *they* are doing the work that is fulfilling the direct goals of the organization. If you are doing your job right, you will shield them from bureaucratic chores to the greatest amount possible. This might be accomplished by having your staff people perform these functions, or it might mean that you do them yourself. It will mean that you complain bitterly to your boss about them, and hope that she or he does the same, perhaps stemming the demand flow. Finally, it will mean that you will initiate and serve on teams to study how these types of demands might be reduced throughout your organization. For if we are going to kill the applicability of Parkinson's law, we have to kill the bureaucratic nonsense that causes it to exist in the first place.

10

The Peter Principle

10.1 Definitions

In 1969, Lawrence Peter dazzled the management world with his book *The Peter Principle* (PETER69). Peter formulated his reasoning in such a convincing way that almost everyone involved with management at the time was left to ponder: Why didn't we see this before? Perhaps many of us did, but mindsets fixed toward external blame prevented us from articulating it. Let us begin by a paraphrase of Peter's great observation, which follows:

> In any large organization, the most competent individuals at a given level will inevitably get promoted to jobs of higher and different (often more complex but less technical) responsibilities, until ultimately they reach a level where they are both nonproductive to the organization and miserable in their job satisfaction.

Before we evaluate the veracity and ramifications of this premise, let us define the terms used. By *large organization* we are referring to one in which there is the possibility of promotion. As with our systems definition, above, however, an organization can be extended or contracted, depending on the desired context. Thus one of the broadest applications of the Peter principle might be a high-school student with the highest grade-point average who is "promoted" (i.e., accepted) to a renowned university where most of the student's new peers are far superior in intellectual ability. The "large organization" in this case is the generic educational system. However, recognize that most of Peter's applications were in business areas where the organization had grown to a point where a bureaucracy had developed.

Sec. 10.1   Definitions

By *most competent individuals* we mean those who are perceived by their management to be performing their tasks in a superior manner. They are generally perceived to be more productive than their peers in accomplishing what the promoting managers perceive to be in their, and possibly the organization's, best interests. (We use the term *promoting managers* to refer to the set of managers who have the authority to promote these individuals.) We emphasized the word *perceived* since two factors are quite critical to successful promotion: (1) the extent to which their perception is consistent with reality, and (2) the extent to which the promoting managers' interests are consistent with overall organizational goals. If the promoting managers' interests are consistent with organizational goals and their perception of reality is perfect (i.e., that this is the most competent person at this level), the Peter principle is most applicable. It is interesting that even in this best-case scenario, the end result, according to Peter, is close to catastrophic. It is an excellent mental exercise to surmise whether a relaxation of our assumptions improves or further exacerbates the situation.

By *given level* we mean the formal level that the relevant individuals hold within the organization chart. The principle was stated to hold for all levels, including the very lowest. In fact, many of the anecdotal proofs given by Peter were drawn from the lowest levels. When we say *promoted*, we may seem to imply that the person is placed at a higher level within the organization. Recognize that this might be formal or informal, depending on how the organization is defined. In the case of our stellar high-school student, there is no formal raising of position, although within society the status that this person has now attained would certainly be at a much more influential level. Similarly, applications could be made to people in industry accepting jobs with other companies or going from one sector of the economy to another (e.g., military to government to industry).

By *higher and different responsibilities* we are referring to the new environment in which this person must function. *Higher* refers to expectations within the organization (i.e., higher management). Note that the emphasis is not on the higher but on the *different*, which introduces the real cause of the problem. We stated (parenthetically) that these were less technical but more complex, which merely mirrors a subtheme of this entire book: that people are much more complicated than physical entities. The responsibilities of higher management are increasingly more people oriented. For example, a first-line manager might have the primary responsibility of dealing with those immediately below, a few immediately above, and several who are on the same level. This is an order of magnitude more complex than that of the lowest-level subordinate, where the primary concern is in pleasing the boss and possibly, getting along with some co-workers. Further, with promotion to the next level, managers must somehow influence how their subordinates deal with the next lower level of subordinates. Indeed, this is a much more difficult task than the one faced by the manager dealing directly with one set of subordinates. (We recognize that we are speaking in a very Theory X way here, and the principles of technimanagement can and should overcome these established paradigms. Please bear with us.)

The term *nonproductive* is not intended to mean worthless or totally incompetent; only much less skilled at the new task at hand. Indeed, newly promoted persons bring with them considerable background and experience that might be crucial to the company's success. But as we shall see, like the courier of gold who is suddenly swept overboard, this valuable baggage contributes further to "getting in over your head."

Finally, the term *miserable* is used as a relative measure of happiness. One of the primary keys to a successful technical organization is the sense of well-being and contribution that every member of the organization possesses. A few miserable people vegetating in key positions can prevent the organization from functioning properly. According to Peter, this condition will ultimately become pervasive, almost universal.

Now that we have explored our interpretation of Peter's observations, we need to consider further the validity of this principle. If it is as extensive as he asserts, it will certainly be valuable to document countermeasures to it. On the other hand, if there are exceptions to it, we need to understand them and recognize them as potential forerunners of effective countermeasure development. In the next section we explore the validity, and in the final section we present countermeasures based on these findings.

10.2 Validity of the Peter Principle

The Peter principle provides a comfortable explanation for many of the problems we see around us in organizations today. Chances are that if it were not generally true, it would have been shot down before the book was even sent to press. We have to respect the response of the general population of managers, although they have been far off base before. (After all, why didn't someone formulate this before 1969?) But let's not just go along with the majority. Peter recognized that his principle was not a basic cause of problems—it was a symptom of underlying bad management practices that go back to the days of the Tower of Babel and continue to this day. By looking at these underlying causes we can validate the principle and perhaps identify countermeasures.

We might begin by asking a series of questions: (1) Do managers tend to promote those who are outperforming their peers, and, if so, why? (2) Does this, in fact, cause the result that Peter observed? (3) Does it do so in all circumstances, and if not, why not? and (4) Can we generalize the exceptions in a useful way? This provides an outline for the remainder of the section.

There can be little doubt that managers do tend to promote those who outperform their peers. Let us put aside the complications given above of perception and reality and whether the organization's or the managers' best interests are under consideration. It is not that these are irrelevant, just that they are highly variable within each organization. To the extent that managers' perceptions are not consistent with reality, they will make worse and worse decisions. And to the extent that managers make decisions in favor of their own agenda and contrary to the organization's goals, these decisions will deteriorate further. Placing those issues aside, let us assume that the manager knows who is performing best and that "best" here is measured consistently with organization goals. Given these assumptions, we might still ask: Why do managers promote those who outperform their peers?

There are several reasons. The reason given by Peter was that management surmised that their success in their current position predicted success in their new position. In the absence of all other data, and all other things being equal, it is very difficult to argue with this reasoning. But Peter produced other arguments that showed that the best tinkerer rarely makes the best manager of tinkerers. And all other things are not equal. For example, there are characteristics displayed by individuals (as opposed to current-level productivity) that indicate which people would make better managers. So one cause of the problem to this point is lazy, simplistic thinking (or possibly ignorance) on the part of promoting management. Although ignorance could have been a cause in 1968, it is hard to imagine too many managers who have not heard of the Peter principle since then. Yet dare we surmise that this problem persists even after it has been exposed? There must be other, deeper underlying reasons.

Perhaps the greatest underlying cause is the fact that promotion is so intertwined with the reward system. The jobs of upper management are recognized to be of greater responsibility and difficulty. The chances of failure are greatly increased. The constituency is enlarged and so is the aggravation of having to deal with a much more diverse set of complaining individuals. Thus greater prestige and monetary rewards are given to those who are promoted. These rewards motivate those at lower levels to seek promotion. The usual methods employed (i.e., the rules of engagement for success), which are encouraged by the upper levels, includes the demonstration of success at the currently assigned task. When a clear "winner" emerges (i.e., someone who is clearly recognized by all levels to be outperforming), it is unthinkable to hold this person back and promote someone else. To do so would clearly be demoralizing to this extremely valuable person, who could get a job with any number of competitors. Peter's observation seems to be devastating: The best tinkerer will inevitably become the manager of the tinkerers.

As if this were not enough, another major self-validating cause is the organizational structure itself. The promotion of persons from the lower-level peer group is a no-win situation to the promoting managers. Only one gets the prize; the others get left behind to look for another job or to fret in their anxiety and concern for their own self-worth, which often finds its resolution in bitter external blame. If they are good, they will move on. If they are not, they will stay around and blame everyone else for their plight. Thus management needs some way out—some easy way that will be acceptable by all to determine promotion or nonpromotion. If that is not possible, the problem is one of damage control. Here, too, the logical trap is to promote the person who everyone recognizes as being the superior performer. So not only is this the thing to do as far as that person is concerned, the short-term negative impact on the rest of the organization is also minimized. Finally, the manager who had the primary responsibility for determining the promotion is also quite safe and secure, organizationally speaking. Who could blame him or her? Again, Peter's perception is overwhelming.

When positions become available, the current organizational structure as well as basic human nature are generally locked into promoting the superperformer from the next lower level (we will ignore for the present the possibility of bringing in someone from the outside). Given this fact, the next question that we need to address is: Is this *always* bad? Again, we can argue from the general acceptance of the Peter principle that there is no doubt that this is sometimes bad and there are strong arguments that this is generally bad. But Peter took the position that it was, allowing a few possible exceptions, almost always bad. Let us first review the reasons for this conclusion (i.e., why is it bad?) and then explore the significant exceptions.

Peter gives several negative effects of promoting the exceptional lower-level employee to a higher level. Of primary importance, especially to the highly technical organization, is that the organization no longer has access to the talents that got this person promoted in the first place. This two-edged sword cuts quite deeply into the quick of the organization's ability to fulfill its mission. The one edge denies people who are promoted the option of being committed to their first love (i.e., the skill that the organization so desperately needs). At the same time, they are called on, again and again, to come to the rescue when these skills are not adequately provided by those left behind. This cuts in the other direction, preventing the person from being an effective manager, which is a full-time job, especially for the novice.

At the same time that the organization is calling for these old skills from behind a shrouded alter, so is the still small voice from within: "Oh, for the good old days when I knew what I was doing. . . ." And just about that time, a crisis hits to which only our poor victim can respond in any meaningful way. Yet, now with an inept sense of authority, there is a grappling to accomplish the task by fiat. "After all, I did it, why can't they?" But the proper use of power and authority must be learned by training and experience, and typically our victim has neither. Trite analogies abound—*a fish out of water*, or perhaps, *out of the frying pan into the fire*. Still, there is the persistent urge to get back and help out, if not outright dominate, the old specialization, a characteristic that is commonly called *dabbling*.

To whom do these poor newly promoted people turn? The promotion has created a dramatic sociological change. They were very strong informal leaders at the lower level (see Chapter 12). The change in formal position will most certainly destroy that, for now they are suspected of speaking from the point of view of their new position. And, whereas they were at the top of their old peer group, they are at the bottom of the new one. As newcomers, they have the lowest status (while previously enjoying the highest). They have exhausted their power by virtue of their promotion. They must prove themselves—to everyone—once again before this informal power can be reinstated. Unfortunately, however, they are the last to know it, and this misperception of reality can be one of the most deadly. Virtually shut out of the informal organization, where does one go?

Introduced above was one of the most critical elements of failure—lack of relevant education and experience. We dare not let it pass by without additional elaboration. What was it, other than technical knowledge, that made this person such a tremendous performer? In many cases this technical knowledge is obtained at the expense of management and people skills. This is not because the technical employee is a bad person or does not care about people. It is a matter of time and attention allocation. Obviously, we are all allocated the same amount of time per year, although clearly some people use it more effectively than others. For most people to get really good technically, there has to be a sacrifice of something else. Although we might reason that there is little time to take a management course, read up on management issues, or discuss management strategy with one's peers, the problem is much deeper than this. We will call it an *attention allocation* problem. Many very skilled technical people just do not believe that management issues are important—and remember that we are talking about the extremely competent (i.e., those exceptional people who are considered for promotion). They trivialize it as being "common sense." Thus, with some notable exceptions, they have not attained the necessary management skills either through formal education or experience at the time of their promotion. (We recognize that this is not as definitive a support of the Peter principle as some of the arguments given above and that it does not have to apply to every skilled technical person.)

The failure to understand principles of good management leads to a chain of management blunders, which we will not detail here. There is one that we will mention, however, because of its interaction with the dabbling instinct. This is the failure to delegate effectively. A common complaint of management is that they tend to delegate everything which they do not understand and dominate everything which they think that they do. Thus the aggravation of upper management over the allocation of parking spaces while the organization is on its way down the sewer. Our stellar performers who have just been promoted are certainly not immune to this, as they dispose of the nasty things which they should be mastering largely by default. After all, Joe has always handled that; let him handle it. You do not have to guess where they fail to delegate. It is just easier to do it yourself than it is to get (which usually means train) someone else to do it. Not only is this dabbling counterproductive to effective management because of the time loss to other vital areas, but it kills morale and the incentive in the old organizational component to establish a new informal leader and get on with the job.

As Peter so effectively points out, the long chains of management blunders, caused by the Peter principle in action, do not effectively argue for the replacement of the people who have been promoted. There are several reasons for this. In the short term, there is the coupling of the "experience" excuse with the reluctance of the promoting managers to admit a mistake on their part. After all, you have to give the new manager some breathing room. This is quite valid. However, when these errors continue and months turn into years, action is clearly warranted. But at this point it might be too late. What is the organization to do with these unfortunate people? It dare not promote them. On the other hand, what signal would be sent if these people, many of whom have attained heroic workaholic status, are demoted, or worse yet, fired (the unthinkable)? So some of the very same short-term simplistic thinking that caused the original problem serves to keep these people in place.

The paragraph above assumes that there are members of the organization who recognize incompetence when they see it. In fact, one of the management blunders caused by the Peter principle in action is propagation of the Peter principle itself. In a blaming organization [i.e., one that almost always shifts the blame from management to some other sector (usually labor)], there is a very good chance that no one (or, at best, very few) will even recognize incompetent management. They are too busy emulating one another to open their eyes to reality: We have found the problem and "it is us."

Peter argues further that if by some stroke of fate, you are one of the exceptions and happen to perform your management task in an exemplary way, you are not immune to the Peter principle. For most assuredly, you will be promoted to yet the next-higher level and to even stranger tasks. Eventually, you will reach your *level of incompetence*, and then the principles of the paragraphs above will kick in. Thus you will be stuck at your level of incompetence and there remain in a purgatory of woe awaiting your ultimate release in retirement. (Where, by the way, an excellent argument could be made that this too is analogous to a promotion, which might explain why so many extremely productive people die shortly after retiring. But let's not get into that.)

We think the paragraphs above argue quite well for the general applicability of the Peter principle. (Perhaps Peter would disagree, and we urge you to read his book.) If you detected some degree of overstatement in parts of our argumentation, we commend you. For, *if we take Peter to the limit, all organizations would collapse of their own weight, nothing would be produced, and our economy would have come to a screeching halt long ago.* Although this has not happened, admittedly we can use the Peter principle to do better, but only if we define those areas where it applies and those where it does not.

10.3   Shortcomings of the Peter Principle

The greatest flaw in the Peter principle is that it does not account for simple human adaption. Just because skills are lacking at one point in time does not mean that they cannot be acquired. While the brute-force method of thrusting competent persons into new and strange environments might not be ideal, there are times when it has its advantages. It brings new and different approaches to the decision-making process, and it ultimately produces a broader perspective in the manager.

[As an interesting "aside," our society is absorbed almost to the point of obsession with the entertainment value of adaption. Perhaps over half of the story lines of our theater entertainment and as many TV stories are based on this very simple premise: Place people in a strange environment and watch how they bring old solutions to bear on new problems. One reason it is so often employed is that it is a simple way to create humorous situations (from the Real McCoys of the 1950s, to the Beverly Hillbillies of the 60s, to Beverly Hills Cop, Mr. Mom, etc.). However, dramatic productions have also drawn on this most common theme, from *Gone With the Wind* to *The Fugitive* (original) to *The Fugitive* (remake). These examples from our pop culture are not intended to disprove the Peter principle, only to bring to light another well-proven principle of humanity: recognition of our ability to adapt. Sure, we struggle along quite humorously for awhile, applying some of the old paradigms to our new environment. But most of us survive change somehow. In most cases this requires a healthy break with the past.]

This is not to say that most excellent technical people who are promoted to management positions make good managers. However, they adapt quite well to the established structures with which they are confronted. In very large part, the structures themselves (i.e., the bureaucracy) were established to protect the organization against bad managers, and they tend to do that quite well, albeit at the expense of sacrificing the original goals of the organization. *The problems are not with incompetent managers in a good management structure, but with competent managers in a poor management structure.* One symptom of this poor management structure is the Peter principle itself. However, the inability or unwillingness of managers to change these structures might still be laid to their charge. Indeed, the changing of these structures is the essence of technimanagement.

So where does this lead us? Is the Peter principle true or false? Does it hold generally or is it an oversimplification based on the false assumption of human inadaptability? Let us state a qualification of the Peter principle that might serve to clear the air:

> The Peter principle is valid to the extent that newly promoted individuals fail to adapt to their new job responsibilities by acquiring the management and human-relations skills necessary to be effective.

The fact is that some people never change! Like the proverbial bull in the china shop, they have no idea (in today's jargon: "not a clue") of the havoc they are provoking in the organization and how many years it will take to clean up the mess they are making.

If we are going to solve the problems that accrue from its applicability, the Peter principle must be viewed as a symptom of a deep-rooted problem of management structures as we know them. Thus to write something off as being caused by the Peter principle is not at all a solution (and we think that Peter would agree). To attack the symptom will not bring about a solution; it might even be counterproductive in masking the true cause. In the next section we present some countermeasures to the problems caused by promoting people to their levels of incompetence. However, it should be recognized that these are not isolated techniques. Unless they are integrated into a transition that generates the mechanisms by which true technimanagement can be implemented, these techniques are like a Band-Aid on an amputated arm.

10.4 Solutions to the "Peter" Problem

For lack of a better term, we define the problems (plural!) that accrue from all of the negative ramification of the Peter principle as the *Peter problem*. (We hope Mr. Peter will forgive us if this is the least bit offensive to him.) It is essential from the outset to identify the making of promotions as a *discrete optimization* problem. First it is discrete in that it consists of a choice from available alternatives, not from a continuum of ideal choices, as the promoting management would wish. It is an optimization problem in that the choice from the various alternatives should be made in light of maximizing the overall benefits to the organization (or minimizing the costs), as opposed to being viewed as totally beneficial or totally destructive. Thus the "right" decision is the one that balances the advantages and disadvantages in the way that most promotes the organization's goals.

This concept of optimization (defined in Chapter 8) is quite important in this application. The downside of any decision will always be seized by critics to further their agendas. In this case there are as many downsides as there are competitors for the position. Thus, in promotion considerations, the very first question should be: Do we need to make this promotion at all? Is there a law or custom requiring it? Is it really a critical need of the organization? Given that a position exists which must be filled, the question is: Do we bring someone in from the outside, promote from within, or open it up to either? Both have their downsides. Once the pool for selection has been established, the problem is one of selecting the best person for the position. Best, in this context, must be measured in terms of the total effect of introducing or reassigning this person to/within the organization.

Unfortunately, quick and simple rules create more problems than they solve. This is a complicated problem with a multitude of ramifications, and it affects the single most important ingredient of success: your people. It is essential not only that the characteristics of this person be taken into consideration but also the potential impact that this will have on everyone else within the organization.

With this preface we will present some considerations for ameliorating problems created by the Peter problem. They logically fall into two categories: the individual and the organization. The first addresses the ways that you, as an individual, can escape the Peter problem and thus live happily ever after. The second addresses the problem from the viewpoint of management countermeasures that can benefit the organization. Finally, we emphasize again the necessity for embedding these considerations within mechanisms for the transformation to technimanagement which are detailed in other parts of the book.

10.4.1 Individual Countermeasures

Peter's advice was primarily to the individual; and indeed, if individuals would avoid the Peter problem, its impact on the organization would probably take care of itself. Peter gave a number of personal initiatives to which we refer the reader. However, one of these bears mentioning because of its applicability for other purposes as well. According to Peter, a direct refusal to be promoted might cause displeasure to key individuals within the organization, potentially eliminating future options and definitely eroding personal power. As an alternative to this, he recommended something which he called *creative incompetence*. This was behavior that would disqualify one for consideration for promotion, but it was not so extreme as to hurt one's intended career. Examples might include "carrying on" at the weekly departmental meeting or singing loudly in the halls. No rules were given. Rather, the person needed to do whatever it took to be different without being too obnoxious.

We recognize and appreciate the humor with which Peter wrote, and realize that he was not trying to write a formal treatise on management behavior. The humor does not in any way diminish the value of his advice. If anything, it promotes the underlying subtheme of not "taking ourselves too seriously," which is good healthy advice. The concept of creative incompetence in this context is quite beneficial if it leads to the type of beneficial nonconformity that is generally necessary to bring about positive change. This type of activity is all too often stifled by the fear of being disqualified from consideration for promotion. In Chapter 9 we referred to ways to utilize Parkinson's law by intentionally deferring selected bureaucratic chores. This kills two birds with one stone: It demonstrates overt resistance to bureaucratic nonsense, and it produces the effects of creative incompetence at the same time.

As far as practicing creative incompetence just to escape promotion, however, we believe that Peter would agree that in this day and age there are much more straightforward ways of going about avoiding promotion than resorting to this type of reverse manipulation. Most large technical companies have adopted what they call a "dual advancement" system, where there is the equivalent of a promotion to middle management for those who wish to remain in their nonmanagerial technical jobs. But even if your company does not have such a plan, there is no reason that an excellent technical person should not be able to negotiate for something of this nature. Surely the reduction of the promotion decision space by one should be welcomed by management, and if they do not understand how your promotion would be counterproductive, you can always read them the Peter principle. A number of promotion-avoidance techniques can be applied; we advise the most straightforward and open possible.

If promotion avoidance is practiced by some people, they should understand and accept its downside. While "equal" promotion paths are touted, clearly some of these are more equal than others. When people choose the option to refrain from management, they give up much of the potential to have a greater role in control of the company. When people avoid promotion, they must be willing to accept this loss of control over their work environment. The alternative to such acceptance is inevitably a bitter attitude toward those who have taken the other option, which can be just as bad personally as reaching your level of incompetence.

While promotion avoidance is advised for many people, we feel that very few of them will be reading this book. We hope that you are reading the book because you want to get prepared well before the dread promotion arrives. This may take several forms: (1) formal education, (2) informal self-study, and (3) continuous job enrichment. The first two are quite straightforward. Take every course that you can that is relevant to your management success (if your immediate supervisor is foolishly suspicious, get advice from your staff or university advisors). Secure books and periodicals that relate to management technique and devour them.

Before defining the third form of preparation (continuous job enrichment), let us consider a major aspect of the problem—discrete promotion. By *discrete promotion* we refer to the act of plunging employees who are totally skilled and absorbed in one function into another for which they have no background. This is the essence of the cause of the Peter problem. This would seem to be a fairly easy problem to circumvent. *Continuous job enrichment* is a term that we use to designate a seamless alternative to discrete promotion. It involves the acquisition of responsibility on a small, incremental basis. It involves sharing management responsibility with your immediate supervisor. However, it goes beyond this to involve service on work teams and task forces that may be established to improve operations and solve specific problems. In brief, it requires that the person develop skills within the informal organization and thereby acquire the preparation that will be essential when the formal position is finally attained.

The purpose of continuous job enrichment for the person seeking promotion is to acquire the management and human-interaction experience that is so crucial at the next level of management. The primary deterrent to this is short-term thinking. There is absolutely no guarantee that the promotion will ever come, and if it does, it will probably take years. Most people do not have the patience to practice continuous job enrichment over such a long period. However, it offers the best alternative to reaching one's level of incompetence, and it is just the opposite of creative incompetence. Astute upper management will recognize those who are contributing by their continuous job enrichment activities, and they will be given the edge in the promotion decisions.

Continuous job enrichment is a key element of the transition to technimanagement, and this transition will require that it ultimately take place throughout the organization (see Chapter 24). However, what do you do if your organizational component is not currently supporting this type of cultural change? Do it anyway! But use good judgment. This takes the ability to "read" your management to determine those additional decisions for which you should take the initiative and those for which you should seek assistance. Demonstrate your willingness to make decisions, but do not do it in a way that is at all threatening. Know when to advance, but also know when to beat a hasty retreat.

In summary, there are two personal approaches to solving the Peter problem. Both will work. Creative incompetence avoids promotion. Continuous job enrichment is an aggressive buildup of your position to accept and train for additional responsibility. The latter approach will ultimately put you into a position where you will need to consider organizational countermeasures.

10.4.2 Organizational Countermeasures

An understanding of the Peter principle is imperative to effective management decision making. Some countermeasures have already been applied in many progressive management organizations, including the dual-path promotion policy that was introduced above. These, however, are not necessarily founded on the principles of technimanagement, which is essential if they are to continue to be effective.

At the end of Section 10.1, we traced the cause of the Peter problem to the absence of effective management mechanisms. Without describing these mechanisms in detail (which we do in Chapters 14 and 24), at this point we consider their effect on countering the Peter problem.

It would seem that it is a mistake for management to promote a person who has demonstrated the greatest competence at the lower level. The obvious countermeasure is to choose someone else—someone who is not the superior tinkerer—to become manager of the tinkerers. But like most simple solutions, this one is wrong. It is based on nothing other than a reaction to the Peter principle. We look for a principle-based approach which will assure that promotions enhance rather than degrade the organization and its capacity to attain its goals.

We have explored the management mechanisms that are currently in place. They consist of a set of what we have called promoting managers who review the available candidates and ultimately decide which one will be promoted. This mechanism fails when it promotes someone who, on promotion, immediately reaches his or her level of incompetence. The problem is one of management ignorance. It is assumed that the successful person will continue to be successful, even in the new sociological environment. However, in this case the causes of success at the lower level might actually cause failure at the next-higher level.

Is it possible that the promoting managers can increase their intelligence and determine which of the people under consideration would not fall into the Peter problem trap? Possibly. But even if this occurred within the current structure, the promotion of the second- or third-place performer could be extremely detrimental to the morale of the entire organization. Clearly, the control system is dysfunctional. Not so clear is the definition of an effective mechanism to replace it.

Let us begin by suggesting that two mechanisms need to be replaced: (1) the decision-making mechanism and (2) the "morale system" (i.e., the reward system). Actually, the two of these rise and fall together. Under the traditional management structure there is a partitioning of tasks, authority, and responsibility. This has created a *belief system* with regard to promotion within the organization given by the first column of Figure 10.1.

Traditional	Technimanagement
Life at the next-higher level is much more rewarding relative to the ability to control, perquisites, and take-home pay; thus a promotion is a sought-after commodity; and If I convince my management of my value to them, I will be more likely to be promoted to this position.	Life at the next-higher level is *somewhat* more rewarding from the views of ability to control, perquisites, and take-home pay; thus a promotion is a *certainly to be considered;* and If I convince my management of my value to *function most effectively in that position*, I am more likely to be promoted to this position.

Figure 10.1 Redefinition of Belief System

In the traditional management structure, which includes most organizations today, these beliefs are a fairly accurate perception of reality. Although they do not consider the personal disadvantages and headaches of increased managerial responsibility, who ever does? Thus we are correct to exclude the downside from the current belief system. (We are not talking about underlying morality or religious beliefs when we use the term *belief system*, only the perceptions that the members of the organization hold toward it.)

The key to high morale within an organizational component is the underlying *belief system* of its members with regard to the organization in general (and the component in particular). For morale to thrive, this belief system must be consistent, accurate, and positive with regard to organizational goals. By *consistent* we mean that everyone (at all levels) within the organization must share the same belief system (the alternative is manipulation, and manipulation is the major killer of morale). By *accurate* we mean that the belief system must be an accurate perception of reality, not a made-up set of slogans for the purpose of artificial motivation. And by *positive* we mean that the feeling which each person has toward the organization is one that does not conflict with the person's perceptions as to what is in his or her own best interests.

It is essential that we recognize the necessity to change reality first. For obvious reasons, all changes in the belief system will be extremely counterproductive to morale unless they are preceded by changes in reality. (We say that this is obvious, but it seems to escape many managers, who think it is quite clever to keep as many of their colleagues and subordinates in the dark as possible.)

We revisit that part of the belief system which motivates the Peter problem and see how it might be modified slightly to promote better morale. The two beliefs with regard to promotion within the organization have been modified for technimanagement as given in the second column of Figure 10.1.

Note first that this is not a revolutionary change. We are not saying that everyone should believe that managers do not enjoy some additional control capabilities by virtue of their positions. However, to improve morale, we propose to change "is much more rewarding" to "is somewhat more rewarding." Similarly, the other changes do not require radical alterations in the belief system. However, these changes are significant, and they will have a major effect on increasing morale. First, there will not be as great an expectation of reward with promotion and thus not nearly the letdown when it is not received (or worse yet, received). Second, it will be perceived that the promotion was not so much a reward as an attempt to *place the most appropriate person in a better position to serve the organization*.

We do not expect this perception to be changed without ample evidence. This requires a change in the way that promotions are viewed by all members of the organization. The problem is actually to change reality such that these perceptions reflect it accurately. A second major problem, which we do not address at this point, is to convince everyone in the organization that this change has been made. However, if reality is not changed, this becomes moot. Changing reality calls for the replacement of those mechanisms that maintain the old with those that support the new philosophy. Although these are discussed further in subsequent chapters, let us suggest the approach given in Figure 10.2. The objective is to establish mechanisms to bring about the modified belief system shown in Figure 10.1.

- Establishment of mechanisms (e.g., task forces and work teams) to distribute decision-making powers to those who are in the best positions to make them.
- Ongoing cooperative development of a joint understanding between managers and their corresponding subordinates as to the decisions that will be made at each level.
- Establishment of a clear direction of development of mechanisms within the organization toward distributed decision making.
- Establishment of pay differentials that reflect the qualifications and preparation for each position, as opposed to the position itself.
- In conjunction with the other items above, movement toward the determination of promotions and rewards by means of a collaborative process that heavily involves self and peer review.

Figure 10.2 Suggested Approach toward Belief System Modification

We recognize that transition to these mechanisms does not instantaneously solve the Peter problem and that it cannot be solved in a vacuum. Rather, the Peter problem must be viewed as just one of the many symptoms of comprehensive organizational problems. Solution of these problems requires a total systems approach involving the evolution of principle-based mechanisms, which are discussed throughout this book. Our objective here is to illustrate how these mechanisms can address this particular problem specifically.

Before concluding this section, some tie-in with Section 10.4.1 is in order. After the smoke cleared, the primary positive recommendation to the person seeking promotion was to pursue aggressively continuous job enrichment. Among other things this involved the expansion of the person's job to acquire a broad range of managerial experience prior to actually assuming the position. Under technimanagement this will be encouraged by management throughout the organization. Notice how the five mechanisms suggested above are consistent with this approach, as shown in Figure 10.3.

- Participation on empowered task forces and work teams generates experience in cooperative resolution of problems, human interaction, and decision making, thus eliminating management incompetence in the participants.
- Dynamic reassignment of responsibilities supports continuous job enrichment by encouraging (1) lower-level employees to participate in decision making in groups, and (2) individual managers to groom key subordinates for higher positions.
- Establishment of a clear direction of development of mechanisms within the organization toward distributed decision making creates a dynamic climate of continued excitement and expectation providing the motivation for involvement.
- The reward system will encourage individuals to acquire a wide variety of useful skills, including management skills; this system will reward all of these skills according to their level of difficulty and demand as opposed to the position that the person holds within the organization chart. That is, salary will reward increased productivity due to job enrichment, without the necessity for formalized promotion.
- Utilization of a peer-review process will encourage cooperative and supportive activity directed throughout the organization, as opposed to that directed toward a single person (i.e., the manager).

Figure 10.3 Technimanagement Countermeasures to the Peter Principle

It should be clear that the organizational (upper-level) response to the Peter problem will not be effective unless it is accompanied by individual efforts on the part of subordinates. Essential to this scheme is the breaking of our sports-culture paradigm (i.e., the *generation of a "winner" does not necessitate that all other competitors are losers.*) We put *winner* in quotes to emphasize the downside of promotion. Technimanagement calls for merging of the continuous job enrichment and the distributed management paradigms, which will ultimately have the effect of blurring the discrete levels of traditional management.

The Peter problem will continue to be a plague to business and industry as long as effective countermeasures to it are not implemented. In this chapter we have presented the problem and some example mechanisms that will be initiated by the transition to technimanagement.

If all of this sounds idealistic and impractical, we are pleased. What is being proposed here is a radical departure from business as usual. A thousand reasons will be posed to explain "why it will not work here." Arguing with this opposition is akin to dealing with a teenager who declares "I can't do math." In fact, the belief of this statement is the only thing that makes it reality! This maps quite well to the major problem that is considered in the next chapter.

11

The Paradigm Problem

11.1 Definition of The Problem

We owe credit for the packaging of many of the thoughts contained in this chapter to Barker (BARKE91), who has produced a videotape on the subject. Technically, the word *paradigm* should appear in quotes in the chapter title since the word is not being used consistently with its original English definition. The word itself carries the meaning of a model, pattern, example, or template. With this definition we can hardly think of paradigms as problems, since they are the tools that we use daily to encapsulate our current knowledge so that we can extend it to solve future problems. Contemporary use of this term, however, has tended toward applying these meanings more toward the way that we view our world, and more specifically, our views toward methods for accomplishing organizational goals. This becomes a problem only if we lock into a paradigm within which only suboptimal solutions exist. Since this does seem to be a fairly common characteristic of human nature, it must be addressed, and we will use the term *paradigm problem* accommodatively to encapsulate this concept.

Let us begin by giving a more formal definition of the paradigm problem:

> Once thought is well established within a given paradigm, optimization tends to take place within this mindset; it is exceedingly difficult for most people to even conceive of another paradigm, and once one is suggested, it is difficult to gain general acceptance of it.

To define our terminology, recognize that the paradigm problem is strictly concerned with *thought* (i.e., that which goes on inside people's heads). It might have very little to do with any reality other than the real-

Sec. 11.1 Definition of The Problem

ity of what is going on there. *Optimization* was defined in Chapter 8; in this context it involves the selection of one from a number of alternatives such that advantages and disadvantages are balanced to maximize some defined objective. By *mindset*, we mean a habitual routine of thinking (and we have used it throughout this book in that sense). We think the name *mindset problem* is more descriptive than *paradigm problem*, since it focuses on the cause of the problem. However, we bow to the communication advantages of sticking with current popular terminology.

The phrase *optimization tends to take place within this mindset* gives credit to the decision maker for understanding optimization and attempting to bring about the best possible solution. However, even if an optimal solution is attained within a given paradigm, it will not be optimal within a better paradigm. The healthiest, best-bred racehorse, in its prime and ridden by the best possible jockey, could not possibly outrun the worst of functional motor vehicles commonly on the road today. What was it that kept human beings from inventing and utilizing the internal combustion engine for transportation prior to the turn of the twentieth century? While a large number of sociological and scientific factors could be mentioned, one thing is quite certain: As long as we were only trying to improve our horses, there is no way that we could invent the automobile. The paradigm problem in this case was the concentration on more and better horses. Today a similar paradigm problem is that we are locked into internal combustion engines and battery-powered motors. Surely there has got to be a much better way!

It helps to understand the difference between *local* and *global* optimums. We might well approach a local optimal point within a given paradigm; and this is commendable. However, this local optimization is restricted to that paradigm. The very act of seeking optimization within a given paradigm is generally counterproductive toward establishing a better paradigm. For example, what would be the likelihood of those who were breeding better riding horses would come up with the internal combustion engine? (About the same chance as one of our oil companies inventing a non-petroleum-burning engine.) Yet the chances of the global optimum (and thus a better local optimum) existing within another paradigm is far greater than it existing within the current one. This is the essence of the paradigm problem and the reason that it causes such havoc in preventing progress.

But let's not get stampeded to hasty conclusions. For some reason we have made progress despite the paradigm problem, so obviously it has its exceptions. It behooves us to study the veracity of the paradigm problem just as we did the Peter principle. Once we understand its scope and causes, we will be in a much better position to apply countermeasures.

11.2 Validity of The Paradigm Problem

It is difficult to conceive of a paradigm that must contain the global optimal (for anything). Surely there is a better way to do everything than what we currently do, and one good reason that we are not seeing the possibility of major improvement is that we are stuck in our current paradigms. Thus the proponents of paradigm change have a distinct advantage over those who wish to maintain the status quo. The problem that the paradigm-change proponents almost always ignore, however, is the distinct possibility that a paradigm change might shift to one in which the local optimal point returns less than that which could be achieved by the current paradigm! In these cases those "poor souls" who are incapable of conceiving or accepting new paradigms would have the advantage. Thus we are safe in the warning that:

> Some paradigm changes are not beneficial.

The decision to move to a new paradigm is an optimization problem within itself, which must be evaluated against the costs and benefits that it will produce in terms of organizational goals.

Recognize that this is beside the point of the major principle conveyed by the paradigm problem. The problem that we refuse to accept new paradigms is eclipsed by the problem that we fail to conceive of them because our heads are buried in the sand of optimizing within our current paradigms. We cannot fail to accept the reality that there are better paradigms out there, but it is difficult for us to conceive of them.

Let us try to cut a bit deeper into this problem. Traditionally, managers have been trained to believe that basic human nature is resistant to change. People just do not like to change, we are told. So any time that you attempt to modify or change anything, you can expect considerable resistance. Further, this resistance is directly proportional to the degree of difference of the paradigm and the length of time that the current paradigm has been in place. So if a change is in order, it needs to be evolutionary. Is this true or is this a paradigm in which we are stuck? (Bear with us; we will revisit this shortly.)

The Hawthorne studies discussed in Chapter 3 tend to contradict this premise. Indeed, change was not resisted, it was accompanied by increased productivity. It might be argued that these were not major paradigm shifts. We counter that the change from the traditional Theory X approach to the kinder management style introduced by the researchers *was* a major paradigm shift, as opposed to the change of light intensity, which was an attempt at optimization within the old paradigm. "Oh," you counter, "but this was a change that was favorable and enjoyed by the subordinates." Exactly! and the conclusions are obvious. (Interestingly, like many other new paradigm discoveries, this resulted when searching for something else.)

Even if we admitted that the majority of people resist change, does this argue in support of the paradigm problem? Again, the paradigm problem starts with the inability to conceive of new paradigms. Is resistance to change (given that it exists at all) the cause that we are not able to conceive new paradigms? We think not. The adjacent display gives a list of factors that tend to create or support the paradigm problem. However, they are not absolute proofs of its general validity. They do indicate that the paradigm problem will persist to the extent that these causes exist.

Factors That Promote the Paradigm Problem

- *The forest-for-the-trees syndrome.* Symptoms of this affliction occur when managers stop looking for another type of forest (or possibly even a field). With all of their other activities, they just do not have time to dream up new paradigms. This leads to the next cause.

- *Only an "elite few" are responsible for conceiving of new paradigms.* While managers are too busy, the workers are generally not considered to be smart enough. So a few staff people or consultants are charged with the responsibility to improve them. Not having a day-to-day working knowledge of the operations, they rarely do more than optimize within the current paradigm.

- *Anticipation of rejection.* The few new paradigm proposals that do get generated have a difficult time surviving because of pressure from those who had no part in conceiving the major change in thinking. Those major paradigm shifts that have taken place have been successful only because of the tenacity and persistence of their originators.

Factors That Promote the Paradigm Problem (Continued)

- *Limitations in our use of language (communications).* The words describing the most radical new paradigms do not yet exist. Example: prior to the "computer paradigm," terminology had to be adapted to explain the concepts until these could be created and accepted (e.g., *boot*, short for "bootstrap," short for "pulling yourself up by your own bootstraps," an analogy to getting a computer *cranked up* (another one). A second problem is one of acceptance and ambiguity, including our inherent tendency to argue about terminology as opposed to issues of substance.

- *Short-range systemic and economic reasons*
    - *Systemic reasons*: those attributed to the system itself, such as (1) having the maximum power within an organization concentrated in a few people, (2) rewards based on short-term metrics, and (3) the general tendency to follow the leader and not make waves
    - *Economic reasons*: market forces that stress the immediate quarter's bottom line

- *Reinforcing factors: the failure, expense, and inconvenience of past changes.* These produce gun-shy managers who would rather stick with anything that works than to risk repeating a painful change. However, just because a change failed in the past does not mean that it (much less others) will fail in the future. The third factor, inconvenience, is probably the least defendable of anything discussed so far. Interestingly, this reaction causes a situation where past mistakes (experiences) actually become a liability.

The above display, entitled *clichés that reinforce fixed-paradigm thought*, show how our culture has tended to perpetuate existing paradigms. This provides adequate justification for the balance provided by Deming's "continuous improvement forever." However, recognize that the notion that "no one will accept such a major change" is itself a self-perpetuating paradigm. In fact, it is just not true, as evidenced by the many major changes that our society and technology have experienced over the past few decades. This is not to say that change does not often meet with resistance just because of natural jealously and pride or comfort attained with the established way of doing things. We will let you determine which types of organizations bring about these changes: those that tend toward bureaucracy or those that tend toward technimanagement.

> **Clichés That Reinforce Fixed-Paradigm Thought**
>
> Certain clichés are so universally accepted that they cannot help but affect policy. Consider the following as they affect the paradigm problem.
>
> - *Let's not reinvent the wheel.* This is very effective in intimidating the advocates of change back into a state of complacency, especially if uttered from a high-enough administrative position. The basic germ of truth on which it is founded argues against disposing of that which has proven to be effective in the past. Although this is quite valid, the "wheel" might need to be reinvented at times (e.g., the trend in mass transportation might be away from wheels).
>
> - *You don't remove a pitcher when he's winning.* Question: What does it mean: *winning*? Was IBM winning when it failed to recognize the customer demand for open systems? Pitchers are as often removed before their teams are losing, at least as far as the scoreboard is concerned. A good manager must anticipate when the positive momentum is shifting. As useless as the strict application of this cliché is in managing baseball, it still seems to have unreasonable impact on business management decisions, perhaps to justify inactivity.
>
> - *Don't throw the baby out with the bathwater.* Who could disagree with this? Recognize, however, that the baby could be a python. Nothing is sacred; everything must continue to justify its existence (even the baby).
>
> All these sayings could be used for either harm or good. However, when employed merely to maintain a state of convenience for management, they devastate the incentive for positive change.

The paradigm problem poses a real threat only in those organizations in which its basic causes predominate. The statement that "people do not want to change" is a *shift of blame* from management to subordinates which accomplishes nothing other than masking the real causes of the paradigm problem. Ordinary people are quite capable of conceiving of and adapting to new paradigms. The problem is that *most organizations are structured so that they cultivate and fertilize the underlying causes discussed above.*

Can modern organizations survive for long without making major paradigm shifts? Conventional wisdom is quite clear in this regard; for example, consider Kast's observation: "Progress depends upon the uncommon man and his unwillingness to perpetuate a given system" (KAST70,317). Those who have been around for over 20 years know that the only thing that does not change is that "nothing ever stays the same." However, these people are often the very ones who have stopped initiating changes and are spending most of their time resisting changes suggested by others. It is indeed the uncommon person who has the following characteristics: (1) a dissatisfaction with the current system of things, (2) the intelligence and motivation to formulate a practicable superior alternative, and (3) the charisma and communication skills to convince the establishment that the new paradigm is superior. It takes all three of these characteristics to counter the paradigm problem. The higher that managers are in the organization, the more likely that they will have the last two of these characteristics. However, in most established organizations, the higher up, the least likely to have the first—dissatisfaction with current position and past decisions. Since changes emanate from people and not organizations, this tends to further propagate the paradigm problem.

Another aspect of the paradigm problem is highlighted by Keith: "We must also recognize that the term 'management' and the role of the manager will change as time goes by because the nature of business will constantly change" (KEITH71,181). Most managers consider themselves progressive to recognize that change is inevitable. However, few apply this progressiveness to their own roles as managers. Yet this is potentially the most critical consequence of the paradigm problem. That is, not only should the system being managed be subject to change, *so should the very nature of the management process itself.*

In summary, the paradigm problem is not the cause of organizational resistance to change, but it does serve to package a concept from which we can identify causes. To the extent that these causes exist, the paradigm problem and its consequences exist. Approaches to eliminating these consequences must reflect elimination of the causes. Some of these are considered in the next section.

11.3   Solving the Paradigm Problem

One of the most important aspects of solving the paradigm problem is to get it in its proper prospective. Those who lay all of the world's problems at the feet of the paradigm problem have clearly exaggerated its potential

Sec. 11.3 Solving the Paradigm Problem

menace. After all, if the paradigm problem were that universal, we would still be riding horses, using our fingers for computation, and telling stories around the campfire. Things do get changed, for better or for worst. We might argue with the rate, the value, or the quality of these changes, but we cannot argue that things do not change.

The problem is not one of selecting the correct paradigm, nor is it one of getting others to accept a better paradigm. Consider Whitehead's statement: "No society or organization is averse to change, provided the initiative for that change takes place at the relevant level—at that level where the daily activities have shown the need. Under those conditions, change will present itself not as an interruption, but as the natural flow of social living . . . "(WHITE36,78–79).

In fact, if the paradigm is beneficial to social living, it will prove itself and be accepted for the benefits that it brings about. So let us concentrate on the problem of conceiving new and better paradigms, and specifically, organizational deterrents to the creation and formulation of better paradigms. Consider the following possible organizational countermeasures to the paradigm problem:

1. Disband the specialists who are currently expected to invent new paradigms and distribute their responsibilities throughout the organization. This will require the formal establishment of horizontally and vertically integrated groups that are dedicated to this purpose, and the dedication of the time that it takes for them to function effectively. Although team efforts are essential, management must recognize that new paradigms are largely the result of individual efforts.

    Comment: The reason that we cannot see the forest for the trees is that we are totally overwhelmed and backlogged with the day-to-day problems of running the organization. Consider the effectiveness of formally requiring all employees to take one hour a week (2.5 percent of employees' time) in which they are not allowed to conduct business as usual; rather, they are supported in determining ways of improving their operations. Since rarely does any employee operate in a vacuum, this would need to involve group activity. The group stimulation of this "thought process" will continue throughout the week on an individual basis, where the really major breakthroughs are made.

2. Remove organizational deterrents toward the creation of new paradigms. This begins with management's assumption of complete responsibility for the ineffectiveness of current paradigms, which disables the subordinates' defense mechanisms. This must be followed by their open statement of the belief that all paradigms can be improved and the solicitation of assistance from all levels. Thus anyone conceiving of anything

from a minor change to a major paradigm shift will not be defensive (at least of management) of the current paradigms, and will communicate their suggestions (as wild as they might seem).

3. Provide real, tangible, and significant incentives to activate the organization to seek new paradigms. This begins with the removal of "guilt for the present" but extends to a positive reaction to suggestions (regardless how seemingly ridiculous) within all meetings. It must also include financial and promotional incentives toward those who are the true source of improved paradigms. This requires proper identification and protection of those persons and groups who not only originate but evaluate, prototype, perform detailed design, and develop applications of the new paradigm. In other words, the incentives toward improvement must be organization-wide.

4. Prove new paradigms with pilot tests and prototypes before implementing them fully. Many new paradigms are so revolutionary that their premature implementation itself causes all kinds of organizational rejection problems. The primary purpose of the pilot test is to establish (or disprove) the economic viability of the new paradigm. Given that it is found to be beneficial, it will also begin to establish its organizational acceptance by developing the essential communication mechanisms. No pilot should ever be declared a failure, since all are learning experiences that accrue to the future benefits of the organization. If only 1 in 10 prove economically feasible, the results could still be overwhelmingly positive.

5. Evolve a financial structure for the organization that will encourage the development and implementation of improved paradigms. This begins with the education of stockholders as to the ultimate value of the "new management paradigm" itself, which includes an emphasis on continual improvement and a view for the longer term.

These changes should not be difficult to bring about, although the paradigm problem is probably more pronounced in management practice than anywhere else. However, organizations that are not oriented toward and supportive of such changes have their days numbered.

By far the greatest incentive to innovation is a realignment of the *organizational culture* in this regard. Often, the formal "orders" to innovate are counterproductive, as also might be financial incentives if they cause animosity between co-workers. This is often not seen by managers since they mistakenly judge the sociology of their employees by the standards of their own social environment. It is essential that the informal organization be totally supportive of new paradigm formulation and organization paradigm shifts. That is, change is not induced by management, it is brought

Sec. 11.3   Solving the Paradigm Problem

about by the presence of a holistic *learning organization*. This is the essence of evolving the organizational culture toward technimanagement.

Probably the most important element in bringing about a learning organization is the recognition that new paradigms are not hatched spontaneously. They must be conceived in an environment conducive to their survival. They must be incubated in a comparable but significantly different environment. They must be nurtured to maturity in yet another. All of these environments do not come about by accident or by the mere presence of "good" people. They must be created and maintained by concerted and continuous management efforts.

How does a manager create organizational involvement and support (i.e., one that will not be attacked by the originator's peers)? Unfortunately, the short-term orientation, which is distributed throughout the entire organization, makes this quite problematic. The primary key is a recognition of *informal organizations* and their effects on acceptance. We consider this in considerable detail in the following chapter.

12

Informal Leadership and Organizations

12.1 Definitions, Methods, and Problems

Informal organizations (note the plural form) exist within all formal organizations. They are mandated by the basic observed principles of the sociological nature of humankind. The formal organization is characterized by its *formal leaders*, who are usually given titles by those "higher up" in the organization. The informal leaders derive their power from peer identification, for which the English language is almost devoid of an effective word. Those familiar with Spanish might better appreciate the adjective *simpático*, which we will use in the sense that one is able to get inside others' true feelings (apart from overt actions). Our term *to have sympathy* is close, but it carries with it the excess baggage of pity. While our word *charisma* is effective in conveying a sense of identification, informal leaders do not have to be charismatic; in fact, some of the most effective remain constantly in the background, exerting power almost subconsciously with little recognition of this fact by anyone. However, the charismatic leader with considerable *simpático* is certainly most visible and probably has the highest potential to motivate action (for good or ill).

We have defined this word to make this point: The peer group is not *simpático* with the informal leader as much as they instinctively recognize that the informal leader is *simpático* with them. Further, to be *simpático* cannot be created: It must be genuine, and any attempts to create it artificially will have the opposite effect with virtual certainty (even though they might seem to work for a while). It is very difficult (we might even say virtually impossible) for managers to be *simpático* with their subordinates just by virtue of their positions. Thus any attempts to eliminate informal organizations by merging them with the formal organization should be abandoned. The best advice to managers who are even toying with this idea is: *Don't even think about it*.

A further humbling observation is that over time, informal leaders tend to dominate social attitudes. Consider the observation of Kazmier: "The emphasis of the formal organization is on the positions in the organization, whereas the emphasis on the informal organization is on people and their relationships" (KAZMI69,160). Which do you think will dominate over time? If yours is an organization in which organizational positions dominate "people and their relationships," you have a very sick organization and it is time to initiate the healing transition introduced in this chapter and detailed in Chapter 24.

In this section we continue discussing the methods for identifying informal organizations and the problems associated with some most commonly applied approaches toward dealing with them. This will lay the groundwork for the remainder of the chapter, which deals with the benefits and empowerment of informal organizations. Finally, summary thoughts on informal communications are presented.

12.1.1 Identification Methods

If you, as a manager, have not been able to identify informal organizations within your organizational component, this will be the first problem in dealing with them. The problem is usually not a failure to identify any evidence of informal organizations as much as it is to oversimplify this very complex and dynamic sociological process. For example, if only one informal leader has been identified, chances are that several other informal organizations have been overlooked. In addition, if the same informal leader has been recognized for a specific capacity over a very long period, perhaps the changing nature of informal organizations has not been considered.

Let us attempt to establish some underlying principles that relate to the characteristics of informal organizations so that we are in a better position to identify them. Consider the following:

1. Informal organizations are formed spontaneously to meet sociological needs which, by definition, are not met by the formal organization. *Spontaneously* implies that they are not planned. So, at least at the outset, there is no intent to accomplish anything with them. Frequently, their existence is not recognized by either management or their membership, due to their informal nature.

2. Generally, there will be many informal organizations of varying levels of maturity and cohesion within any formal organization, and several may exist within any formal organizational component.

3. Since informal organizations have little, if any, respect for the formal organization, organizational boundaries are quite easily traversed, and membership can come from any component.

4. Although informal organizations tend to be most predominant at the lowest level (due to the numbers), managers often form informal organizations, especially with their peer managers.

5. The synergy that makes an informal organization viable and identifiable comprises the common interests that the members share.

6. Informal organizations either grow or diminish in strength, depending on the perceived needs from which the cohesion of the group is derived. The more intense these interests are perceived to be, the stronger the informal structure that will hold them together. Potentially the most intense of these is a perceived common threat from the formal organization. However, the effects of informal organizations can be quite beneficial to the goals of the formal organization. (The positive aspects of the formal organization are usually deemphasized, for reasons discussed below.)

7. Informal leaders are defined by their respective informal organizations, not vice versa. While the views of the informal leader will most certainly affect the attitudes and beliefs of all members of an informal organization, it is a grave mistake to view the informal leader to be the cause of that informal organization. *The informal leader is an effect, not a cause.*

8. No informal leader is capable of leadership under all circumstances or conditions. There are two dimensions to the formation, strength, and longevity of informal leadership: functional and temporal. Functionally, a single person is not able to supply all informal organizations with leadership, since each informal organization generally has its own set of needs which caused its formation in the first place. Temporally, informal leaders might fail to change as the needs of their constituencies change, thus automatically generating a new informal leader who is more responsive to these needs.

9. Members of a given formal organization might be members of, and even exercise leadership within, several informal organizations.

10. Informal organizations tend to communicate with each other to the

extent that they have a commonality of sociological needs; this communication is facilitated by people who hold membership in multiple informal groups, and the ultimate of such communication is a merging of two or more informal organizations.

11. Informal organizations tend to become formalized, either by (a) the formal recognition of management, (b) the generation of an internal structure (e.g., a "worker's association"), or (c) unionization, regardless of strength.

12. By definition, when an informal organization becomes formalized, *it ceases to function under the principles of informal organizations*, and other informal organizations arise to address those issues that are still not perceived to be resolved adequately by the formal organizations.

The understanding of these principles allows managers to identify informal organizations by direct observation. This last principle is significant, for it implies that informal organizations operate on a completely different set of principles than do formal organizations. They are quite flexible and completely voluntary, which gives them their inherent power.

The bounds of an informal organization might be impossible to identify. However, informal leaders usually become quite apparent. In most cases they will be the group spokespersons, but this can be exceedingly deceptive, due to the informal organizations' dynamic nature. If, in fact, managers believe that they are dealing with a given informal leader when in fact they are not, this could create a major breakdown in the control system (loss of the measurement component). On the other hand, if the proper management mechanisms are in effect, it might not be so important who the informal leaders are at any point in time. This is the ideal situation, analogous to a thermostat working regardless of the temperature. However, the concepts of control do not apply to informal organizations as they do to formal organization, as we shall see in the next section.

12.1.2 Problems with Current Approaches

It is impossible to address all the approaches that might currently be applied by the formal organization toward informal organizations. We attempt to discuss those that seem to us to be the most intuitive and possibly the most common. By far the worst approach is one of *denial* that they exist, which results in no approach at all. Actually, this is the default approach most commonly taken, but it is based on ignorance that the informal organizations even exist, and thus it cannot possibly be effective because of its total disregard for reality. This can be a rather grievous error. The adjacent display explores the potential causes of denial.

Since informal organizations are a reality in all but the most infantile of formal organizations, their effects will become quite apparent to the manager in due time. These effects can be both positive and negative. The positive aspects are rarely attributed to informal organizations; instead, the manager takes personal credit for them as "just good management practice" (a further form of denial). However, when a perceived negative effect is forthcoming, it is blamed on insurrection. At this point a whole host of countermeasures might be employed to deal with this perceived foe.

> **Causes of Denial of the Informal Organization's Existence**
>
> - *Threat to the manager's ego.* Consider the common attitude revealed by the following quote:
>
>   After all, I have been designated as the manager of this department. That should be the end of the issue. What further leadership is required? If I only had total allegiance [i.e., control] over every aspect of my people, things would work out fine. But X, Y, and Z seem to have their own agendas. Why don't they understand that this is all for their own good?
>
> - Managers need to learn that the very last thing that they want is total control—no one is smart enough to be all the brains within an organizational component; and it is only the manager's own ego that prevents this recognition. This is a weakness to which the novice manager is particularly vulnerable.
>
> - *Threat to managerial authority.* This arises from a misunderstanding of the basis of authority (see Chapter 5). Although counterintuitive, informal organizations can greatly increase the authority of effective managers, since their authority is measured by the power of their organizational components, not by their power to control individuals.
>
> - *Oblivious ignorance.* This is expected in technical organizations, where the managers are basically schooled in their technical specialty as opposed to the sociological aspects of management. However, this is the most forgivable of the potential causes of denial—and the easiest to overcome.

Before considering these fallacious countermeasures, let us observe that like the alcoholic person who causes a fatal car crash, there is a sudden recognition of a problem by everyone. The logical person would reason that this should precipitate a proper assignment of fault on the part of the alcoholic, who would then initiate an effective rehabilitation process. Unfortunately, this is the exception rather than the rule. Instead, there is a tendency on the part of the alcoholic to blame anyone (everyone) else, which includes the blaming of just bad luck (transfers the guilt to the ultimate higher power). As strange as this might seem, this irrational behavior is often accompanied by what is called a codependency, where most everyone who is close to the offending person participates in reinforcing the blaming behavior by apparently agreeing with the offender's assessment.

In our analogy, the manager refuses to recognize informal organizations as an effective approach to solving organizational problems and, instead, views problems as isolated incidents caused by troublemakers. The codependents are the other members of the organization, who in their ignorance or reluctance to intervene, reinforce these perceptions. Some methods employed under this influence include:

1. *Fire the informal leader.* Informal leaders frequently become the focal point, since they typically surface as spokespersons. It would be simple if this were the cause of the problem. Question: Has assassination ever been effective in stomping out any cause (whether worthy or not)? On the contrary, it usually accomplishes just the opposite over the long term. By making a martyr of a person, those in authority display their inability to lead and the total moral corruption of their cause. Remember, the informal leader is a symptom, not a cause. (This does not mean that firing an unfit or incompetent person is always counterproductive to the formal organization; it does mean that when such action is taken it should be recognized as an essential action by the vast majority of the organization. This would very rarely be the case where someone has risen to the position of leadership within one of the informal organizations.)

2. *Fire everyone causing the problem.* The same comments apply as those given immediately above. It should be recognized that this is a multiplication of the consequences, and even the thought of such action is probably an admission of poor management philosophy.

3. *Divide and conquer.* Find a champion at the lower level who is willing to stand up to the informal leader and let him or her do the dirty work. When this countermeasure is employed, the issues in which they disagree are usually quite irrelevant. This is the epitome of manipulation, and those who employ it will not have long to wait for the consequences that this will wreak within the entire organization.

Although we have not explored the gamut of all possible attempted countermeasures in this regard, we need go no further to see the folly of any actions taken which continue to deny the rightful existence of (or to destroy) the informal organizations. It is just impossible to make good decisions when they are not based on at least a fairly accurate perception of reality. In this case the perception is about as far from reality as a manager can get.

Allow us to postscript our analogy by taking it a few steps further. The practicing alcoholic in denial often does great harm to the very people who are attempting to help. In this case it might be the members of the informal organizations themselves. Further, like alcoholics, managers should never believe that they are ever "cured." The concept of *recovering* is quite analogous to that of *continuously improving forever*, and it also carries with it the sense that there is always the danger of regression. To deal with reality, we must first recognize it.

So let us turn our attention to those actions that might be taken when there is an accurate perception of the presence of informal organizations. Let us suggest that there are several potential actions which are commonly employed that might cause problems even if the existence of informal organizations is accepted. Consider the following temptations:

1. *Promote the informal leader.* This would seem to be a logical thing to do, and sometimes it might be required (this was discussed in detail in conjunction with the Peter principle). It is very tempting, especially if the informal leader is both intelligent and highly productive. However, it should never be done to disable an informal organization. This is true simply because the leader is the effect of the creation of an informal organization, not its cause. Thus the promotion of a competent person within the formal organization will have the effect of creating a vacuum within the affected informal organization that may be filled by someone who you would not promote in your wildest dreams. This action is often an attempt to take advantage of an informal organization by merging it with the formal organization. The theory is that an informal organization will become (more) loyal to the formal organization if its informal leader becomes the formal leader. The only problem with this theory is that it is wrong. This was explained above: When informal leaders are given formal recognition by management, they cease to be the informal leaders and others will arise to take their places. In addition, the temporary dual role that the informal/formal leader will assume will create a role conflict that will disable her or his ability to function effectively in either role until it is resolved.

2. *Become the informal leader.* If the reader has believed even half of the principles discussed above, this should seem laughable. Unfortunately, however, the concept of "leadership" espoused by many management philosophies reveals a complete misunderstanding of the nature of

informal organizations. Even if workable, this approach would not be effective, for the same reasons as those described in action 1 above. However, rather than promoting the informal leader, this is an attempt to usurp the authority of the informal leader. By definition, however, this is an unworkable approach in that the formal leader is not qualified to be an informal leader. (This is referring only to becoming the leader of an informal organization within the manager's organizational component. There is no reason that a manager would not become the leader of an informal organization that includes people at other levels in the organization. In fact, this will become more widespread as the transition to technimanagement progresses.)

3. *Manipulate the informal organizations.* (There's that nasty word again.) Once informal organizations are understood, this will certainly be a temptation. Recall that the primary ingredient of the manipulative approach is that of deception. Thus any attempt to use an informal organization that is not totally and candidly open to every person within it is destined to fail.

These approaches display a misunderstanding of informal organizations even though they admit that it exists. There is an underlying implication above that management views the informal organizations with suspicion. To fully understand the proper approach toward informal organizations, this must be overcome and the benefits of informal organizations must be appreciated.

12.2 Benefits of Informal Organizations

Much of the following was inspired by thoughts of Keith (KEITH71). However, because these might not totally reflect his opinions, we urge readers to review his excellent work on this subject. To formulate positive and productive approaches toward interacting with informal organizations, it is essential that their benefits be understood. Management tends to see only the bad side of informal organizations because they seem to surface only when there is a problem. We clearly allow for the potential downside (see Section 12.2.6). However, we defer the discussion until the benefits are understood, since the major downside is the failure to obtain these benefits.

12.2.1 Filling the Decision-Making Gaps

One of the most valuable benefits of informal organizations is their role in filling the gaps in communication and decision making. To illustrate the decision-making potential, consider the following example scenarios:

Scenario 1. An emergency arises in department X which requires that a special gadget be produced in department Y or else several weeks of experimentation will be lost. Without consulting his supervisor, Joe calls on his friend Harriet in department Y, who works through lunch, produces the part, and gets it to Joe for installation so that only a minimum number of data are lost.

Scenario 2. Joe had been telling his supervisor about the squeaking gadget in the olgunameter for weeks. Just at the most critical time, it breaks and threatens to interrupt the data gathering such that several weeks of experimentation will be lost. Joe complains to his supervisor, who takes the broken gadget to the supervisor in department Y. However, the manager of department Y is in the middle of a firefight himself, and he complains about the fact that they are backlogged and will not have time to fit the gadget in. He proposes going to an outside source. Joe's supervisor believes that this is unreasonable and goes to his manager about this. His manager has no direct authority over department Y, but he contacts his boss, a vice-president who is playing golf with a multimillion-dollar client. (Anyway, you get the idea, so you can finish the story.)

Postmortem. A high-level meeting of all divisional managers is held in that it is decided that in the future, broken gadgets (or such) which threaten to cause a significant shutdown and loss of data in other departments will be given top priority over other production items unless critical external customers are affected. A memorandum to this effect is written at the vice-presidential level and distributed to all departments. However, additional guidelines will be forthcoming to determine what constitutes a "significant shutdown" and what constitutes a "critical external customer." They will hold another meeting in the near future to resolve these issues. *Question*: To which scenario should we assign the postmortem?

Scenario 1 presents informal organizations in their finest hour. These types of transactions occur in all productive organizations, mainly because of the inadequacy of formal organizations to cover all eventualities. In scenario 1, Joe took the leadership role. Harriet courageously followed his lead, as did several other "accomplices." Did they know that they were possibly going against some company policy—and certainly against the wishes of one of their supervisors? Quite likely; maybe they even enjoyed that aspect of it.

With these examples as a backdrop, let us put forth this principle for further consideration:

> In all effective formal organizations, informal organizations are encouraged to exist to solve those problems that are not handled explicitly by the formal organization.

Unfortunately, the effectively functioning informal organization is often seen as a threat to the authority of the affected managers. While the supervisor of department X probably appreciated the actions of the informal organization in scenario 1, it is clear that Harriet's supervisor might have had other concerns.

The only difference between scenario 1 and scenario 2 was the effective functioning of an informal organization. By *effective functioning* we mean that it is conditioned to respond to the needs of the organization. Will formal management always bungle the problem, as indicated in scenario 2? Of course not; but that is not the point. We used some exaggeration to demonstrate what typically happens when informal organizations do not function effectively and thus the formal organization has to solve all problems. Note that informal organizations still exist; they are just not responding to this particular problem.

In reality, the formal organization is just not capable of anticipating every eventuality, and even if it could, the likelihood that it would be able to draw up policies to respond properly to all of these is quite remote. If the formal organization is fortunate, an informal organization fills the gap, as in scenario 1. But good fortune is not just a matter of random luck. Rather, it is the formal organization (i.e., management) that creates the environment for informal organizations to be productive or not.

The postmortem above is an example of an attempt by a formal organization to formulate a future policy, after the fact, to address similar problems in the future. Note that all of this effort would be skipped under scenario 1; thus all of this valuable upper-administrative time could be employed on more productive endeavors (such as entertaining valuable clients). However, the example illustrates the attempts of the formal organization to resolve issues that properly belong to informal organizations of intelligent people.

We recognize that our examples are not even historical anecdotes, and therefore they certainly do not prove anything. They are given to illustrate one potential value of informal organizations. We appeal to the reader for proof. Consider how organizations would work without such informal mechanisms. Consider what would happen if everything were done strictly by the book (assuming that there is a book). In fact, in all functioning organizations there are many informal mechanisms that enable work to get done in a more efficient manner. These are not to be feared; they are to be respected and rewarded for the benefits they produce.

Let us conclude by stating once again that informal organizations are a reality. They can either produce benefits for the formal organization, as exemplified by scenario 1, or they can be passive to the formal organization's needs, as in scenario 2. Where was the appropriate informal organization in scenario 2? We do not know. However, given that the members did not respond to the problem at hand, there is a good chance that they were preoccupied with nonproductive, possibly counterproductive, endeavors.

12.2.2 Personnel Development

We were emphatic that the informal leader not be promoted just to disable informal organizations by merging them with the formal organization. However, there is a function of personnel development that informal organizations play. This is primarily one of sociological development, especially with regard to gaining an understanding of the needs of the members within a given informal group.

There are a number of ways that the leadership characteristics developed or identified by informal organizations may serve the goals of the organization. While the promotion of an informal leader can be quite detrimental, there are times when it is appropriate. Two such instances might include (1) assuming mutual approval, the relocation of the person who has been promoted to a division of the organization where there will be minimal, if any, contact with informal organizations over which this person exercised leadership, and (2) situations in which the informal leader is strongly supported for promotion by most of the other members of the informal organizations. Neither of these should be, nor should they be perceived to be, attempts to disassemble the informal organizations. Although we see some obvious problems with the second of these, clearly exceptions must be made to any rule which totally prohibits the promotion of an informal leader when it is clearly in the best interests of the organization and the individual to do so.

Aside from promotion, personnel development within the informal organizations should be recognized as a beneficial result of the process of informal group formation and functioning. Individuals at all levels should be encouraged to develop (or not discouraged from developing) leadership skills through their respective informal organizations. These key leaders should be recognized as critical to the information gathering essential for effective decision making. They should be considered heavily and given broad discretion and empowerment (see Section 12.3) as they are assigned to work teams and task forces which distribute the decision making throughout the organization. This brings us to our next topic.

12.2.3 Useful Channels of Employee Communication

The value of the informal leadership in providing more effective communications with management was introduced above. This can be extended to include the value of communications between employees at all levels, especially within informal organizations. These communications are essential to organizational morale and harmony, which are the primary keys to productivity. While it is impossible to increase morale by forcing people to communicate with each other, nothing kills morale faster than a breakdown in communications. *Formal communications* (i.e., meetings, memos, and other interactions necessitated by the formal organization) are just not sufficient to bring about increased morale. If anything, they tend to be demoralizing, usually communicating the necessity for efforts which are outside those which would naturally be performed to satisfy organizational goals. Something is necessary to counterbalance these rather depressing communications.

Informal organizations provide new channels of communications, and this is one of the main reasons that they are created. Of course, we recognize that communications can be used to ruin morale as well as to improve it. But the absence of communication is no more an option than the absence of informal organizations. Recognition of this enables the manager to attempt to provide an environment in which healthy and morale-building informal communications will take place. We resume our discussion of informal communications in Section 12.3.

12.2.4 Improving Managers and Management

Just recognition of the existence of informal organizations will improve management. Further recognition of their various manifestations and leaders within the informal organizations will further strengthen managers in establishing effective relationships with them. A final major advance comes from the recognition of informal organizations as positive forces for improving the capability of meeting organizational goals. These factors certainly improve overall management capability. However, in this section we are concerned primarily with the mechanisms by which managers and management are improved by the functioning of informal organizations, recognizing that better managers are recognized only by their ability to create better management practices.

One factor mentioned in the literature is the pressure that it puts on managers to exercise effective leadership. This pressure is induced from the leadership exerted by the informal leaders. There is a temptation to think that this would best be realized if formal managers view their informal counterparts as competition. However, this as a very unhealthy viewpoint by the manager, for two major reasons. First, the manager and the informal leaders function in two relatively independent realms. Although they clearly overlap geographically, and at times functionally, the primary difference in these realms is one of perceived authority. It is impossible for the formal leaders to get within the informal leaders' spheres of influence as long as they remain "managers." Second, the idea of "competitiveness" arises from a jealousy that the manager has toward the informal leader. This is one of the most dangerous of passions, since it leads to a disregard for objective decision making. Thus we conclude that it is quite unhealthy for managers to see their informal counterparts as competition.

Even if managers have the proper noncompetitive view of their informal counterparts, their recognized presence can spur managers to greater managerial excellence. For the power of managers to achieve organizational goals comes from the determination of their subordinates to overcome all obstacles in reaching those goals. Informal organizations can be the most contributing factor to this (and, if so, they are also potentially the most destructive). This contribution will be positive only if it is preceded by a realization on the part of managers that they are not in full control of all aspects of the workplace. This requires a certain sense of humility that is quite difficult for most managers to attain, regardless of experience or educational background. However, once attained, it establishes a management atmosphere in which initiatives can be taken to evolve mechanisms by which informal organizations can unleash their potential.

Thus managers are improved by informal organizations primarily by the establishment of effective communication and alliance mechanisms. This is just the opposite of the competitive spirit that pervades many organizations. It is essential that both the formal and informal leadership agree on, if nothing else, their mutual need to work together if their organization is going to prosper to everyone's benefit. In Section 12.4 we will see that this is achieved by allowing informal organizations certain decision-making powers that have traditionally been considered to be management prerogatives.

12.2.5 Improving Worker Morale

The fact that informal organizations are created spontaneously out of a social need implies that their creation and growth would increase worker morale. However, while the existence of informal organizations is not in question, poor morale persists in many organizations. Thus we must formulate some basis for the relationship between informal organizations and morale. Consider the premises in our line of reasoning given in Figure 12.1. It goes without saying that one of the major objectives of management should be to polarize informal organizations in a positive way. But recall that the formation and operation of informal organizations is reactive, not proactive. Thus management cannot directly polarize the informal organizations in a positive way (the negative polarization potential is obvious). Positive polarization must be done by the lengthy process of establishing the proper environment and relationships.

---

1. An informal organization creates itself spontaneously in response to a perceived sociological need to satisfy objectives that are not satisfied by the formal organization.
2. These objectives are usually a mixture of two extreme poles:
   • *Positive:* those in complete harmony with organizational goals
   • *Negative:* those counterproductive to organizational goals
3. The position of the objectives with regard to these two extreme poles is almost completely correlated with worker morale; the more the objectives are toward the positive pole, the greater the organizational morale.
4. Management can polarize informal organizations in either a positive or a negative way.

Figure 12.1 Factors in Management Polarization of Informal Organizations

This presents a dilemma, since we are expecting the formal organization to have an influence on an organism that came into being solely to satisfy needs that were not being satisfied by the formal organization. However, these fall into two categories: (1) those that could not be satisfied by the formal organization, and (2) those which were due to deficiencies in the formal organization. These map almost perfectly to the positive and negative subdivisions in premise 2 of Figure 12.1. Let us consider these in order.

- *Needs that cannot be satisfied.* There are certain social as well as functional needs that cannot be satisfied by the formal organization. There is probably no argument as to the social needs. Even though some formal efforts are made to facilitate social activities, no one begins to believe that all social needs could be thus satisfied. Functional needs of the organization, however, are much more controversial and thus subject to denial. We have expressed the fact clearly above, however, that no formal organization can possibly consider and address all the scenarios that will occur in the future. Too much is going on simultaneously in any meaningful organization for the formal machine to keep up with it all. Given that informal organizations are attuned to organizational goals, they will fill in these gaps, and when this happens it moves the organization toward the positive pole and higher morale. However, when this urge to further organizational goals is stifled by management, and when management threatens the existence of an informal organization in this regard, we can expect a reaction toward the other pole and a radical decrease in morale.

- *Needs due to deficiency in the formal organization.* Neither the inability of the formal organization to provide total social satisfaction, nor its inability to cover all organizational functional needs should be viewed as deficiencies of the organization. They are facts of life. In contrast, the following examples are considered to be deficiencies of the formal organization: (a) outright incompetence on the part of management to provide the basic functional organizational structures, (b) unfairness, (c) guile, and (d) manipulation. These types of deficiencies will polarize the informal organizations in the negative direction and produce low morale. In the extreme case, they can create a situation in which the informal organizations act in ways that are counterproductive to the goals of the organization. Depending on the passions produced, this could either be due to a countercontrol situation or it could be intentional.

The surest way to bring about the negative type of dissatisfaction is to try to stomp out an informal organization's legitimate attempts to solve problems in meeting organizational goals. This is one of the gravest of management errors, but we feel it to be the rule rather than the exception. (More on this in Section 12.3.) This leads us naturally to the next section.

12.2.6 Downside of Informal Organizations

The purpose of this section is to provide some balance to the discussion above, which stressed the positive aspects of informal organizations. This viewpoint was taken because we feel that ignorance of these positive aspects causes far more problems than does ignorance of the downside. In fact, the perception of informal organizations as a threat to the formal organization creates the management-by-fear philosophy expected of fearful management.

Clearly, every positive aspect of informal organizations corresponds to a downside if it is absent. Like anything else that has the potential to bring about good, it can be used in a way to produce ill effects as well. The most benign of these is the mere absence of its good effects. In the worst case the potential good effects are turned to produce much more destruction than good (analogous to atomic energy).

Let us delve a bit deeper into this dark side of informal organizations (we dare not be naive). In the extreme case we have total insurrection—a situation in which everything that the formal organization wants to accomplish is opposed by the informal organizations, and vice versa. Even in the worst-case scenario within our society (e.g., the strike) we usually see a disagreement in the methods, not overall goals. While isolated events occur in which direct damage to the formal organization is sought, it is hard to conceive of a situation where total insurrection rules. In such an environment, there seems little that could be done to salvage the situation.

[We might interject here that unions fall under the category of informal organizations only in their infancy. Once they are formalized they become an integral part of the formal organization (for better or for worse). Indeed, unions will have their own informal organizations, which only further complicates our model of reality. Fortunately, this is not that relevant to technical management.]

Given that the extreme is not realistic, let us withdraw somewhat from this view. Most organizations within our society function with a general agreement between the formal and informal organizations as to the overall organizational goals. However, there might be considerable disagreement as to the methods that should be employed in accomplishing these goals (i.e., the objectives). These disagreements will vary considerably between organizations, reflecting the management strategies that have been employed in the past as well as other sociological aspects of the workforce. Few organizations have a totally harmonious relationship between the formal and informal organizations, and even fewer are in a state of total insurrection.

Ideally, management will have a realistic view of where they stand with regard to informal organizations. The informal organizations will be operating somewhere along the spectrum from total insurrection (totally dysfunctional) to total harmony (fully functional). The question that management should address is: Given our current state, how do we go about creating an environment in which we can move continuously and forever in the direction of greater harmony. This recognizes that we will never "get there." It also recognizes that informal organizations can and often do have a very formidable downside. It is for this very reason that they cannot be ignored as one of the major factors in effective management.

12.3 Empowerment of Informal Organizations

At the end of Chapter 11 we attempted to provide some solutions to the paradigm problem. This was somewhat difficult to do in the absence of a consideration of informal organizations. We have given some of the potential benefits of informal organizations above. Conspicuously absent from these is a solution of the paradigm problem. (The failure to recognize informal organizations is itself an example of the paradigm problem—but that is another issue.)

We propose that there is a close relationship between the organizational paradigm lock and the lack of proper involvement of informal organizations. Let us revisit Whitehead's statement which we quoted in Chapter 11: "No society or organization is averse to change, provided the initiative for that change takes place at the relevant level—at that level where the daily activities have shown the need. Under those conditions, change will present itself not as an interruption, but as the natural flow of social living..."(WHITE36,78—79). Given our understanding of the cause of spontaneous informal-organization generation, we should be able to see that informal organizations could well develop in response to the paradigm problem. When this happens and the formal organization does not respond to it adequately, an informal organization might transform itself into an entirely new formal organization (i.e., a new company based on the new paradigm). Numerous examples could be cited of this in contemporary industry.

An alternative application of the foregoing quote gives us a practical way of dealing with the paradigm problem. The lack of involvement of informal organizations is one of the primary causes of the paradigm problem, since it prevents "the initiative for . . . change" to take "place at the relevant level." Management-imposed changes are almost certain to be opposed by the informal organizations, even if managers were smart enough to conceive, develop, and implement them (which they are generally not, making the resistance even worse). We might note that in a majority of the cases in which resistance is forthcoming, it is not the fact that it did not originate within the organization as much as the perception by the informal organizations that the paradigm shift is not one that will lead to optimality. There is nothing that the workforce hates more than the discomfort of a paradigm shift which they know beforehand will not result in improved organizational performance.

We concluded Chapter 11 with the question: How does a manager create organizational involvement and support (i.e., one which will not be attacked by the originator's peers)?, and we suggested that recognition of the informal organizations was a primary key. Now that we understand what informal organizations are and how they operate, we are in a position to discuss management strategies with regard to them. This is closely connected to the paradigm problem, since their involvement requires the continuous creation and acceptance of new paradigms.

If the primary key is the recognition of informal organizations, the essential secondary keys are harmonious relationships that are established between the formal and informal organizations. This process of moving toward a positive, functional working relationship between the formal and informal organizations we call *empowerment*.

Probably no management term carries such an excess baggage of fear and misunderstanding. It is essential that we recognize that empowerment is to be viewed in the context of optimization and what is sought is a degree of distributed decision making that is optimal for the particular organization under consideration at this particular point in time. At the same time, we want to put organizational mechanisms into place, which will evolve a different, more effective empowerment structure as the organization matures and learns with respect to its various responsibilities.

Recognize that this concept is difficult to understand in the abstract and that the paradigm problem is taking its toll here. The most predominant current management paradigm is heavily oriented toward Theory X. Management establishes the rules and delegates the various responsibilities as they see fit. Once the organizational chart is set up, it is expected to stay in place ad infinitum. In other words, the formal organization is assumed to be fixed—unlike the informal organizations, which are the essence of fluidity in both their existence and form. When problems arise, management establishes rules which they expect will solve the problems once and for all (as in our example postmortem). Further, the informal organizations which inevitably form generally operate totally independently from the formal organization, even though they have the same basic goals. Typically, these covert operations are perceived to be essential because of the fear that management will desire to control and ultimately disable the effective informal mechanisms which have evolved—a fear that is quite well founded.

It is extremely difficult for managers in the middle of their operations to recognize this (forest/trees). But ask any of the people who are actually getting the job done in the organization, and this is exactly the situation they will describe. (We recognize that there are a few notable exceptions, in which case the principles of empowerment have already begun to be applied.)

Returning to our subject, we have two major objectives at this point: (1) to define the concept of empowerment as it applies to the effective and harmonious relationship between the formal and informal organizations, and (2) to define the mechanisms that will bring these relationships into existence.

12.3.1 Effective Empowerment

At this point we wish to distinguish between effective empowerment and ineffective empowerment. The fear of empowerment is the fear of ineffective empowerment, for there is nothing to fear in effective empowerment. Unfortunately, the formal organization has a tendency to resist all empowerment of informal organizations to the point where they often deny their very existence. This denial of reality is the most effective way of disempowering themselves. This leads to the first principle of effective empowerment:

> **Shared power principle**. Effective empowerment increases the power of managers in the same proportion that it increases the power to their subordinates.

Thus managers should seek empowerment for their subordinates as opposed to trying to isolate them from it. However, we qualify this by stating that this must be *effective*. It is essential that we differentiate between effective and ineffective empowerment. For like any other power for good, ineffective empowerment can produce its ill effects.

Figure 12.2 demonstrates the difference between effective and ineffective empowerment. Given these potential misinterpretations, it is no wonder that "empowerment" is suspect by both the formal and the informal organizations. In the remainder of this chapter and this book, when unqualified the word *empowerment* will refer to *effective empowerment*. At times we will qualify the word for clarification and emphasis.

We began this section by presenting a version of the shared power principle, which views empowerment from the point of view of management. The characteristics of effective empowerment given above present a symmetry that responds to the question: Where does this power come from? It must be recognized that tremendous power, possibly exceeding that of management, exists distributed throughout the various informal organizations. As the organization that understands the principles of empowerment evolves, not only is the power of management transferred to the informal organizations, *but much of the tremendous power of the informal organizations is transferred to management*. This is probably the most misunderstood aspect of effective empowerment on the part of manage-

Ineffective Empowerment	Effective Empowerment
Motivated by a desire (either on the part of the formal or the informal organizations) to control people	Motivated by an understanding of human needs and organizational optimization and control
Manipulative gimmick either by the formal organization (to get the lower levels to "buy in" to their methods) or by an informal organization (to wrest power from the formal organization)	Requires that both the formal and informal organizations understand the purpose of the transition
Seen by management as a loss of their authority; seen by informal organizations as a tool for further manipulation	Threat to neither the formal nor the informal organizations
Either mandated by management or the result of a power grab by the informal organizations	The result of a plan that is developed jointly by the formal and informal organizations
The precipitous result of arbitrary delegation or collective bargaining	Evolves as the organization matures true self-healing and learning structures
Constant struggle for more and more decision-making power at all formal and informal levels	Collaboration at all levels to determine which person or group is in the best position to make each decision that must be made to meet organizational goals

Figure 12.2 Contrast between Effective and Ineffective Empowerment

ment, although its understanding by the informal organizations probably accounts for much of their resistance to empowerment. This resistance is unfounded and comes from a misunderstanding of the true nature of effective empowerment. To allay these fears we present the following equivalent restatement of the shared power principle:

> **Shared power principle.** Effective empowerment increases the power of informal organizations in the same proportion as it increases the power to management.

Thus this is a clear win–win situation. Both management and the informal organizations gain power because in working together the entire organization prospers, grows, and becomes more powerful. Also, by optimally sharing power as opposed to competing for it, management and the informal organizations unleash the tremendous power to accomplish organizational goals which is usually spent in disabling and ultimately destroying the organization (e.g., gridlock and internal energy dissipation).

(This is another example of where some of the major influences on our culture—our sports and gaming paradigms—break down so terribly. With the ubiquitousness of sporting and gaming events, and the religious zeal with which they are pursued, it is very difficult for us to think of the episodes of our lives as other than games in which there must be a winner and a loser. With effective empowerment, if there are any losers, everyone loses; and if there are any winners, everyone wins. Oh, that our sporting and gaming events could reinforce anything but the opposite of this principle!)

12.3.2 Empowerment Mechanisms

Given that we have arrived at a healthy view of empowerment, the next question is: How do we establish the mechanisms that will bring this about? In the ideal sense, we know where we want to get—the difficulty is in getting there. Indeed, many have attempted to embark on this pathway before, and most have failed to reach the goal. To introduce empowerment mechanisms, it is first of interest to explore the reasons that these attempts have failed and then address them in terms of sound principles. We do this in this section; however, this subject will be revisited in much greater detail in Chapter 24 after several other sociological principles of techni-management have been introduced.

Consider three primary reasons that past attempts at empowerment have failed:

1. They have been seen as mere organizational assignment and delegation changes and have not been viewed as the initiation of an evolutionary process requiring continuous effort and improvement.
2. When suffering the inevitable setbacks of any major paradigm shift, they regressed over and over again to Theory X solutions,

Sec. 12.3 Empowerment of Informal Organizations   159

ultimately cycling back to a predominantly Theory X organization.
3. They did not have a proper understanding of effective and ineffective empowerment.

These three are certainly not independent of each other.

[We were tempted to use communism as an example of failed empowerment, but we recognized that its failure resulted primarily from its major (hidden) goal, which was to manipulate the masses. Thus the method for its implementation was irrelevant; although that too was flawed by all three deficiencies listed above. A perfectly excellent bad example!]

It must be recognized that persons who have lived predominately under Theory X organizations all their lives (beginning with their educational institutions at about the fourth grade) are not going to be able to transform to a collective empowerment organization overnight. The process must evolve, and there will be a considerable number of setbacks along the way. Whether the reaction to these setbacks is positive or negative will go a long way to determining the success of the transformation (*positive*, meaning continued efforts toward making empowerment work, as opposed to the regression toward Theory X solutions). New control mechanisms must evolve in which the development and implementation of all three elements of control are optimally distributed throughout the organization. If the traditional goal-setting, measurement, and correction techniques are applied (i.e., imposed by management), the organization cannot evolve toward one in which effective empowerment is possible.

Recognition of informal organizations is the first step toward collective empowerment. By *recognition* we mean in the minds of managers, not formal recognition. (We have stated that formal recognition would eliminate existence of an informal organization.) Although recognition is essential, clearly it is not sufficient. To recognition must be added respect, and to respect must be added consideration, and to consideration must be added trust, and to trust must be added reliance. All of these characteristics must be mutual: mutual respect, mutual consideration, mutual trust, and mutual reliance. This is an evolutionary process, the process of establishing a working relationship between you, the manager, and those informal organizations with which you interact. This does not take place overnight. Given the expected setbacks, it is a process that may involve years before it finally comes to major fruition. However, the benefits of just setting the proper direction are immediate. The detailed steps in this process are given in the adjacent display.

> **Process of Evolving Empowerment**
>
> 1. *Respect.* Beyond recognition, this is the knowledge that informal organizations already have the power to make the formal organization succeed or fail.
>
> 2. *Consideration.* Based on this respect, this initiates communication to assure that the relevant informal organizations are appropriately integrated into all aspects of decision making. This needs further definition, since it could be rationalized that "appropriate" might be minimal or zero. All modern philosophies of management promote the idea of getting everyone to *feel* involved. *Consideration* mandates that they will not just feel involved—they will *be* involved. Integration requires much more than mere subordinate involvement; they must become intricate and necessary elements of the organization. This can result only if the formal organization realizes that they are essential to effective decision making. Finally, *consideration* conveys not only that the opinions of others be considered but that they be given *consideration* in the legal sense (i.e., rewarded for their contribution).
>
> 3. *Trust.* This cannot be mandated or imposed; it must proceed out of the successful application of consideration. That is, as the mutual involvements of the formal and informal organizations move in the direction of optimality, the results should become self-reinforcing, to the point that integrated decision making becomes a natural part of the functioning of the organization.
>
> 4. *Reliance.* This represents the final stage of the development of the relationship between the formal and informal organizations. It is here that both the informal and formal organizations have come to a point where they not only recognize their interdependence but thrive on it. In effect, it is a final surrender to this reality. Reliance implies a fully integrated, controlled, and reliable organization

The ultimate legal authority of management to mandate action is not disputed, although the wisdom of mandating things that do not have informal-organization support is highly questionable. Management's ultimate authority places squarely on their shoulders the responsibility to put in place those mechanisms that will initiate the evolutionary process toward the optimal shared decision-making structure.

Although all of the platitudes above seem fine, they will not succeed without implementing the *mechanisms* by which they can be realized. This is initiated when managers establish task forces and work teams that fully integrate the informal leadership. This is a very tedious process calling for optimization considerations to balance the destructiveness that formalization can have on the informal organizations with the mutual empowerment that comes from this full partnership. The system must be finely tuned to assure that the informal leaders continue to enjoy the full benefits of communication with their informal organizations so that they can bring the *pulse of the organization* to the table (i.e., they must remain informal leaders). It falls on the manager to see that the tediousness of this process does not cause this to become another exercise in manipulation. However, full integration coupled with openness serves as an adequate barrier to this danger.

We revisit the mechanisms for creating and evolving these transitions in Chapter 24. There is no substitute, however, for individual initiatives on the part of managers to establish the proper relationships with the informal organizations with which they interact. The formal mechanisms for empowerment can easily be built on this foundation; however, in its absence these formal mechanisms dissolve into mere tools of manipulation.

By far the most idealistic situation is for each member of the organization to be an equal shareholder of the corporation, and this may ultimately evolve in many organizations. It is already a reality in many multilevel sales organizations. Consider the situation, for example, where half of one's income is made from salary and the other half is based strictly on the success of the company. In this idealized situation, everyone is an owner of the company and has an equal interest in its success. The decision as to "who makes the decision" will automatically evolve over time to its optimal point, since everyone has the interests of the organization at the same level as their other personal interests. While we state this as an ideal, some employee-owned and profit-sharing organizations are moving in this direction. Even if this ideal cannot be fully achieved due to the legal and financial structure of an organization, this model gives us a theoretical goal for which to strive. (In reality, many employees have much more than half of their fortune tied up in the company, especially if it goes under.)

In conclusion, let us revisit the primary premise on which the entire reformation of the organizational structure is based:

> The transition weakens no one; it empowers everyone.

It empowers management because it enables them to implement decisions by means of a highly motivated workforce. It empowers the informal organizations because it gives them direct input into all of these decisions to the point where many of them are implicitly turned over to the informal organizations. Most important, it empowers the entire organization because it eliminates the internal stress that consumes most of the traditional organization's energies. Thus, by more effectively accomplishing its goals, the empowered organization works to all of its members' benefits.

12.4 Informal Communication

This section on informal communication is largely a postscript to the chapter. Its importance was introduced above with little elaboration to keep from interrupting the flow of the intent of the section. To further define its significance, we might ask: What does an informal organization do other than communicate? The answer is nothing! All of its positive and/or negative effects are brought about by individuals. Its sole function is to provide the mechanism for informal communications between its members. To the extent that these communications evolve leaders and cohesiveness of action, informal organizations have their impact on the accomplishment of the goals of the organization.

Cannot the same thing be said for the formal organization? True, a very large part of the formal organization is involved with communication, but it has many other functions as well. It is in the formal organization that policies are developed, decisions are made, funds are expended, and so on. The informal organizations have none of these prerogatives, and thus the responsibility of the formal organization to enlist their involvement is further clarified. While the most oppressive of organizations have attempted to suppress informal communications in order to eliminate them, the very idea of this is repugnant to our cherished concepts of freedom. (And besides, it never works!)

It is very important that this be understood, because this is the heart of understanding the nature of informal organizations. Managers often resent informal communications, especially when they lead to individual actions contrary to that which would have arisen out of the policies of the formal organization. Hopefully, this chapter has served to demonstrate the destructiveness of this attitude, mainly because informal communications are not going to be stamped out. This being the case, they need to be appreciated for the benefits that they can bring to the organization. Keith has delved somewhat deeper into the problem, as he states: "Part of the problem stems from the fact that you and I, occupying positions of authority, have too little comprehension of the things people one, two, or three levels below need and want to know. Too often we assume that because we are aware of events others are also aware. Informal lines of communication may also develop simply because the formal line acts too slowly" (KEITH71).

We agree with Keith that the problem is one of incapacity to communicate all of the things to all of the people. It is not just a technological and time problem, it is a problem of insurmountable ignorance. Insurmountable, just because of perspective: It is impossible for managers to understand thoroughly the information needs of their subordinates, and quite often we assume that others already have our understanding and perspective. This is only human, and it is one of those factors that *reliance* is supposed to get us to admit and accept.

This does not argue that the formal organization should not take additional initiatives to counter the misinformation that is bound to be generated by some of the informal organizations. Most of the problems between the formal and informal organizations accrue from the lack of effective and credible communications, and the mechanisms that management must put in place to evolve optimally integrated organizations are almost solely mechanisms of communication. However, the purpose of more effective communication is to replace misinformation with a better perception of reality on the part of everyone. It is not manipulative as in the case of the dispensation of propaganda (perhaps even misinformation) on management's part. Most important, it is not to supplant informal communication, for not only is this impossible, any such attempt will be viewed as a threat to disrupt normal social activity.

Informal communications inevitably become the most trusted source of information within the organization, mainly because there is no impeachment of the source. Management is always suspect for ulterior motives, at least at this end of the transformation. Subordinates do not necessarily distrust their managers integrity, they just recognize that their perspective is different and that their perception of reality is therefore not the same. Even in the best case of totally honest management, this is true, and no amount of management training or subordinate stroking will change this reality. Again, it is management's job to minimize the negative effect of this by presenting communications that are as accurate and candid as possible.

The problem is compounded greatly if there is a perception that the formal lines of communication are not reliable or are intentionally deceptive. Feedback from informal organizations themselves can indicate to management if this is the case. The process of healing this type of wound must be initiated by management, and it can succeed only if the only medicine is major doses of honesty and demonstrated actions applied consistently over long periods of time.

If management is open and candid, constantly attempting to communicate the truth, they have no reason to fear informal communications which also convey the truth. In fact, informal organizations can contribute heavily to the goals of the organization by providing essential communication functions. Informal organizations that consistently fail to communicate the truth will soon lose their credibility and their ability to affect people's behavior. As the formal and informal communications become more consistent, the organizations will, of necessity, develop better relationships.

Now that we have a grasp of both the formal and the informal organizations, let us continue our discussion of the nature of organizations in terms of a concept called organizational entropy.

13

Organizational Entropy

13.1 Definitions

We continue our discussion of organizational behavior, but now from a macro point of view. We will begin by defining terms and stating some principles that apply to organizations as they mature. After this we return to a consideration of how this affects, and is affected by, informal organizations. Finally, group formation and disintegration are discussed.

The principle of organizational entropy is borrowed from the concept of *entropy*, which is fundamental to the field of thermodynamics. This application was suggested by Miller: "The disorder, disorganization, lack of patterning, or randomness of organization of a system is known as its entropy" (MILLE65,195). Entropy is quantified in thermodynamics; the greater the disorder, disorganization, and so on, the greater the entropy.

The *second law of thermodynamics* in its most general form states that *left to themselves, all physical systems tend to increase in entropy*. This should be intuitively obvious from general observation. Automobiles break down, iron rusts, wood rots, nothing works for long without maintenance. In short, our entire world is in a state of decay.

The second law of thermodynamics does not explain how certain components of the universe move into states of higher entropy. While every human body decays, babies are born. The analogies in all other areas of endeavor are readily apparent. Components of this universe are placed at levels of decreased entropy through the consumption of energy, usually massive amounts of energy compared to the energy potential that is created. As this energy is released, these components increase in their entropy levels as they decrease the amount of potential energy that can be turned into useful power. Now let's see how this applies to organizations.

13.2 Principle of Organizational Entropy

The *principle of organizational entropy* states that:

> The second law of thermodynamics applies to organizations.

The analogy is practically perfect. An organization in its infancy is like a physical component to which a relatively large amount of energy has been applied to create an entity of relatively low entropy. It starts out as being highly organized, not in the traditional management sense but in the sense of organizational morale and cohesion—the key ingredients of organizational productivity. However, as time goes by, this organization will fly apart of its own volition, given that large amounts of "energy" are not constantly consumed to maintain the original potential energy of the organization.

As the organization matures, its entropy fluctuates around a steady-state median, depending on the "natural" tendency for it to increase as opposed to the effective application of management energy to decrease its entropy. Two terms need additional definition. By *natural* we mean management's disregard for, or incompetence toward, the maintenance process. The phrase *effective application of energy* implies that there are ineffective applications of energy. Like short-circuiting a battery, counterproductive management activity, especially countercontrol, is more detrimental to organization entropy than no management action at all.

To visualize the process, recognize that all successful organizations begin with a small, very idealistic, zealous, self-motivated membership. This group usually starts as an informal organization within a larger, more mature formal organization. The informal organization breaks out of the original formal organization by creating a formal organization of its own (e.g., a new company). Having no momentum of its own, the organization's very survival depends on the satisfaction of its customers, which include both the users of its product and possibly the stockholders who are financing the young operation. This, together with a tight and highly motivated organization, leads to an initial spurt of success. (Of course, if these ingredients are not present, the enterprise might crash and burn before it gets off the ground.)

Over time, successful organizations grow. There are a number of reasons, which are discussed below, that the nature of this organization changes. As it matures, layers of management are added as well as specialized staff positions. Over a longer period of time it becomes clear that the organization will survive, and the original cause of its success begins to be taken for granted. This is especially true as second and third generations of management and workers outnumber the original membership. As a bureaucracy develops, concern within the organization shifts radically from satisfying the customer to personal and departmental survival. Ultimately, people within the organization work harder at making things easier and more secure for themselves than in their concerns for the customer. Often this takes the form of infighting or meaningless exercises that serve only the bureaucracy, and tremendous organizational power and resources are consumed in these nonproductive endeavors.

This is the organization increasing in entropy. Although entropy increases can be countered, the number of organizations cited ten years ago as models of ideal organizational behavior which have subsequently declared huge financial losses demonstrates a failure to do so. As long as these organizations blame the economy, foreign competition, or anything else, they will never be able to address their real problems. The meta-control systems of these organizations failed because they did not observe the following principle:

> To prevent the second law of thermodynamics from destroying an organization, the measurement element of meta-control must concentrate on the direction, velocity, and acceleration of its entropy.

Organizations that have all the earmarks of increasing organizational entropy need to take countermeasures before they reach a point of no return.

Since organizational entropy is solely a sociological phenomenon, its control is rooted in the attitudes of the members of the organization. Avoiding increases in organizational entropy is mainly a matter of avoiding or countering its causes. Consider the causes given in the adjacent display. These items are highly interdependent (i.e., they cause each other) and it is difficult to differentiate causes from effects. Further, we reemphasize that these causes do not have to occur; countermeasures were described for most of them in earlier chapters.

Causes for Increases in Organizational Entropy

- *Attraction to success.* The original organization was formed by an informal leader who hand-chose the membership when the organization was first formalized. Since the organization was successful, its original members had to be both competent and team players. Those attracted to the maturing organization may not have the ideals or motivations of those in the original group. Volunteer organizations are particularly vulnerable to the attraction-to-success syndrome. Although it could be reasoned that the original members of nonvoluntary organizations choose the other members, two problems remain: (1) each succeeding generation loses some of the characteristics that made the original group so successful, and (2) even the first generation may not be competent in selecting others like themselves.

Causes for Increases in Organizational Entropy (Continued)

- *The Peter principle.* As the organization changes, the original group functions in ways for which their experience and training have not prepared them. This is particularly true of rapidly growing organizations. As opposed to being promoted, members of the original organization (and their successors) are thrust into new and different situations. Although this arises differently from the Peter problem, the net effects are the same.

- *Informal group formation.* Even though the original core group arose from an informal organization, they might not understand or tolerate similar formations within their own formal organization. Without proper consideration of informal organizations, their formation will probably result from negatively polarized dissatisfaction, further increasing organizational entropy.

- *Paradigm problem.* The informal leader who sets up a new formal organization has the responsibility to provide a vision for the future which will prevent the organization from becoming like the one from which it emanated. We would expect that if anyone could relate to the deficiencies of the bureaucracy, it would be the original leader. However, in our discussions with these new captains of industry, the focus always seems to turn to the reasons that "it cannot happen here" as opposed to "What actions are we taking to prevent it?" Perhaps this is a lack of concern for the longer term, perhaps it is pure ignorance, or perhaps this is what happens to people when they are suddenly thrust onto the other end of the authority/responsibility spectrum. In any event, it is clearly a blindness that fits perfectly within the definition of a paradigm problem.

Observation of the frequency with which the major reputable organizations (e.g., Chrysler, IBM, etc.) have effectively "come apart" in recent years shows that no organization is immune to the curse of increasing organizational entropy. The major root cause might be quite different in various organizations, and the current state of the organization determines to a large extent its potential for entropy increases.

Volunteer organizations are interesting because they tend to accelerate the process greatly. An excellent recent example involves a group that advocated extremely tough penalties for drunk drivers. It grew far too quickly, attracting many followers largely because of its success. It quickly rejected its founder and established a bureaucracy that consumes a significant proportion of its income. Much of its central activity is currently spent in determining policy and its original focus on severe penalties for offenders has been abandoned for the broad-based set of countermeasures which are not significantly different from those that have generally been accepted within the mainstream of the traffic safety community. When first formed this organization had low entropy and tremendous potential power, and it was a major cause of the large reduction in the proportion of alcohol-related traffic fatalities which occurred at that time. Only time will tell if it can regain its original power or if it will follow the typical pattern toward mediocrity, if not total self-destruction.

This relates to technimanagement only in that it provides a microcosm which can be studied, such as the fruit fly provides for those studying genetics. We know that the process of increasing entropy occurs merely by the examination of most large organizations. First, we must commend them on the fact that they have survived to this point. They must be doing something right! But it is internal confidence in this fact that is a major cause of the paradigm problem. On the other hand, good managers seem intuitively to take counterbureaucratic actions that lower the organization's entropy before things get critical. In other organizations, disaster is imminent—like a nuclear reactor that reaches a state of no return, a self-destructive organizational *meltdown* occurs. Since this is countercontrol at its worst, let us formulate the following:

> The vulnerability of an organization to organizational meltdown is measured by the direction and speed with which its entropy is changing; it is not measured at all by its bottom line.

By the time the bottom line is affected, it is far too late to reverse the entropy change! In fact, the perceived success of the organization makes such a change virtually impossible; the paradigm problem is never so vicious as when things are apparently going well. Thus measurements must be made on the presence of the symptoms of increasing organizational entropy, which we have been describing throughout this section.

The problem with an organization that is perceived to be in steady state is that it is very difficult to measure the change in entropy. An inability to measure the entropy change might lead to as much complacency as would an ignorance of its existence. It might be reasoned that it makes no difference as to the current state or the change of entropy, management should do everything in their power to decrease organization entropy. However, we are dealing with dynamic human forces that react to different stimuli in different ways at different times. To lower the entropy level, it is essential that a control system exist in which countermeasures are monitored and adjusted as the current environment dictates. This is the difference between a one-shot management approach and the creation of *mechanisms* to maintain the organization in a state of low entropy.

Clearly, countermeasures to increasing organizational entropy include avoiding the causes described above. From a more holistic point of view, however, consider the following principle, which might guide the entropy management process:

> As an organization exerts greater control over its members, its organizational entropy increases.

We feel that this is counterintuitive to most managers merely because of our observation that the response to most organizational problems is an attempt to increase, rather than decrease, organizational control over its members (we have referred to this approach to organizational problem solving as the "knee-jerk regression to Theory X").

This, of course, brings us back to the underlying theme of the entire book: The solution lies in establishing mechanisms for the empowerment of the individual and, in turn, the informal organizations which they choose to create in support of organizational goals. With this established, let us consider further the methods and reasons by which these groups are established.

13.3 Spontaneous Group Formation

Most new formal organizations are formed as the result of the maturing and negative polarization of informal organizations. We saw in Chapter 12 that informal organizations were formed as the result of the failure of existing formal organizations. *The failure of the existing organization does not have to have a tangible cause; it can exist merely in the minds of those who are dissatisfied.* The perception of the organization membership is reality when it comes to the formation of informal organizations. Since the organization is destined to fail if its members reach a certain level of dissatisfaction, management can ignore this perception (however false) only at its own risk.

Of course, the reason that we spend so much time discussing perception versus reality is that managers (at all levels) often have a completely different perception of the "success" of the organization from that of their subordinates. This would be fine if (1) this perception alone determined the satisfaction of the remainder of the organization, and (2) this perception was, in fact, accurate. Neither one of these factors are generally true. Thus, when informal organizations create themselves spontaneously, it tends to come as a rather shocking surprise to many managers. We saw in Chapter 12 that these informal organizations, at varying degrees of cohesion, can either help or hinder the organization from reaching its goals. However, unlike our discussion in Chapter 12, here we are concerned only with informal organizations which become so negatively polarized to the original organization that they break out and form a formal organization of their own to compete with the original organization.

For the moment, let us turn our attention to some very basic motivators of human behavior in an attempt to grapple more effectively with this somewhat universal problem. Consider what we will call the *favored motivator principle*, which states:

> It is easier to motivate most people with dissatisfaction than with satisfaction.

Most implies that there are exceptions. Although they are so rare that their overall impact on organizational behavior is minimal, overgeneralization of the principle stated above can lead to extremely damaging errors when these special people do appear on the scene. For now, let us stick with the ramifications of the rule rather than its exceptions.

The cases in which people are motivated by satisfaction are not exceptions. We are not stating that people are never motivated by their satisfaction with their current situation or with their dedication to another person or a cause. We are stating that they are motivated much more easily by dissatisfaction rather than satisfaction. The easiest way to manipulate people is to motivate them with dissatisfaction, as opposed to loyalty to a positive cause or principle. Aside from its inherent dishonesty, the primary problem with this approach is that it motivates behavior that is negative toward the organization as opposed to continuing positive activity toward its goals. The fact that we do not endorse manipulation should be ample evidence that we are not recommending this as an approach to motivation. However, many charismatic leaders use this principle (in some cases subconsciously) to manipulate their followers. For proof, try to find counterexamples.

Evidence of the validity of the favored motivator principle surrounds us in almost every human endeavor. In the most extreme case we might ask: What drives the mob to such strange behavior? Is it ever satisfaction? What drives the continuation of hostility between races and religions to the point of war? This unreasonable state consumes the very resources that would enable us to live together in peace while spawning another generation of hatred. Is the motivation for this ever satisfaction?

This general nature of the motivation behind the formation of new organizations within society has been recognized for some time. One relatively recent quotation that certainly relates to our society today emanated out of the 1930s observations. According to Mayo: "It is unfortunately completely characteristic of industrial societies we know that various groups when formed are not eager to cooperate wholeheartedly with other groups. On the contrary, their attitude is usually that of wariness or hostility. It is by this road that a society sinks into a condition of stasis—a confused struggle of pressure groups, power blocs, which, Casson claims, heralds the approach of disaster" (MAYO45,7–8). (*Note*: The reference to Casson is: S. Casson, *Progress and Catastrophe*, Harper & Brothers, Publishers, New York, 1937.) The contemporary word for *stasis* is *gridlock*. It is very clear to all objective observers that our governmental organizations have followed the classic patterns of increasing entropy to the point that the infighting, even within political parties, has led to a condition of stasis.

Whether this is for good or evil depends on one's political leanings, but none can deny this reality. That most political energies are spent at infighting really does not disable the economic side of our society, provided that things are already operating close to optimal. However, if this is not the case, and laws need to be passed to address continuing inequities, this process contributes to higher entropy, and ultimately a meltdown situation might result, as has recently occurred in the Soviet Union. Coming to the country's salvation are external situations that have tended to decrease entropy by bringing about a temporary unity and cohesiveness. Popular wars, the launching of *Sputnik*, and the fuel crisis have served to decrease entropy at the expense of massive efforts and great consumptions of organizational energy. It is interesting to note that these motivating factors were in all cases reactions to intruding outside forces (hatred or discontent) as opposed to the accomplishment of positive national goals.

How does this relate to technimanagement? We speak of governmental organizations, mobs, and military actions just to establish the general nature of human motivation and prove the validity of the favored motivator principle. It provides an understanding for the societal structure in which we must operate. Its application to the management of technical organizations is quite direct. First, it should be recognized that informal organizations per se are not the problem. The problem becomes quite apparent when these informal organizations become so negatively polarized that they break out of the original organization. Management must recognize the extreme motivating effect that negative polarization, as defined in Chapter 12, can produce. This is because of the inherent animosity that can tend to build between subordinates and management—it does not have to be nurtured by external forces.

It should be apparent that the cause of this problem originated long before this symptom is perceived. It is a failure on the part of the organization's management to give all of its membership adequate opportunities to realize their creative energies. This is accomplished by making the formal organization as effective as possible, and the primary key to this is integration of the informal organizations (including but not limited to their leaders) into the decision-making process.

To summarize the countermeasures discussed in Chapter 12, there are two things that informal organizations might motivate their members to do in response to their discontent: (1) positive actions, which complement management activity and further the goals of the organization, or (2) negative actions, which preoccupy both management and their subordinates to fight each other. Unfortunately, it is management's negative response to the first of these that so often leads to the second, and when this happens, organizational entropy increases dramatically. We consider this further in the next section.

13.4 Management Response to Group Formation

In the ideal world the benevolent all-knowing dictator would have complete knowledge of all organizational operations and thereby dispense orders down a perfectly understanding chain of command to everyone's satisfaction. In the real world of (at best) imperfect (and, at worst, egocentric) managers, this never happens. Thus dissatisfaction with management decisions must be expected, and those who do not expect criticism are inherently ignorant of human nature. We will deal with the necessity for and use of criticism in Chapter 20. Here we want to deal with the common symptom of the inability of managers to deal with their own shortcomings, which usually manifests itself in reactive criticism of their subordinates and fear of the informal organizations. This, in turn, is one of the major causes of increased organizational entropy.

Again, we go back to the early formative days of "scientific management," 1936. Consider the statement by Whitehead: "What is feared of integration within a small group is that it may organize itself in opposition to the larger whole—and this it certainly will do if its existence be threatened; but equally, a protected group will endeavor to satisfy its wider interests by collaborating with the organization of which it is a logical part. In this way, its loyalty will extend to the firm as a whole" (WHITE36,98–99).

The problem according to Whitehead is management's fear of informal group formation (integration) because it might organize itself in opposition to the formal organization. He states that this is a valid fear if their response to this is one of repression. On the other hand, if it is "protected," it will collaborate with the formal organization in satisfying its goals.

The concept of *protection* needs further elaboration. We interpret this to be the provision of an environment of support. This would include involvement in the decision-making process and rewards when informal organizations contribute to the goals of the organization. The terminology used most often in the literature is *goal integration* (i.e., doing that which is necessary to assure that the goals of the informal organizations are kept consistent with the goals of the formal organization). To be effective, however, *this must fall far short of formal recognition*.

An essential ingredient of goal integration is effective communication. This gets back to the necessity for informal leaders to know and understand the goals of the formal organization and the methods employed to achieve them. This is often regarded as a selling of the subordinates, or, according to contemporary terminology, getting them to *buy in*. However, there is an implicit credibility problem with this arm's-length perspective. The very fact of being management makes such a sales job next to impossible. The other problem lies in the problems associated with distancing subordinates from the decision-making process itself. To formalize this into a recommendation, let us state the following:

> Effective goal integration requires that subordinates be in the decision-making process, not just to buy in to it.

If, in fact, they are in the decision-making process to the point of goal definition, it is obvious that they will support the goals of the organization. Although this might not be attainable immediately, the clear establishment of this as a direction of the organization is attainable immediately. This is accomplished by establishing mechanisms that include the informal organizations with the promise to extend and expand their influence as they prove successful. At this point we want to conclude our discussion of organizational entropy by discussing group decomposition.

13.5 Spontaneous Group Decomposition

By *decomposition*, we mean just the opposite from formation. Both formal and informal organizations have finite lives. We saw that the formal organization emanates from an informal organization, and the members of

Sec. 13.5   Spontaneous Group Decomposition

informal organizations create it spontaneously (i.e., without planning or design). This accounts for their formation. The formal organization terminates when there is sufficient reorganization to cause it to lose its identity, even though it might keep its same legal name. This might be difficult to identify, but when its entropy expansion begins accelerating toward meltdown, the end is not far behind. The demise of the informal organization is much more difficult to determine, since its formation was never officially identified, and it may not even have been recognized.

Although the termination points of these organizations might be difficult to identify, it is important that we understand that they do have a finite life. This is especially important from management's approach toward informal organizations. We have discussed organizational entropy to this point purely in terms of the formal organization. Informal organizations (and indeed even individuals, as we discuss in Chapter 19) also follow the second law of thermodynamics. The entropy of informal organizations increases for much the same reason that it increases in formal organizations: namely, neglect.

This implies that the manager need not move directly against an informal organization which is causing negative effects. Our original recommendation, of course, was for management to eliminate the conditions that give rise to a counterproductive group. However, this is not always possible because of inevitable shortcomings of management judgment and the possibility of misunderstandings. Thus managers might have to deal with a negatively oriented informal organization to prevent them from getting further polarized. With an understanding of the nature of organizational entropy, however, the following can be formulated:

> Proactive efforts are not required to dismantle an organization; without a continuing reason to exist it will dissipate of its own increase in entropy.

Therefore, rather than direct action, which will tend to decrease its entropy, a twofold approach is recommended: (1) a cessation of the deficiency that gave rise to the problem, and (2) continued effort at goal integration as described above. As opposed to direct action against the informal organization, this action will either increase the entropy of the negatively oriented informal organization, or it will reorient it toward the support of organizational goals. In either event, it will disable its negative effectiveness. This is not manipulation since it is a direct, open, and honest attempt to rectify the problem, not an attempt to repress this informal organization.

At the same time, management should recognize the benefits, support, and thereby decrease the entropy of those informal organizations that are contributing to the goals of the organization. This is essential to the effective operations of the organization, since there will always be gaps that management cannot possibly cover.

As a final thought on organizational life and death, consider the following:

> No organization is any stronger than the loyalty of the membership to the principles on which it was founded.

This holds for both formal and informal organizations. It serves to explain the reason that organizations based on discontent and reaction are easy to create but difficult to maintain. Entropy in these organizations is maintained at a low level only by continuously fabricating enemies. Unless there are strong positive principles that can serve to unify the organization and mitigate the natural increase in organizational entropy, these organizations will not survive for long. (However, this does not mean that others will not spring up to replace them.)

As an example, if the *only* goal of an organization is to make money, its members will probably end up either in the poorhouse or in jail. If this is the only motivator within its members, they will jump ship at the first offer for more money. The entropy of such an organization will grow quite quickly to a meltdown state. On the other hand, the goal of making a profit by contributing to society in producing a useful, high-quality product is one that serves to unify and motivate both the formal organizations and its supportive informal organizations as well. Genuine management commitment to these principles is as essential to maintaining low entropy levels as are the overall goals of the organization itself.

The principle stated above also explains the power of informal organizations. Since they are purely voluntary and based on the unwritten principles of friendship and mutual trust, their cohesion is bound to be stronger than that which can legitimately be imposed by the formal organizational structure.

A major step in the direction of effective integration of all levels comes with a common understanding of the principles of technimanagement. If all members of the organization will commit to the common principles of technimanagement, a transition to the continuous improvement of the management process will result. In the chapters to come, we will be suggesting more definitive mechanisms to evolve organizations that liberate both the individual and informal organizations to unleash their full potential productivity.

Finally, the management engine must be continuously maintained, oiled, and tuned for efficiency. Management cannot be an activity that is completed. It is essential that you, as a manager, visualize and proactively deal with those organizational problems that will surface one, two, and six months from now by establishing the mechanisms for eliminating them before they significantly increase the entropy of your organizational component.

14

Group Dynamics

14.1 Definitions

In Chapters 12 and 13 we discussed many of the ramifications of informal organizations and their impact on the capability of the formal organization to meet its goals. We saw that this effect is going to occur, regardless of management actions, but that it could as easily accrue to the benefits of the formal organization as it could to its detriment. Finally, we saw that although this was not entirely within their control, management has more control on the work environment than anyone else.

In this chapter we shift gears in two different dimensions. First, we discuss an aspect of the formal organization as opposed to the informal organizations. Second, we move from a macro to a micro view. That is, in the previous chapters we considered the general effects of informal organizations and formulated very general strategies that could be applied to set up the environment in which they could most effectively contribute to the goals of the organization. In this chapter we deal with specific local strategies of establishing and managing the most basic of organizational components.

Very little is accomplished by technical organizations solely on an individual basis. This is because our technology is so advanced that single individuals rarely have the capacity to fully understand and then design and develop entire systems. Even most systems components require a combination of skills for their design and development (e.g., computer hardware, software, electronics, telemetry, robotics, etc.). For this reason *project groups* are created to address these issues. It is not that individual effort will not be required—very few technical tasks take simultaneous effort as might be true with physical tasks. In fact, without purely individual efforts, very little *measurable* progress can be made (i.e., lines of code generated, design documentation, etc.). Even progress at meetings can only be measured by the minutes, which are written by a single person. However, all of these tasks are generally preceded by considerable project group interaction, including training, coordination, and mutual assistance.

We define several terms for the purpose of our communications here. It is not our intention to create or to try to impose general use of these

Sec. 14.1 Definitions

terms. However, to communicate new concepts we need to clarify the terminology we are using. The two adjacent displays contrast traditional groups with those proposed under technimanagement. We refer to these structures collectively as *groups*. The collective functioning of these groups together with the meta-control structures to integrate them into the organization are the primary *mechanisms* of technimanagement.

Traditional Groups

- *Project group:* an organizational unit at the lowest *line* level that works in direct pursuit of organizational goals; a set of people who are recognized by the formal organization to be assigned to and responsible for a given set of tasks, which are objectives of the organization. This might be one project, or it might consist of a series of related projects (related by task type or the group's qualifications).
    - *Project group leader:* generally established and recognized by middle and upper management, this term refers to this lowest level of management. This assignment is mainly for its functional convenience and control (i.e., delegation, responsibility, and reporting). Commonly, the terms *first line* and *lowest level* are used synonymously with *project group leader* in technical organizations.
- *Committee:* traditional approach to creating organizational structures outside the line organization when additional interactions between line and staff organizations are required. It is appointed and charged by upper management, and usually has a finite life, although ongoing committees are not uncommon.
    - *Committee chairperson (or chair):* formal leadership, which applies only to the context and charge of the committee. The chair does not automatically endow any line position or authority.
- *Staff group:* another traditional alternative to the line organization, established in the traditional organization to deal with issues that are generally common to many, if not all, of the line components of the organization. Usually, there will be a *staff group leader* appointed for any given project, although a given staff group usually has a permanent manager as well.

Technimanagement Organizational Support Groups

- *Work team:* nontraditional structure which has the following characteristics that distinguish it from a committee or staff group:
    - Deals with a specific aspect of continuous improvement within a project group or common to several project groups.
    - Its charge evolves continuously depending on external circumstances, and ideally it is almost totally self-directed.
    - Depending on its charge, its membership content might require:
        - *Horizontal integration:* the membership of individuals representative of all affected line organizations
        - *Vertical integration:* the membership of all employees at pertinent worker and managerial levels necessary to accomplish the work team's objectives cooperating on an equal basis
        - Customers, who are given preeminence on the team
        - Suppliers
        - Customers and suppliers might be external or internal to the organization.
    - Its life is not specified, and it is expected to last as long as the process that it is charged to improve, although obviously both the people and the type of membership will evolve.
    - The formal leadership, called its *chair*, is analogous to the committee chair as far as scope of authority is concerned; however, the chair tends to be a facilitator of this group as opposed to a traditional manager.
- *Task force*: equivalent in all respects to the work team, with two exceptions: (1) it is charged to address a specific problem as opposed to the constant improvement of a process, and (2) it is expected to have a finite life due to its charge.

We have been using the term *organizational component* to refer to the subset of the total organization over which one manager has the span of authority. In our list of groups given above, the only one that fits this definition is the project group. Yet they do not map directly, since one manager might be responsible for several project groups. Further, a manager over several project leaders would have all of these project groups as her or his organizational component.

The other four groups do not have line authority, and they are not directly responsible for the accomplishment of the goals of the organization. (Thus great care must be exercised in their creation.) The two traditional structures, the committee and staff group, have typically been used to advise management and to bring specialists to bear to "fix things," respectively. The performance of these functions will be revised dramatically under technimanagement by the use of work teams and task forces, as we have defined them. To distinguish these from the (line) project groups and the traditional structures, we refer collectively to the new organizational structures as *organizational support groups*, which have a general objective as follows:

> The objective of organizational support groups is to distribute the decision-making functions throughout the organization.

The committee and staff group are traditional organizational structures that have violated this principle consistently, and therefore we expect that only a few remnants of them will remain as technimanagement evolves. The differences between these approaches and technimanagement reflects an entirely new paradigm of management. Thus the key to understanding the difference between staff groups and committees (on the one hand) and work teams and task forces (on the other) is empowerment, which we consider further shortly.

A major goal of this chapter is to prescribe methods for keeping informal organizations consistent in goals with the formal structure (i.e., goal integration). Since this cannot be achieved by interfering with informal group formation or functioning, it must be accomplished by the effective functioning of the formal organization. In this chapter we begin this process by further establishing the concept of empowerment. We continue with a consideration of group selection followed by a discussion of group sizing. The major determinants of group productivity—harmony, compatibility and morale—are considered next. Finally, some techniques are given for bringing about group consensus.

14.2 Group Empowerment

We anticipate that many managers will respond that the "new" groups above are already in place (possibly by other names). Giving things new names and putting them in place, however, does not make them functional. If groups such as task forces and work teams are in place, and if there is the type of empowerment described below, your organization is well ahead of the trend. We would expect to find some aspects of these structures in all organizations, although many of them may be accomplished by effective informal organizations.

We are not talking absolutes here. While the normal practices of good management may have implemented some technimanagement structures, that does not mean that the entire organization has embraced the philosophy or that an effective transition has begun. Every organization is at some stage of maturity in this regard: We have come a long way from slavery and the sweatshops, but we still have a long way to go.

Vertical integration begins to approach the issue of empowerment. To make this point, we emphasized *equal* in the definition of vertical integration within the context of organizational support groups. This requires that the various managerial levels come together as equal members of the team. Thus persons of higher rank will not dominate the team; they will not be feared. But this is only the tip of the iceberg. Let us delve into the concept of empowerment in general before seeing how these new structures produce this intangible commodity.

Although the concept itself is not new, we believe that it is generally grossly misunderstood. Consider the following: "The *power-equalization* approaches, in their early stages at least, shared a normative belief that power in organizations should be more equally distributed than in most existent 'authoritarian' hierarchies" (LEAVI65,1154). The problem with this concept is that there is a perception that "power" resides at the top and can thus be doled out as the CEO sees fit. Maybe only the CEO realizes how inaccurate this perception is. (We see this as one reason that CEOs accept the principles of technimanagement much quicker than middle managers do.) In reality, the CEO's power is determined almost totally by the performance of the organization. As a person without the productivity of the organization, the CEO is powerless. Thus anything that the CEO does to increase the performance of the organization will increase the CEO's power base. If this means allowing customers to come in and tell the employees what to do and how to do it, why not? (We are not saying that this is always optimal, only that if this were known to be the case, the CEO would be a fool not to allow it.)

Make no mistake: We recognize that ultimate decision-making authority within an organization is with the ownership, and the ownership retains the CEO to act in its best interests. Thus any higher level of management can legally overturn decisions made at lower levels. But is this power? On the very opposite end of the spectrum, the lowest levels have the legal right to walk off the job at the most inappropriate of times (or, what might be even more devastating, stay on the job and practice malicious compliance by doing everything "by the book"). But, is this power? Indeed, both are types of power, but they have little to do with what we are calling *empowerment*. Do not proceed unless you have this firmly in mind; if not, reread Chapter 5.

We get into the principles and details of the power game in Chapter 19. For now, let us summarize by stating that ultimate empowerment is attained by all members of the organization when they all understand and support the *distributed decision-making principle*:

> For each decision that must be made to attain the goals of an organization, there is a unique person or organizational support group who/which is in the best position to make that decision.

One major objective of empowerment within technimanagement is to dynamically create the organizational support groups to improve on the decisions made by traditional management practice. It should be anticipated that internal organizational stress will occur in cases where, even though the distributed decision-making principle is understood, there is still a difference of opinion as to whether a person or an organizational support group should make the decision. [Even if this is resolved, the individual participant(s) are still in question.] Ultimate resolution of these questions remains with the relevant manager—and therein lies the necessity for leadership. For a tremendous amount of power can be lost if their resolution is perceived to be arbitrary. One of the major power-building techniques of the manager is to anticipate such conflicts and resolve them well before they reach the stage of confrontation.

The important thing to recognize is that the type of empowerment required by technimanagement is not a transfer of power from a manager to a subordinate as is the case in traditional delegation. Since every member of the organization will ultimately evolve this understanding of empowerment, there is as much transfer of power to managers (when the subordinates recognize that managers are in the best position to make the decision) as there is a transfer of power to the organizational support groups, and ultimately to the subordinates. However, since the entire organization benefits from the improvements in the quality of the decisions that are made, everyone's power is increased dramatically. This contrasts greatly with Leavitt's traditional concept of power equalization and spells one of the most significant differences between technimanagement and traditional management concepts.

Turning our attention to how this translates into holistic organizational empowerment, consider the attitude of many, if not most, members of the organization toward the traditional organizational structure known as the committee. Kast covers this quite eloquently, as follows: "Typical quips are . . . 'The purpose of a committee is to: (1) reduce tranquility, (2) increase dissatisfaction, (3) divide responsibility, and (4) stave off action'" (KAST70,289). Does anyone disagree? (With a few notable exceptions, of course.) With such general animosity against this structure at all levels, is there any wonder that the organization itself is disempowered?

In implementing technimanagement, the most important element that must be communicated to all members of the organization is that the work teams and the task forces are not committees. For this reason, we strongly recommend that they not be called committees and that as much distance between the two be made as possible. (One way to do this is to abolish all committees summarily and set up a task force to establish effective organizational support groups. This was done at a level above the author's organizational component with absolutely no discernible downside! However, Chapter 24 presents more evolutionary methods for initiating the transition.)

This requires that the new decision-making structure be understood and its potential for mutual empowerment be appreciated throughout the organization. Lower-level managers who have implemented technimanagement within their organizational components can attest to the fact that the distribution of the decision-making function throughout their components leads to their own personal empowerment. However, the radical change in the management paradigm that this requires in most organizations cannot be attained overnight. Thus, if this is to be accomplished, it is essential that the mechanisms be put in place that will set a direction from which this can ultimately evolve. Again, we refer the reader to Chapter 24 for details.

Given that the concept of empowerment is understood, the importance of getting the right combination of people on any given (project or management support) group cannot be overemphasized. This is the topic of the next section.

14.3 Group Selection

While addressing the issues involved in group selection, two closely related issues arise: (1) Who should select the members? and (2) who should select the formal leader? Consideration of these issues could be quite convoluted, since, for example, the question might arise as to whether the leader should be chosen first and then be involved in the selection process. Often these issues are self-resolving, based on the particular circumstances at hand. For example, if a given person took the leadership in working on the successful proposal that obtained funding for a project, this person would be a logical choice to be group leader.

A consideration of alternatives in the abstract enables some principles of group selection to be introduced. Consider the alternatives given in Figure 14.1, which proceed from the more Theory X approach to those which are oriented toward technimanagement. Alternatives 1 and 2 are almost equivalent; however, the first generally gives the formal group leader more of a Theory X orientation. There is something about formulating things in private and then hatching them on everyone else which gets things off to a nice Theory X start. Alternative 3 is a step in the right direction in that the group does not feel that their manager is being imposed on them. There is still a danger in potentially negatively polarizing the informal organizations, although it is the informal organization itself that is thrusting the leader into a formal position. Thus although this might alter an informal organization over the long term, it can hardly be considered to be the result of management manipulation. (It should be recognized that everything that management does has some effect on the informal organizations.)

Alternative	Description
1. Centralized	Managers select the group leader first and then, collectively, select the people in the group.
2.	Managers select the people in the group as well as the one who is to be the group leader.
3.	Managers select the group and allow the group to select its own leader.
4.	Managers select the leader but allow the "organization" to create the group.
5. Distributed	The "organization" selects the group members and the group selects its leader.

Figure 14.1 Alternative Approaches toward Group and Leader Selection

Alternatives 4 and 5 are also almost equivalent. While the pure distributed decision-making alternative might be considered ideal for a mature, highly motivated technical organization, alternative 4 is sometimes unavoidable due to the history of the project. There are a number of mechanisms by which the "organization" might create its own assignments. Many companies have formal mechanisms by which individuals "bid" on jobs in a variety of ways. It is generally understood that every person within the organization will not always be totally satisfied. Whatever method is adopted, the most important aspect is that it is perceived by everyone to be totally fair and open to all. Further, neither of these methods can function without the involvement of the informal organizations and the accompanying communication by which they hash-out acceptable assignments.

It might be asked if a group leader is really needed, and this is an intriguing question that should be addressed. We recognize that the group will evolve one or more informal leaders (depending on its size and functions) regardless of whether or not a formal leader exists. Recall, however, that the moment that management recognizes and treats informal leaders as formal managers, a significant portion of their informal leadership will be disabled. Since management will inevitably deal with the de facto formal leader, there is little choice but to make it official. The formalization of this at the outset will avoid confusion on the part of both management and subordinates. The informal organizations that will arise can contribute to success if the principles of the former chapters are applied. Finally, there might be an advantage to allowing a group to evolve without a formal leader for a short period of time, but in this case the manager to whom all of the group reports will be the de facto formal leader.

The optimal alternative is dependent on the maturity of the organization and the situation. Between the unavailability of most members of the organization due to commitments on other projects and the necessity for particular expertise on the project, there might not be very much flexibility at all. Sufficient familiarity of management with those who are available, and vice versa, will contribute heavily to the chances for actually applying the optimal approach. This optimal approach is inherently determined by management–subordinate negotiations. That is, when both management and the subordinates together agree that one of these five alternatives is the best approach, then, by definition, it is. When management communication with the informal organizations determines that a given method is generally accepted, it should be implemented. At that point, contention about the methods employed signals that changes might be required in the meta-control system.

Once the method of selection is determined, those actually selecting group members will quickly recognize that the choices must be obtained from a very limited set of alternatives. However, this choice can be the most critical decision that is made with regard to accomplishing the group's charge, and even one error can cause serious consequences. In essence, the formula for a successful group is a combination of balanced technical competence and collective group harmony. Some of the major considerations are presented in the display below. If these factors are given due consideration, the choice between the available alternatives should be greatly improved.

Considerations in Group Selection

- *Charge and group type.* Very different decisions will be made in the establishment of project groups as opposed to work teams or task forces, and the charge to the particular group will also have a major impact. For example, project groups rarely have need for vertical integration as opposed to organizational support groups, which almost always require it. These overriding factors influence each of the considerations that follow.

- *Technical and leadership contributions.* This might be the overriding consideration if a highly technical, cutting-edge project is to be performed. Individuals should be selected (hired) for their future potential as opposed to current skills. There must be a balance such that one skill does not become redundant and another relatively absent (e.g., originality versus documentation). In addition, nontechnical characteristics, including diversity, must be considered, especially as it might relate to group harmony.

- *Group harmony.* A balance between organizational and technical needs should be sought. More than this, however, is the ability of each person to recognize the need for each other member of the group and to respect their capabilities to satisfy those needs. It will do no good at all to assemble the most technically competent set of group members if they spend all of their time in infighting. Certain people might play a minor technical role, but they more than make up for it by bringing harmony to a group, since harmony might be as important to any project group as technical ability (i.e., technical ability might be acquired, which is unlikely of harmony once lost).

Considerations in Group Selection (Continued)

- *Fragmentation and disruption.* The advantages of a single person being a member of multiple groups includes the integration obtained by the diversification of experience of the person as well as the contribution made simultaneously to both groups. The clear downside is the fragmentation of individuals to the point where they become ineffective on all projects. Effectiveness might rise quite quickly as the marginal advantage of the second or third assignment of a given person is made. However, it levels off quickly and heads south when the person becomes overassigned. Global priorities have to be given major consideration in multiple assignments.

- *Formal and informal leadership roles.* The formal leaders are restricted to the group over which the formal organization places them. This is not at all true for informal organizations. Expect that informal leaders will arise out of informal organizations which form both within and across several groups, simultaneously. It is not expected that managers (or anyone else, for that matter) can keep up with this complex, dynamic situation. It is relevant to the assignment of people to groups only if these assignments cause significant disruptions in the informal organizations.

We anticipate the question to arise at this point: With all of these interacting and complicating factors, how do you expect the organization, as opposed to a single manager, to consider all of them and be able to establish optimal team membership? Our response is that it is the very complexity of this that requires the organization to do it—the manager will botch it every time. (Even in the perfect call there will be dissatisfaction just because of the perception of arbitrariness.) Recall that it is the managers' jobs to establish the mechanisms by which these difficult decisions can be made, not to make them. We will not prescribe a solution to this problem at this time, only suggest that it lies within the realm of a task force that totally involves management and the informal leadership.

A final key concept is limited to the selection of individuals on organizational support groups. This is the *principle of volunteerism*:

> The membership of all work teams and task forces should come only from the subset of members of the organization who have enough confidence in empowerment to be fully motivated to serve enthusiastically in the group.

That is, they should volunteer, not be coerced. This principle is essential to the successful functioning of all groups, but especially those that have management support as their objective. If the only way to obtain the membership of these groups is management or peer pressure, the organization has not matured far enough to warrant their establishment. Much more education and "piloting" in technimanagement is required (see Chapter 24). Ideally, we would also apply the principle of volunteerism to project groups. However, the inherent motivation of "the product" makes this less important than with organizational support groups. Organizational constraints might also make it infeasible.

At this point we diverge slightly from the choice of specific individuals for the group to consider something that needs to be established before these people are chosen—group sizing. After that we consider the subject of group harmony in much more depth.

14.4 Group Sizing Theory

Decisions as to the size and number of groups that need to be assembled to fulfill the goals of the organization are some of the most critical in that they tend to define a major aspect of the structure of the organization itself. Because of the significant differences between project groups and organizational support groups in this regard, we treat them separately in the next two subsections.

14.4.1 Project Group Size

Project group size is often dictated by the ongoing operational limitations of the organization itself. Many design and development contracts are government driven, where project size is almost totally determined by the request for proposal and ultimate contract. Private-industry initiatives are much more flexible in overall project sizing. Even with this overall project size constraint, however, there is still considerable latitude to determine the number and size of groups within an overall project.

For purposes of this discussion, suppose that we are concerned with a moderately sized project that requires a total of 30 person-years over three years. One option, of course, is to set up a 10-person project group (on the average). But the other alternatives number in the hundreds (e.g., two 4-person groups and one 2-person group, two 3-person groups and one 4-person group, etc.). This problem space is greatly reduced by the functional subdivisions within the project itself and practicality (e.g., five

2-person groups). Thus if there were three uniquely different functions that had to be developed, it would seem obvious to have three groups.

It is still informative for us to consider this problem in the abstract. To do this, we might pose the question: If we had total flexibility, what is the optimal group size? To address this question, consider the larger-group advantages: The larger the group, the more expertise and the more it will be expected to produce. It is a fairly intuitive conclusion that the productivity per person will diminish above a certain (optimal) group size. However, the fact that absolute productivity will diminish when group size is increased above a given point, although not intuitive, is a well-documented fact. So this part of the upside of larger group size must be qualified. To do this, consider the following premise:

> The potential for problems within a group increases with the number of possible communication links and personality problems between group members.

Since this might explain both the relative and absolute changes in productivity, let us consider the validity of each component of this compound premise.

Number of possible communication links. The purpose of the group is primarily that of communication. Individuals push buttons (or computer keys), produce design drawings, and most of the other tasks for technical design and development. Why, then, organize them into groups? Primarily to "help each other out," which requires communication. We hope that the strengths of each will make up for the shortcomings in another, and vice versa. However, for this to occur, it is essential that the group coordinate its efforts. If not, its actions reduce to that of individual efforts (or worse). Although there are many models of restrictive communications that could be applied, in most technical organizations, every group member "talks" with every other group member. (If this is not the formal model that is applied, the informal communications will make up the difference.) Coordination requires that an updated knowledge of reality be communicated (directly or indirectly) from every knowledgeable group member to every one who lacks and needs that knowledge.

Even when there is every intent to keep communications open and accurate, every communication link increases the potential for inaccurate or inconsistent communications to take place. This leads to the following minor premise:

> For open communication groups, the addition of one member to a team increases the number of communication links within the group by at least three links.

This can be seen easily by drawing all of the possible communication links in a three-, four-, five- and six-person group, as is done in Figure 14.2. For groups of size seven or higher, the number of links increases by much more. As shown by the examples in the lower half of Figure 14.2, it is possible to reduce the number of links by restricting communication through certain individuals. However, this is done at the expense of enforcing restrictions that might seem unreasonable to professional personnel.

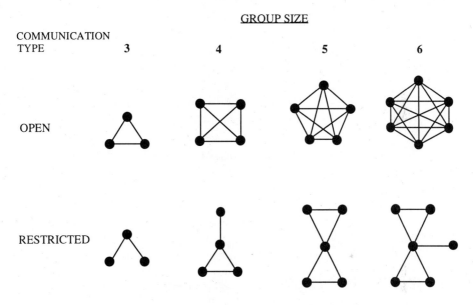

Figure 14.2 Growth of Communication Links

Number of personality problems. This is not as easy to measure as communication links. However, it seems reasonable that as the size of the group grows, so does the number of potential personality problems, if for no other reason than many such problems are caused by miscommunications. Exceptions occur when people are added to a group who by their very nature reduce the potential for these types of problems, as discussed above. If this is not the case, it would seem reasonable that the larger the number in the group, the greater the potential for personality conflicts. In small groups, personality conflicts are usually restricted to individual conflicts. However, in larger groups, informal organizations that greatly amplify personality conflicts (cliques) can easily form.

Recognize that the theory given above ignores the action of proactive management. Both improved communications and reduced personality conflicts can be created by effective management techniques, as we will see in subsequent chapters. These must be factored into the problem of determining optimal group size. Clearly, however, management effort will increase considerably when a group gets much larger than three.

The considerations given above are complicated further by the necessity for more project groups when the teams are smaller. This requires a systems approach to this problem, since the potential barriers to communication between groups is much greater than within groups. Thus a balance must be made between reducing the difficulty of communication within a group by having smaller groups as opposed to reducing the difficulty of intergroup communication by having larger groups. This can be approached by having *cross-group representation* (i.e., the same individuals on several relevant groups). Although this person provides the formal communication link, horizontal integration can also be expected via informal communications, and in most cases it should be encouraged.

In summary, it is clear that smaller groups have a distinct advantage. Above four, the number of communication channels increases dramatically, as does the potential for personality conflicts. This does not mean that larger group sizes might not be optimal, since this depends totally on the needs of the project. It does mean that additional management effort is going to be required as group size grows. For that reason it might be expedient to increase the number of groups in order to reduce group size. A second, less desirable option is to restrict communication through certain key individuals.

14.4.2 Sizing of Organizational Support Groups

Virtually all of the considerations given above for project group sizing apply equally to task forces and work teams. However, the decisions made with a project group typically affect only the particular organizational objective for which they have been charged, and possibly those other project groups that have closely related responsibilities. On the other hand, organizational support groups affect decision making over a span that can include the entire organization. This is the reason that both vertical and horizontal integration are explicit requirements of the makeup of these groups.

The most common problem is to establish organizational support groups which are far too large in an effort to get *representation* from every level and component of the organization. These groups become dysfunctional for a variety of reasons: (1) informal groups (cliques) arise which tend to promote their own hidden agendas, thus neutralizing the power of the group; (2) each, or most, individuals feel relatively helpless (disempowered) against the large group, so there is a tendency to accept the inevitable; and (3) the entire concept of representation is one that carries with it the necessity to act in behalf of one's constituency as opposed to the best interests of the organization as a whole. Thus, with regard to organizational support groups, these factors as well as the communication/personalty factors discussed above argue very strongly for smaller groups.

One solution is to seek members who are respected by several levels and organizational components. This requires a type of hierarchical structure to be developed where several different constituencies perceive that their input is being given appropriate consideration by their representative and the group as a whole. This perception can only occur when, in fact, the members of the group utilize the informal communication channels to assure that it is reality. It behooves management to assure that the members of any such group are therefore well "wired" in a horizontal and/or vertical way. This will come out of establishing effective mechanisms for the organization to establish the memberships of these groups. This highlights the importance of the manager either knowing or having an informal mechanism in place for determining these individuals.

From the point of view of size, the idea is to establish a group that will be truly functional. Traditional organizational support groups (i.e., committees) which have included a representative from every organizational component are suspect, since most managers know that most of these are dysfunctional. Thus a question arises: Are they established just so management can say "we gave you a chance, but you were unable to come up with anything"? This might seem clever, but this manipulation is disempowerment of the organization at its worst. We cannot give an optimal size for all circumstances, but the principles given above certainly argue for some optimal point between dictatorial arrogance and stasis (gridlock). Above all, management must have faith in these groups to act in the best interests of the organization.

14.5 Group Harmony

The subject of group harmony has been an implicit consideration in all of the discussions above. Its importance in moving the organization forward and meeting its goals should be quite clear. There is very little that a cooperative technical organization cannot accomplish if it is unified in the understanding and pursuit of its goals. However, this momentum of success can quickly be reversed if individuals, formal organizational units, or informal organizations do not understand and agree with organizational goals and the methods to accomplish them. This results in a state of internal stress, where major creative efforts and energies are applied to defeating the *internal enemies* of the group rather than accomplishing its part in attaining organizational goals.

At this point we wish to concentrate solely on the group. Many of the principles discussed will also apply to other aspects of organizational unity and harmony (e.g., between groups). We begin this section with a discussion of small-group compatibility, after which the more general considerations of morale are presented. In the final section of the chapter we cover a very closely related topic: group consensus building.

14.5.1 Group Synergism

The word *synergism* as we will use it here goes far beyond its dictionary definition (i.e., the idea that a functioning whole is far more productive than the collection of functioning parts). We want to encapsulate within it the concepts of team jell, collective soul and spirit, cohesion, unity of purpose and approach, esprit de corps, and collective essence. It goes beyond collective productivity to the root cause of this productivity—collective enthusiasm. It is the reason that human beings are social animals: They have experienced synergism (as we are defining it), and they know what it has done for them in the past and what it can do in the future. As one further definition, let us relate synergism to the concept of entropy defined in Chapter 13:

> The entropy level of an organization, an organizational component, or a group is inversely proportional to its synergism.

Synergism is the inherent goal of technimanagement, for not only will it increase the productivity of the organization but will also make being a member of that organization a fulfilling experience for all its members. (The question might be asked: If entropy and synergism are so closely related, why is there the need for two different terms? The reason is that entropy best describes the effect in terms of the thermodynamic analogy, whereas synergism best describes the underlying sociological cause.)

The role of the informal leaders in bringing about synergism cannot be underestimated, for it might go so far as to eclipse that of the formal leader. This reemphasizes the necessity for the group leader to establish cordial alliances with the informal leaders. It also highlights a very inter-

Sec. 14.5   Group Harmony

esting phenomenon, which is summarized by what we call the *principle of synergism*:

> Informal leaders will ultimately determine the synergism of groups both internally and with respect to each other. Formal leaders have a very great capacity to destroy synergism, but they have little capacity to create it directly.

We saw in Chapter 13 that management action can bring about extreme emotions of discontent (hatred, in some people). This is an example of synergism destruction. The role of management is primarily to stave off such bumblings on their part—get out of the way and let the informal leaders get on with the job of making the work environment a fun place to be. They need no incentives to initiate this; management's role is to reward all efforts that improve synergism *in any way that will not inadvertently destroy it.*

Rather than create, managers must promote synergism. Since everyone exercises informal leadership at one time or another (knowingly or not), it is important that the potential power of each member of the organization be recognized. In many cases the same power that can produce synergism is squandered by the normal complaining and flailing at management which seems to occur within all organizations. At the other extreme, if everyone accepts the "management line" on everything (the Theory X manager's dream), synergism dissolves into a group-speak brain-dead organism devoid of informal organizations altogether.

This is not a dilemma if we recognize it as an optimization problem as opposed to a quest for an extreme solution. Further, the normal and natural arrangement of the organization, and human nature in general, facilitates an open and honest solution. We have formulated this into the following principle:

> Optimal synergism can be accomplished by obtaining a balance between the promotion of a cooperative attitude to attain organizational goals with a measured resistance to those management practices that are felt to be detrimental to those same goals.

Fearless management should have no problem with this principle since the only thing that they are concerned with is the accomplishment of organizational goals (which we assume include the humanitarian considerations for all its members and society in general). When a member of the organization sees these goals threatened by the policies and practices of management itself (which is virtually inevitable), resistance is in order. We emphasized the word *measured*, since, again, there is an optimal point between passivity and all-out war. Generally, an optimal point can be found which is most conducive to positive change within the organization without getting expelled from it. This optimal point is the most beneficial to the accomplishment of both individual and organizational goals, although we do not expect many traditional managers to agree readily with this.

This surfaces an excellent management precept: *Disagreement is usually seen as being detrimental to synergism, but it is not!* In fact, there is a measured amount of disagreement that is totally essential to synergism. This affects managers directly, since if they react to criticism by trying to squelch, or otherwise calm it totally, a range of possible results can accrue between the extremes of total success and total failure. Total success could bring about the situation in which the informal leaders are so disabled that they are no longer considered as leaders by their peer groups. Other informal leaders, who will have another set of criticisms, will then arise to take their places, and the process will continue until ultimately an entropy meltdown occurs. Thus squelching disagreement is a lose–lose proposition.

This cycle is to be avoided if synergism is to be attained and then maintained. In this regard, let us define the concept of *healthy friction*, which should exist between informal and formal leadership.

> Healthy friction is that state of relationship between the formal and informal organizations, and also between individual subordinates, in which constructive criticism is expected and encouraged to flow in both directions. Healthy friction is essential to synergism.

The key to attaining an optimal degree of healthy friction is for everyone to agree to and strive for the goals of the organization. *The presence of an open set of principles for which the organization stands is essential to bringing about this unity of purpose.*

This set of principles must include methods for resolving differences of opinion with regard to the attainment of organizational goals. This sets up a win–win situation in which the highest levels of morale and synergism can be maintained. For it gives every member of the organization a legitimate positive way of contributing even though his or her ideas might not be accepted by the organization. (We will return to this very important concept in Section 14.6.)

Disputes between individuals should be fun for all involved. If two subordinates square off, each should enjoy the battle of wits, but there should also be the enjoyment of the first one who is willing to "give it up" for the sake of the organization if not the inferiority of her or his own logic. If the ability to reach agreement is rewarded, there will be a bonding which results from a complete airing of opinions; and this bonding will further synergism. (Not to be anecdotal, but the person retaining the author in his most recent consulting job stated: "The reason that I want you on the job is that I saw how dedicated you were to your position, but when you were presented with a better perspective, you were willing to change your opinion." No one loses when better positions are accepted.)

Once again we see the manager walking the tightrope, balancing the best (perceived) interests of the organization with demands, and possibly opinions, of individual subordinates. However, we urge that manager to adopt the same position as we recommended for the subordinate: Keep everything open, aboveboard, and fun. React positively to those criticisms that are valid without giving in to unreasonable demands. Finally, internalize the fact that a certain level of ongoing criticism and friction are beneficial to the organization. Go beyond its acceptance—encourage healthy friction with management as long as it continues to benefit the goals of the organization. And by all means, produce some yourself! But do not overreact and take things personally. Get your ego out of it!

14.5.2 Group Membership Selection Revisited

While actions can be taken by management to promote synergism once a group is established, major gains can be attained by considering group harmony when choosing the group or replacing a member (an ounce of prevention . . .). Most of the principles for this selection were covered in detail above. However, in terms of synergism, consider the following quote from Kazmier: "Research evidence indicates that small work groups assembled on the basis of sociometric choice are more productive than those assembled on an arbitrary basis" (KAZMI69,165). Here, sociometric choice heavily involves the informal organizations, and it strongly favors organizational self-selection, as prescribed further by DeMarco and Lister (DEMAR87,148). Although self-selection is certainly to be preferred as being consistent with distributed management, it introduces a problematic situation. Consider the following, which would certainly upset the sociometric apple cart: (1) the need for a balance of expertise; (2) the need for diversity, especially as it relates to marketing and the creative process; and (3) the "odd person out" syndrome, which often finds some of the most creative (and possibly obnoxious) people to be sociometric outcasts. (Recall what happened to the perennial last person chosen for the schoolyard team? Many times this person turns out to be one of the most successful.)

So although it is quite true that every attempt should be made to get "friends" on the same group, galvanizing the informal organization is not necessarily optimal. We use the term *friends* to include those who have demonstrated their compatibility at the workplace. It is generally unwise to assign close personal friends, and certainly not relatives, to the same project group, since their personal lives will almost certainly affect, and be affected by, this work relationship. While there is no doubt an abundance of anecdotal evidence illustrating exceptions to this rule, our concern here is more for the personal relationships and the legal ramifications than it is for mere short-term productivity.

Once the selection process has formalized the group, it is the responsibility of management to overcome the incompatibilities that other considerations have necessitated. This requires a careful clinical approach, first to assure that there is a problem and then to solve it. It is surprising in how many cases that the "odd person out" becomes the very unifying force that brings about synergism. This is because the socially unacceptable person is first perceived as a threat to the norms (i.e., the accepted and enjoyable work environment) of the rest of the group. However, when the contribution of this person to the group is perceived, and when the goals of the organization predominate, there is a rallying to support this person, who might even become a type of functional informal leader. Our fairy tales are replete with stories to this effect (e.g., Rudolph the Red-Nosed Reindeer), as are many of our contemporary legends. A healthy organization will make this come true in those cases where it is justified.

We might search our selves for examples. The author polled a recent class for examples where people's first reaction to an acquaintance was quite negative even though the person developed into one of their closest or most respected associates. There was not a requirement that they become close personal friends, only that their relationship turned into one that was unexpectedly amiable and productive. All of these young people came up with examples, all of which are counterexamples to the "sociometric choice" described by Kazmier above. Formalizing this into a principle:

> For there to be maximum synergism, each person in the group must see each other member of the group as an essential contributor to the group's success.

To create and maintain this perception is possible only if it is a reality. Thus the first step, which is the major goal of group selection, is to assure that each member will perceive the other members as essential. In certain circumstances after this, adjustments might be made to group membership to maintain this. However, this would be the worst-case scenario, and this type of intervention is not likely to succeed unless the underlying causes are eliminated. While scapegoats present a tempting remedy, their banishment rarely has a lasting effect (and the power sacrifice is tremendous).

14.5.3 Maintaining Good Morale

We have used the term *good morale* almost synonymously with synergism. However, recalling all that synergism encapsulates, morale must be seen as an effect of synergism. Managers can maintain (not create) good morale by not ruining the synergism of their organizational component. We address some of the ways that this can be done in this section.

In an attempt to build power, managers often complain about members of the group in front of other members, thus creating the perception that some members are detrimental to the group. Not only should this behavior be anathema, but a proactive stance toward group harmony must be adopted. Before recommending the attitudes that contribute toward its establishment, however, consider one further breeder of discontent: the concept of "doing your share."

The principle given above mandated that everyone see everyone else as essential. Inevitably there will be those who do not see this, because we all appreciate our own hard efforts (and alternative sacrifices) but are not in a position to see the sacrifices made by others. Everyone with any self-esteem at all will believe that they are doing their share. Thus when a setback occurs (and it most certainly will!), the tendency is to blame someone else. This leads to considerable frustration and a lack of creative energy because of displaced aggression. In the final analysis the only person whose behavior I can thoroughly control is my own, and you, your own. Displaced blame leads to the deferral of even considering our own behavioral modifications, much less making modifications. So the tendency is for the group to increase in entropy and decompose, as was discussed in Chapter 13.

The countermeasure to this is a restoration of control. In this case the solution comes from mutual support, which emanates from mutual trust. Managers should recognize that this is not something that they can just instill by making statements to this effect. Rather, it is evolved by setting in place mechanisms whereby the group itself can accomplish this. The astute manager will build on success, recognizing that every situation is unique and requires the application of certain principles as opposed to rules. Thus the suggestions given in the adjacent display are labeled as example rules which we invite the reader to extend. These can be generalized to get the message across: *If you are going to keep from crashing the synergism of the group, the major person who must be controlled is yourself.*

> **Example Rules for Increasing Morale**
>
> - Never take credit for anything; give all the credit to the other members of the group and do not leave anyone out. If you are good, you won't need to state it. Echo credit between group members when your leadership in this regard is followed.
>
> - As a manager, do not give credit to one member of the group when it is perceived to be undeserved by some of the others; this only aggravates the group against yourself and this (potentially deserving) person. Like criticism, most praise should be given in private because of its counterproductive potential.
>
> - Have a definitive method for handling the inevitable functional and personal conflicts that are inherent in all close relationships. (Conflict resolution techniques are discussed in Chapter 18.)
>
> - Recognize and project your belief that nonpersonal conflicts (i.e., functional conflicts) can be quite healthy to the organization; thus their existence should surface so that a resolution can be developed that is beneficial to everyone.
>
> - Know how to win and lose, and the inherent value of each. (This relates to personal power and is considered in depth in Chapter 19.)
>
> - Never be jealous of informal leaders.

14.6 Consensus-building Techniques

First let us state our reluctance to put any techniques at all in this book for fear that the principles will be replaced by technique and technimanagement become the management gimmick of the year. We are convinced that this is exactly what is happening with the "quality movement" that is currently sweeping the country. That is, techniques are being substituted for principles and then laid on the same old Theory X philosophy under a different label. Perhaps they are hoping for a "Hawthorne effect" for which there will be ample bottom-line credit to go around before going on to next year's gimmick.

(We hope we are misperceiving the current situation; but we are not. Major paradigm shifts must evolve over long periods of time, and this will happen only if technical organizations provide the leadership. This is not to say that there are not several technical organizations which have not reset their direction and are progressing effectively down the right path. But these are the exceptions.)

With this qualifier we present the following technique-oriented subsections as mere examples of what can and should be implemented as an integral part of effective technical management. We have chosen three areas to stimulate thinking in this direction: (1) meeting conduct, (2) brainstorming, and (3) methods for building consensus within groups. These were chosen because they are the simplest and most needful from the point of view of the person who has just been thrust into the role of group leader, and they will be presented to that audience. The literature describing these and a multitude of other complementary techniques is legion, and we dare not even attempt to cite it for fear of diluting the other more philosophical references and omitting excellent works that even now are coming onto the market.

14.6.1 Meeting Conduct

A major responsibility that you will assume as a group leader is to chair group meetings. Most people hate formal meetings because they are generally conducted so badly that they are a complete waste of time. This is counterproductive of both time and morale. Yet this is one area where techniques can easily be applied to relieve the problem greatly.

The primary reason that we describe this as a major responsibility is that the chair is in control of the time of all the meeting participants for the duration of the meeting, and thus whether this time is wasted or used effectively is almost totally in your hands. Second, the results of the meeting can either be one of consensus building, which brings about increased group synergism, or disharmony, which leads to greater entropy.

To counter these potential problems, consider the following rules for the conduct of meetings:

1. *Don't have the meeting at all.* This is the first and most important rule: The burden of proof is on the necessity for the meeting. If there is not an overriding reason for it, cancel it or do not schedule it in the first place. There is nothing wrong with regularly scheduled meetings since they tend to clear the decks at regular intervals and let everyone know when

to be there. However, they should be canceled whenever there is insufficient reason to have them. Remember the default downside of every meeting is the time that you take away and the disruption that you interject to the normal work [i.e., that activity which (unlike the meeting) actually produces something tangible]. However, that said, some meetings are inevitable, so let's go on.

2. *Never have a meeting without a published agenda.* This is one of the key evidences in the burden of proof that you even need a meeting. Agenda items are things that must be discussed collectively by all the members of the group that you have called together. Things that should never be on the agenda include (a) items that should be considered by a subset of the group before going to the entire group; (b) items that have not matured enough to be worthy of group consideration and time consumption; (c) items that are already decided, for which the presentation to the group is merely a sham to get them to feel that they have "bought in"; (d) items of no significance, such as parking lot violations; (e) items that should be determined by individuals; (f) items that will be resolved by an informal organization; and (g) announcements that can be handled by written communications or e-mail. (Some complain that they need to make announcements in their meetings for reiteration. That may be fine, but if your people are not reading your written communications to them, seek, find, and address the underlying problem. If it is anything other than the fact that they cannot read, you are not going to solve that problem by reading announcements to them in a meeting.) In summary, the written agenda should turn the group on to making decisions that will have a real impact on their lives. Finally, any group member should be able to add to the agenda.

3. *Do not entertain items not on the agenda.* Everyone has a chance to add items at any time except in the middle of a meeting. Obvious exceptions include emergencies.

4. *Never have a meeting without having distributed the minutes to the previous meeting.* In fact, for efficiency, you might distribute the minutes of the previous meeting and the agenda for the next one at the same time. Distribution of the minutes implies that minutes were taken. Take direct responsibility for the quality and the accuracy of the minutes. If someone else can do it better, delegate it—but recognize that its completeness and accuracy are your responsibility. Minutes should be brief and contain only the decisions that were made at the previous meeting. This would include any decision to table an item (i.e., defer its consideration to a later meeting) or to send it to another subset of the group for further refinement.

5. *Take optimal control of the meeting.* Most managers conduct formal meetings much as they conduct their informal communications (i.e., exercise informal leadership). However, over time they might evolve to a very dictatorial, ram-rodding style that discourages discussion. The optimal point is between these two extremes. Its determination is highly group dependent, and it varies with the agenda items and the urgency of the issues. Watch the group. If they are having fun and enjoying the discussions, let it roll. However, when some are getting edgy and impatient, move it along. Attain resolution by summarizing the opinions expressed and asking if anyone objects to a given action. Then move on to the next agenda item. Use voting and parliamentary procedure only as a last resort.

6. *Stay on the agenda and keep it moving without offending anyone.* This requires a balancing of the need to "get on with it" and the need to listen to everyone. As chair, you are responsible for making this decision, which might, at times, be offensive to someone. However, the default—to allow the meeting to meander endlessly—is offensive to everyone. If a consensus cannot readily be reached, the following alternatives might be employed (in order of preference): (a) get everyone to agree to delegate it to a subset of the group, that being the end of the matter, (b) use one of the techniques of Section 14.6.3 to obtain resolution, (c) defer consideration given a specific bit of additional information that is known to be forthcoming at a given point in time, or (d) table it until the next meeting. The problem with the third alternative is that there is no guarantee that there will be a resolution at the next meeting unless you know that an informal organization will handle it properly in the interim (in which case, great).

7. *Exploit group creativity.* Recognize that the primary value of the meeting is to formulate alternatives. This is where group effort surpasses individual effort and the holistic value of your group can best be realized. Recognize that being a manager does not suddenly endow you with all of the answers. They are buried in the creativity of your people if you have the courage to stimulate your group to dig them out and lay them on the table.

8. *Never put down an honorable suggestion.* If you must argue with it, commend it first by stating all you can that is positive about it. This increases your strength in argumentation. However, be careful, for the objective is to do what is best to accomplish organizational goals, not to win an argument.

9. *Avoid confrontation.* Stay away from lose–lose and win–lose situations.

Try to formulate win–win situations by remaining objective. If you have preconceived ideas that you wish to force on the group (which is Theory X at its finest), you should not bring them to the meeting, which by its very nature is a place where you obtain the consensus of the group and do not attempt to dictate your methods. Objectivity demands that you override your own opinions when they conflict with the group, and your power as a leader will be greatly strengthened by it. Eliminate "always being right" from your list of needs. Freely and gracefully admit when you held a wrong position, and others are much more likely to follow suit.

This list could go on and on, but as you can see, it is merely the principles of technimanagement in action. In Section 14.6.2 we focus on item 7, and in Section 14.6.3 we are concerned primarily with item 9.

14.6.2 Brainstorming

Brainstorming is a very simple technique that is difficult to implement because of the paradoxical nature of its single rule:

> The only thing disallowed is the criticism of another person's suggestion.

This is paradoxical because you, as the leader, must intervene to prevent criticism, and thus you must engage in criticism yourself. However, recognize that your criticism is against those who violate the process of brainstorming, it is not against the ideas that they might be presenting. (We recognize that in a sense all alternative ideas are, in fact, implied criticisms of all others. So we could qualify the definition further by stating that no direct criticism of other's ideas is allowed.)

Brainstorming provides a jump-start to group communication and stimulation of group thought. This is necessary because of the reserved nature of formal interpersonal communications. We hesitate to talk because we are afraid of making fools of ourselves. Let's get all of that out on the table. It is true of all of us, from the very extroverted to the most introverted. In fact, the most extroverted individuals are often driven by fear, and they tend to practice consistently absurd behavior to cover this up. The actions of the introverted speak for themselves.

It is imperative that you, as the leader, establish these guidelines and remove these inhibitions from the outset and that you continue to do this throughout the session. As a consequence, you must be *more concerned with the process than with the outcome.* (But this is true of most aspects of management, is it not? I once started a brainstorming session by correcting the way that one of my students used a term. Needless to say, all that I could do to salvage the situation was to use my poor conduct as an illustration of a bad example. Absolutely no criticism means exactly that—no more and no less.)

One other, rather trivial responsibility of the group leader is to formulate the subject or subjects that will be under consideration. Brainstorming is recommended in the early stages of problem solving—possibly even before problems are thoroughly identified. It can also be quite useful in the formulation of countermeasures once problems are defined thoroughly. As the group leader it is your responsibility to spell out the subject of the brainstorming session in sufficient detail so that the "problems of the whole world" are not addressed.

Having fun is an integral part of brainstorming. The idea is to encourage wild thought—the wilder the better. The principle is that such wild thoughts tend to stimulate the considerations of entirely new paradigms in others. Also, the enjoyment of synergism produces a heightened sense of exhibition. This is another example of the holistic approach—where the whole is producing more than they could functioning separately. The rationale for this principle is summarized as follows:

1. Incorrect ideas stimulate creativity and provoke dormant thoughts which might have merit.
2. Those who potentially have the best ideas are often reluctant to reveal them due to their own insecurities.
3. Being in the company of others who are allowing themselves free reign produces the same effect even in the inhibited, since this is the expected norm.

Thus by enjoying the process of making fools of ourselves, new paradigms can be created.

Recognize that brainstorming is for the creation and formulation of alternatives, not for a choice between them. The minute that alternatives begin to be compared, the brainstorming session has terminated. Choosing among alternatives generally requires prioritization, which is considered next.

14.6.3 Prioritization Techniques

Virtually all complex decisions involve the establishment of priorities. Rarely is sufficient information available to resolve all decisions beforehand. For example, many budgets have to be allocated before the entire amount to be spent is determined. If decisions are simply a matter of a choice between a relatively small number of alternatives, a simple vote might be in order, which is merely a simplification of the Delphi method, presented below.

Before we get into the two techniques that we present to exemplify the multitude of others that exist, recognize that the purpose of all of these techniques is not to *resolve differences* but to *bring about a consensus*. Let us explore the difference so that these tools can be properly applied.

We define the term *resolve differences* to mean that differences exist and we need some way to terminate our discussion and take some action so that the organization is not deadlocked (assuming that a decision has to be made before we proceed and recognizing that no decision is a decision for the status quo). Difference resolution does not change anyone's mind; it simply provides a way to proceed. For example, the traditional application of parliamentary procedure fits our definition of a difference resolution technique, and for this reason we do not recommend its use—it only gets in the way of consensus building. So it is with majority rule or any other difference resolution technique. After the vote is taken, the disagreement may still linger on.

The alternative, but not necessarily opposite concept is one of *consensus building*. Here there is an a priori agreement within the organization (in this case, the group) to come to agreement and to use these techniques for arriving at a consensus. It is not that everyone will be satisfied with the outcome. However, there needs to be an agreement from the outset that the resolution will have the full support of everyone within the organization once the technique is applied. This is a holistic approach—the recognition that, by definition, the decision of the group is superior to the decision of any one person. This accrues from a mutual respect of all members of the group. Let us summarize the essence of this concept in what we will call the *principle of consensus*:

> To build consensus effectively requires an attitude on the part of every member of the organization that they would rather yield themselves to the group than to "have their own way" if this is against the will of the group.

This attitude cannot be enforced—it must be internalized by each member of the organization.

The principle of consensus is central to all of the mechanisms of technimanagement. All of the recommendations that call for vertically integrated teams and negotiations between managers and subordinates are based on this principle. However, it requires the development of a level of trust that is not inherent in our current culture. This is the reason that technimanagement must evolve over time, the time depending on the current maturity of the organization in this regard and the individual capacities of the membership to internalize this principle.

If a manager can arrive at a *spirit of consensus*, the problem is 99 percent solved and the particular procedure might not matter. However, it is essential that (1) there be a fair procedure which gives everyone an equal and ample chance to influence all of the other group members, and (2) there be someone, in this case the manager, who is *looking out primarily for the process and is concerned only secondarily with the outcome*. With this in mind the two displays below introduce two consensus-building techniques. One that has been around for a long time is commonly called the *Delphi technique*. This technique is quite effective but (if not computerized) it takes a considerable amount of time to compile and report back the lists. A more time-effective method is patterned after the way that the National Football League (NFL) allocates new players (or, as they are called, draftees); hence we choose to call it the *NFL-draft method*. Both are quite simple and understandable, and both require as a prerequisite some list of alternatives that is in need of prioritization.

The Delphi Technique

- Requires that each participant vote on each of the items on the list by assigning it a number (say between 1 and 5).
    - Lower numbers to items with highest priority enables a prioritized list by arranging the list in the numerical order of the sum of the votes.

- There might be several rounds of voting.
    - After each round all participants have an opportunity to see how their votes compared with all other participants' votes by rearranging the list and presenting the total vote summary for each item.
    - Once that is determined, there is a round of discussion in which each person has the opportunity to influence all others. This is followed by the next round of voting.

- Usually, a consensus is developed quite quickly, especially in larger groups.

The NFL Draft Method

- Starting with a prearranged list, the participants take their turn around the table of taking "one shot" at the list. (It seems appropriate to start the process by flipping a coin to determine if the turns go clockwise or counterclockwise.)

- The "one shot" consists of any one (and only one) of the following:

  - The raising of an item on the list to any other position on the list, at which point all other items below it get bumped down one position,

  - The lowering of an item on the list to another target position, at which point all items above the target position get raised one position, or

  - Pass, in which the participant is satisfied with the arrangement.

- After each "shot" the participant has the opportunity to explain and discuss the reason for the change and any objections to other changes. Critical discussion can be kept to a minimum by disallowing any criticism. However, this is not necessary (as it was with brainstorming) and can be encouraged or discouraged according to the application.

- The procedure continues until there is either:

  - A resolution in which no further changes are desired, which is ideal, since a true consensus has been reached, or

  - A cycling of changes, which reveals the two (or more) persons who cannot agree while the rest are indifferent to these changes.

- There is a possibility of all kinds of convoluted cycling and horse trading; however, once cycling is detected, which is usually quite quickly, a vote can be taken to break the deadlock.

To prepare this, it is recommended that someone, or a smaller group, do some work up front and anticipate the obvious—that is, generally

Sec. 14.6 Consensus-building Techniques

arrange the list in the order in which the group is anticipated to prioritize it. It should be made clear that this ordering is made not to influence them in any way, but to facilitate their consideration of it. Most group members appreciate the great time savings which that brings about, especially when they recognize that the group can rearrange things in any way they want for the final priority list.

In this chapter we have dealt with the lowest and probably the most important level of management—that of the group. In the next chapter we compare the principles of technimanagement presented to this point with those typical of the current "quality movement."

15

Total Quality Management

15.1 Introduction

It is with great trepidation that we enter into a discussion on a subject that seems to squirm like a fish out of water. Indeed, the very name is fraught in controversy, to the point that new terms seem to be proposed at each application. Those who have any familiarity with the "quality movement" will recognize that many of the principles documented in this book are quite similar to those espoused by W. Edwards Deming (DEMIN85), who was unquestionably the father of this movement in Japan and the United States.

One of the secondary purposes of this book was to demonstrate that most of the principles on which total quality management (TQM) is based are not new. They are ageless principles of sociology, psychology, control, and basic human nature. This is not at all an attempt to demean Deming in any way. On the contrary, we establish the validity of his work. While Mayo and others may have formulated some of the underlying principles of TQM well before Deming came on the scene, they were not able to bring these principles to bear on the industrial establishments of their time in any revolutionary way.

This cannot be said of Deming. Whether Deming reinvented the basic principles or whether he just studied and applied them is of no practical consequence (we suspect a combination of both). He was an idealistic and unrelenting person, but much more—he was right! His accomplishments speak for themselves, and without his great work it is doubtful that books such as this one would even be conceived.

It is our intent throughout this book to present sound management principles independent of any movement or personalities. For it is our contention that only by applying sound management principles can positive change have an enduring effect. Through no fault of Deming's, TQM has become the management gimmick of the year to many, and the avoidance of this name by others manifests the problems inherent in following trends.

The purpose of this chapter is to present what we believe to be the essence of TQM so that any new principles, concepts, or guidelines introduced therein can be incorporated within technimanagement. At the same time, it is to demonstrate that most aspects of TQM are consistent with established principles which have been presented in earlier chapters. We do this so that organizations who have begun the transition in terms of applying TQM principles will continue and extend these toward a full technimanagement implementation. As for a comparison between technimanagement and TQM, we defer this until the end of this chapter, once TQM has been presented.

Our purpose in presenting the underlying principles of technimanagement first (Chapters 1 to 14) was an attempt to separate these from personalities and gimmicks and thereby convince the reader that since these principles are universally true, they can and should be applied to the management of all technical organizations. Indeed, we went one step further in declaring that the productivity obtained by applying these principles is so overwhelming that those who ignore them will suffer the same fate as the slavemasters of old. Now that these principles have been presented in earlier chapters, we can study TQM in this context.

Since the very title of this chapter creates chaotic visions in the minds of many, let us begin by giving our definition of TQM. We emphasize that this is the author's definition, as is the interpretation of Deming's philosophy throughout this chapter. We emphasize this so that none will mistakenly believe that we think there should be a unique TQM interpretation. *We are convinced that Deming would, if he were able, argue with many of the interpretations that we will make*, and we do not claim to be experts on the inner workings of Deming's mind. However, recognize that our goal is to apply the applicable precepts of TQM to technical organizations, which was only a subset of Deming's goals.

We chose to focus on TQM in this chapter because it can best serve to encapsulate the philosophy set forth by Deming. In this context, *our* definition of TQM follows:

> Total quality management (TQM) is a philosophy that has as its intent to provide guidance and motivation for organizations to realize fully the benefits of McGregor's Theory Y.

Admittedly, this definition does not give all the details of the content of TQM. Books have been written with that goal, and we invite the reader to consider the remainder of this chapter as a start and to delve into as many of these books as possible.

Consider the ramifications of this simple definition. By *philosophy* we mean that it provides the doctrine and the incentives. The doctrine is interwoven with motivation for management to make major changes in the way in which they conduct their activities. We couch our definition in terms of *McGregor's Theory Y*, not to take anything away from Deming, but to show that he was in the best of company. We find nothing in Deming's philosophy that is inconsistent with Theory Y. However, TQM and Theory Y approach management in completely different ways. Theory Y provides the underlying foundation in terms of basic human nature. TQM is oriented toward what managers should do to use this theory for the betterment of humankind. Thus we say that Deming's major contribution was to provide directives (what he called "obligations") and incentives so that Theory Y could become a reality in many work environments.

Our definition also separates TQM from Deming, and we believe that he would have had it no other way. It is a grievous error to believe that sound principles depend on personalities; in fact, just the opposite is true. Deming's instructional style contained as many questions as answers. His intent was to get people to *think for themselves* as opposed to giving them "answers," which in turn would foster laziness on their parts. This took tremendous self-control on his part, for in his position virtually anything that he said would have been believed. Such would have been inconsistent with the spirit of initiative that he was trying to cultivate. The alternative, cultism, is not healthy, and Deming was humble enough to know it. Unless managers can internalize principles for themselves, quick answers are of little value.

We begin our review of TQM by looking at its general effects first. This answers the question: What has come about in organizations that have moved in the TQM direction? We then explore one of the primary concerns of Deming with current management practice which motivated most of his doctrine: the cost of short-term thinking. Then we present Deming's 14 obligations of management, which form the essence of his management philosophy. We then review two aspects that have been instrumental in implementation of the TQM philosophy: quality management approaches and the necessity for assuming a customer orientation. In the final section we relate all of this to technimanagement, which will set the stage for its more advanced principles in the following chapters.

15.2 General Characteristics of TQM

While TQM is a philosophy and not a set of procedures, there are general characteristics that tend to evolve once its precepts begin to be practiced. These are summarized in Figure 15.1. These items should be viewed as end results of Deming's obligations of management, not as ends in themselves.

- *Flatter organization.* There is a redistribution of responsibility and authority throughout the organization. In other words, a more holistic approach to unification is sought, modeled after the bee swarm, anthill, or even the lowly earthworm (as opposed to the mammal with a single head).

- *Reorientation of motivation.* This evolves from fear-based to peer-based, and it is accompanied by a liberation of latent internal motivation. This enables the establishment of effective group and individual controls, which heavily stress measurement. Critical to this concept is the use of assessments for individual and group feedback as opposed to rewards and punishments.

- *Emphasis on cooperative efforts.* A primary goal of TQM is the formalization of cooperative efforts by the formation of management-recognized, self-directing teams.

- *Group incentives.* Rewards are attained, at least in part, by methods of group sharing. This might be called profit sharing or gain sharing, but the objective is to eliminate competitiveness among co-workers to encourage cooperative efforts.

- *Empowerment.* This is an ultimate end result, not a primary cause. However, it does provide a tremendous incentive throughout the organization, and it tends to keep the momentum of the organization at a high level. Although not defined precisely in Deming's work, it is closely akin to that described in Chapter 14.

Figure 15.1 General Characteristics of TQM

Figure 15.1 illustrates the difference between fad and principle. We could take any of these characteristics and resolve to reverse-engineer it into our organization to provide a formula for success. Unfortunately, such efforts are doomed before they start. Even worse fortune befalls the perpetrators in that these efforts often show some immediate success. That is, they bring about a short-term benefit (akin to the Hawthorne effect). However, they fail because they are procedural in nature as opposed to being a reflection of an ongoing commitment to principle. Procedures are static, while principles apply regardless of the changes in the work environment, society, technology, or personnel.

A major difference between fads and principles is the duration of the commitment. Without a long-term vision and commitment, the implementation of TQM is impossible. The next section is dedicated to this concept, which also provides a primary motivation for its implementation.

15.3 Cost of Short-term Gains

The price paid for short-term thinking is covered in several places in this book. Life itself is a grand optimization problem wherein each of us makes day-to-day decisions as to whether we will take our short-term gains and pleasures or whether we will defer these and invest this time and money in our longer-term interests. Inevitably, it seems that the balance tends to shift to favor the short term. This tendency of human nature in the personal realm is greatly amplified in the corporate world, where short-term gains are almost the sole measure of a CEO's success. This institutionalized impatience was clearly recognized as a key deficiency of management by Deming: "For years, price tag and not total cost of use governed the purchase of materials and equipment. Work standards, quotas, exhortations, numerical goals devoid of methods to achieve them, failure to invest in knowledge, failures of training and supervision, have added their contribution to the decline" (DEMIN85,7). If managers, at all levels, are assessed, promoted, and rewarded solely on this quarter's or even this year's bottom line, how can we expect them to act in the best long-term interests of the organization? Indeed, the very survival of many CEOs depends on their quickly turning a profit. It is no wonder that some of our finest minds are employed in the purchase, cashing-in, and destruction of entire companies—a business that the archaic structure of our government is in no position to control.

Are these problems beyond our scope of control? We think not. However, these are cultural issues that have accrued over a long period of time. If improvement is to be forthcoming, it must begin with a change of direction. Since sound management principles impose a longer-term view on individuals, it is only reasonable to think that they might accomplish the same throughout the culture, including those who compose the company ownership. Or alternatively, the self-destructive nature of short-term thinking might make them obsolete.

Traditional management is static. It seeks an immediate solution which, when put in place, can be "run." It does not look ahead in anticipating but merely attempts to maximize current gains. Twenty-five years ago the common expression to justify short-term thinking was: *When you* are up to your neck in alligators, it is hard to remember that the original objective was to drain the swamp.* (It probably originated with a cartoon; think of the imagery.) It was a nearly perfect analogy that conveyed the picture so well. Everyone wants you to think ahead, but you have your problems now. But what are these problems, these alligators? Your customers, your subordinates, your boss—constantly taking bites out of your hide.

The longer-term approach requires that we investigate and determine the original cause of the swamp. It is interesting to extend this analogy a bit further. Most drainable swamps inhabited by alligators are caused by beavers, and no matter how much you dynamite their dams, they just come right on back the next year. How do you get rid of beavers? The most effective natural control is (you guessed it): alligators. Alligators eat them. So the very things that are eating your hide are not really the problem—they are the solution (your customers and subordinates, that is). Just get out of the swamp and let them do their jobs! But, then, that is a long-term solution and we need resolution today. So let the dynamiting begin!

As we go through Deming's Fourteen Obligations, be alert to the fact that none of these can be accomplished overnight. They can be initiated immediately, but their full realization take the time and effort that we have called *the transition*. Even if all management understood it (which they won't) and totally believe in it (which they don't), it would still take years for their subordinates to accept the philosophy and obtain the training in its application. Does this mean forget it? No. It merely implies that success is going to go to the patient and persistent. But, then, that is really nothing new.

15.4 Deming's Fourteen Obligations of Management

The most concise statement of Deming's philosophy can be found in his Fourteen Obligations of Management (DEMIN85,10), which we review in this section. Interestingly, Deming did not take these any further and project them into methodology or set of organizational structures. Rather than showing his limitations, however, this really demonstrated his wisdom. We believe that he recognized that every implementation would be different. It would have a completely different starting point, deal with a completely different corporate culture, and probably target a completely different optimal end (which, by definition, would never be fully attained). So instead of prescribing a specific methodology, he gave some directives that reorient the organization and begin its evolution toward excellence. The essence of successful implementation is to adopt the new philosophy and set the direction, as opposed to following a formula.

We hastily add that this is only our assessment; we do not know, nor will we try to discern, what was in the mind of Deming. This is far too complex a task, and we are not sure that even those who were quite close to him could accomplish it. In addition, this is not nearly as important as our being able to use what he said to make our organizations better and more productive.

Our approach in this regard will be, first, to map each *obligation* to one or more of the principles given in previous chapters, to validate its basis in principle. Then we explain how the precept of the obligation applies specifically to the technical organization. Our goal is to provide a basis in principle for TQM to demonstrate thoroughly that it is not a mere cult of personality, but a philosophy that is supported by innumerable years of research.

[In the following sections we restrict our use of quotation marks to direct quotes from Deming (DEMIN85,10). The commentaries on these quotes are oriented toward their application within technimanagement. Although we cannot guarantee that they are totally consistent with Deming's thought, they do demonstrate our belief that TQM is generally consistent with technimanagement.]

15.4.1 Constancy of Purpose

> "Create constancy of purpose toward improvement..."

This obligation speaks directly of goal integration, which is clearly one of the most important elements of keeping the organizational entropy low and its synergism high. We saw that this was one of the most important factors in meeting the goals of the organization. Essentially, this produces a generic goal for the organization, which is quite simple easy for everyone to understand.

The role of management is to "create" this constancy of purpose. This cannot be done by waving a magic wand or promulgating an edict. Indeed, past management practice actually favored the quick-and-dirty over the high-quality product. It is essential to change this orientation by 180 degrees, and this must begin with a change in management attitudes.

Practically speaking, there is no such thing as too much quality. In theory we can conceive of a cost of quality. However, in reality we very rarely approach that optimal quality point above which quality is going to cost more than the value it produces for us. By concentrating the entire organization on improvement, there is very little danger that management will "go too far" in that direction, and the goal integration that results is more than worth the risk.

The application of this to technical organizations is quite direct. Consider the product: research, design, prototypes, and possibly development. In each case there is a tangible product. The documentation can either be readable or a pile of garbage, and so it is with all of these products. Measuring the quality is admittedly not as easy as it is in a production organization. However, it certainly can be assessed—if by no one else, then by the user.

If the only unifying factor to bring about goal integration throughout the organization were the quality of the final product, we might take issue with the adequacy of this obligation. What about the product itself as an accomplishment? This is certainly relevant to the technical organization, which might be bringing something into existence for the first time. This should certainly be an exciting unifying factor. However, it should be recognized that this is not excluded, since "improvement" here goes far beyond improvement of the final product. In a research organization it might apply to the improvement of new concepts to propose and new ways to accomplish existing functions. If anything, the concept applies more to the technical organization than it does to any other, since generally their overall goals are already aligned with doing existing activities in new and better ways.

Most important, improvement applies to the process of management itself. Thus improvement, by definition, can never be static. It can never result in a pat answer to solve a problem, for better solutions are always known to exist. Thus the best that can be done by management to meet this obligation is to establish an evolutionary mechanism within the organization that turns it around and then keeps it accelerating in the direction of constant improvement.

15.4.2 Major Paradigm Change

> " ... adopt the new philosophy... "

The key word here is *new*. This was emphasized above in terms of a radically different orientation for every member of the organization. The underestimation of the impact of this change is the major deterrent to its success. Management takes shortcuts in its quest for a quick-and-dirty solution. Viewing these management obligations as mere add-ons to the established management approaches is a perfect application of the paradigm problem.

There cannot be business as usual. There cannot be the normal knee-jerk reaction to Theory X solutions when things go wrong. This is an entirely new philosophy based on principles that are counter to those currently in effect, and it *must* be *adopted*. This is a major commitment of management that will change forever the way in which they view management.

15.4.3 Process Improvement, not Inspection

> " . . . cease dependence upon inspection. . . "

This is what we have been calling proactive management, and it applies much more in the technical field than it does to the production line. Deming complained that it was totally ineffective to have a quality control function at the end of the line looking at the product after it was produced. He likened it to trying to drive a car by looking out the back window to see where you have been. The concept of proactivity is to anticipate from a review of the process itself.

This obligation requires managers to fix their defective control systems. Often, these control systems break down (or do not exist) and no one even recognizes it! There is no concept of meta-control and no one is responsible for monitoring and improving the control systems themselves. For the most part, current systems are, at best, unresponsive. At worst, they produce countercontrol as their disruption of the social aspects of the workplace ruins morale. This is further compounded by a breakdown in the authority and responsibility chain caused by having one staff group responsible for quality while the line organization is responsible for production (generally emphasizing quantity).

Although clearly this obligation was directed at production organizations, inspection is characteristic of traditional technical organizations. Activities are so intangible and technical that it is quite difficult for managers to supervise the work. Essentially, they "get lost" in the early stages of the project and hope everything is proceeding on schedule. There are few, if any, metrics to tell. However, when some tangible products (e.g., prototypes) do start to emerge, they jump all over them with requests, often for major changes. But the project has usually exhausted well over 70 percent of its resources by this point, so major efforts of management from this point forward are dedicated to damage control.

Efforts have been made to give substance to these intangibles as they are developed by requiring deliverables. For example, detailed government standards have been imposed for software developed for the military. However, this is also inspection after the fact, and it leads to a dual approach: one to do the real job and the other to meet the red tape requirements. Although some documentation is certainly required at the end of each phase of a project, the changing technology dictates that one static standard applied to all future software designs and developments will almost certainly become obsolete before it is even published.

The objective of this obligation is to get the process under constant review by a group whose members are primarily from the process itself. This group is charged not only with the assessment of current efforts but also with developing new methods that can make those efforts more productive. This establishes a completely different control system which is capable of intervening at any point if the project gets off course. Measurement is not performed just on the final product but on the process itself by people who understand it because they are intimately involved with it. The manager's role is to respond to this group with whatever support is necessary to assist primarily in the correction component of the control system.

(Interestingly, the primary means of quality control is inspection in the institutions that have the most impact on the major new supply of expertise into the workforce: the educational system.)

15.4.4 Price Tag Independence

> "End the practice of awarding business on the basis of price tag."

Here we are dealing with another major breakdown in the control system. This simple statement is an attack on just one of the symptoms of the problem, which is a lack of linkage between the measure component of the control system, which monitors the process to assure outgoing quality, and the correction component, which might alter the purchasing or recruiting policies. While addressing a symptom, this obligation can be generalized to require the creation of an inherent control system capable of addressing problems and implementing countermeasures throughout the organization.

The cause of the breakdown has to do with the maturity of most organizations and their progression through the stages of organizational entropy and authority degeneration. We can see Deming's concern if we look to the source of the problem, which, in most large organizations, is the purchasing bureaucracy. They have been charged by their upper management to keep the price down but are generally given no responsibility for the quality of the product produced. Also, they have little, if any, technical expertise on which to judge incoming materials quality. Thus, in their minds all other things are equal, and they have every incentive to make their selections based on low bid.

There are many obvious applications of this obligation in the area of managing technical organizations, the most obvious being the awarding of subcontracts. However, there are others in all areas of supply. Since the major raw material supplied to a technical organization is expertise, decisions with regard to the acquisition of this expertise should not be based solely on price tag. Recently, there has been a temptation to flush out the dead wood and at the same time swap two or three employees for the price of one by the implementation of early retirement incentives. If people were interchangeable parts, this might not be bad policy. But the impact of removing one good person who has acquired every ounce of 40 years of experience and thinking that the same productivity can be obtained by two, or even three, recent graduates is ludicrous, but it is another example of acquisition based on lowest price.

(If nothing else, these early retirement incentive plans reveal just how bad our organizations have been managed over the past 20 or 30 years. If the most mature and most experienced are better cashed in by the organization, something is woefully sick. Either upper management does not understand the value of experience, or these older folks are, in fact, incompetent. Either way, the blame is laid definitively at the feet of the very managers who are creating the early retirement policies. Finally, those who are essential to the organizations usually reappear within a few months at consultant fees of several times their former salaries. This and the cost of golden parachutes calls into question the economics of these approaches to shaping an organization's future.)

15.4.5 Constant Improvement Forever

> "Improve constantly and forever every activity..."

The major premise of this obligation is that our products, our processes, our management procedures, and (yes!) we as human beings *are in constant need of improvement*.

Our discussions in previous chapters of human nature in terms of blame and credit weigh heavily into the basis for this management obligation. Deming was generally quite emphatic in placing virtually all of the blame for any problems within the organization on its management. While, theoretically, we might imagine certain circumstances in which this could be taken too far, self-blame occurs so rarely that those who have these tendencies are often given weird psychological labels. It is an ingrained human characteristic (not just a tendency) to blame anyone else but yourself, your family, your race, your class, and so on. Even when we blame ourselves as management, we rarely place the blame on ourselves personally.

Yet self-blame is a primary characteristic of a good manager, and it should be honest and sincerely believed (internalized). If something goes wrong in the part of the organization to which you have had management jurisdiction long enough for a steady state to develop, it is caused by (1) your bad decisions, (2) your lack of leadership, (3) your lack of influence on your management, (4) your lack of anticipation of future events or aspects of human nature, or (5) your inability to communicate. If secure managers internalize this properly, they will have no problems announcing it to their subordinates and management in order to seek their help in moving toward their own continuous improvement.

This might seem like a radical step, but it is actually a test of one's self-confidence. Only those who are insecure deny the obvious in their attempts to cover inadequacies. Everyone else is aware of your shortcomings; why not admit it to them? Most important, if managers freely accept their shortcomings then, and only then, will they be able to get their subordinates to accept their shortcomings. Thus there is a fundamental tenet of leadership that is motivated by this obligation.

Acceptance of responsibility throughout the organization is a primary and essential element, as without it continuous improvement cannot occur. Without it, corrective action is deferred as each person spends an inordinate amount of time fixing blame and complaining because others do not address the problems that are obvious to everyone. We address this further when we discuss the fear factor in Section 15.4.8.

15.4.6 Immediate Training and Education

> "Institute training and education on the job..."

Before sound management principles can be applied, they must be appreciated, and before they can be appreciated they must be understood. This cannot be accomplished by sending everyone away to a bonding session for a couple of days. It is not accomplished by memorizing buzzwords and other types of indoctrination. It must be integrated into the job itself. In addition, to instill the new culture that will result, it is essential that nonmanagers also be trained in these principles as well as the techniques that have been chosen to implement them. It is a major obligation of management to recognize that everyone within the organization must evolve a common understanding and appreciation of these principles such that they can bring about a common approach toward decision making.

Given this recognition, the next obligation is to implement the training and educational activities by which this understanding will occur. If the educational effort is effective, the appreciation will follow. The principles themselves do not need to be sold, since the benefits of their application should appeal to everyone. However, what does need considerable selling is the fact that management is really serious about it. A major commitment to training will establish this credibility.

A major training and educational effort is required throughout our economy. In technical organizations it is understood that the company will quickly be left behind if all the technical people in the organization do not remain current with the technology. Many organizations of this type have as their primary purpose the generation of new knowledge with regard to their particular niche within the technology. Thus we would not expect that this obligation of management would be neglected in most technical organizations (and we have given it little attention to this point for this reason).

The one area in which training is generally lacking in technical organizations is in the management process itself. As a result, management culture and practices are generally inherited from past experiences as a subordinate. For the younger members of the organization, this is derived primarily from the models within the educational institutions, which are almost totally Theory X in style. (Some of this might be tempered by actual courses taken in management; however, this is almost negligible for personnel who are primarily technically trained.) More experienced personnel who have worked in other organizations will largely be influenced by the management practices there. Those whose primarily experience is with their current organization will be most influenced by the management practices of their own current managers. By the time that a person has the chance to become a formal part of management, a fairly rigid paradigm of management has been established.

The relationship between creative activity and management of this activity is quite complex. Because the job is so nebulous and the results so intangible, everyone is managing something or someone (e.g., themselves). Personal time management is one of the greatest problems since most people have been assigned to several projects concurrently, and even within a project there are usually several competing activities. Resource limitations (e.g., computer access) compound this problem further. Certainly, improvements in productivity would follow if improvements could be made in personal management. However, this expands rapidly beyond the person into the process itself. The advancing nature of tools that aid design and development efforts mandate that training take place if the organization is to stay current.

This discussion is continued in terms of educational programs in Section 15.4.13.

15.4.7 Supportive Management

> "The aim of supervision should be to help people... do a better job."

This obligation highlights current defects in both the control and reward systems. The primary customer of the employee in the traditional organization is, in reality, the manager. Virtually all rewards—promotions, raises, privileges, and so on—are obtained through the manager. Thus the subordinate has (almost) every incentive to do whatever it takes to please the manager. (Alternative incentives are almost always forthcoming, however, from informal organizations.)

Sound management principles require that managers assume the opposite role to that perceived traditionally. In our experience, most effective managers have recognized the necessity for this aspect of the job. To illustrate, we recall interviewing a candidate for a dean's position and were extremely impressed when, pointing to the department heads, he stated: "If these guys are doing their jobs like they should, they will make my life really miserable." This off-the-cuff remark revealed that his perception of our jobs was to lean on him for whatever we needed. However, it is not a question that exceptional individual mangers can and have had attitudes consistent with good management principles. (Those who do not quickly lose the support of their subordinates, and their days are numbered.) Rather, it is a question of degree. The question is: Do current organizational philosophies encourage or discourage this attitude?

This problem is complicated further in technical organizations because most technical managers have other competing activities for which they are rewarded which detract their attention from their subordinates. Examples: the preparation of new proposals, liaison with other peer managers, and reporting requirements. This leads to a situation where there is a good chance that a request for "help" on the part of a subordinate could be viewed as a real annoyance. It really does not matter if this is communicated or not; if a subordinate perceives this to be the case, calls for assistance will be avoided. On the other hand, if a manager does show signs of annoyance or believes that this is not even part of the job, the effect is much more devastating.

Once this obligation is understood, the manager's job becomes more than just responding to the problems that subordinates surface. It is to establish mechanisms that will proactively anticipate these problems and then to support and improve these mechanisms. The manager becomes an integral part of this process and supports the subordinates by establishing formal communication links by which information on this support can be obtained. Subordinates are explicitly encouraged and expected to provide this information through these links. Thus the reward system promotes soliciting the assistance of the manager as opposed to virtually preventing it.

As a final note, recall that we mentioned (parenthetically) above that subordinates obtain alternative incentives from their informal organizations which at times lead them to do other than cater totally to their manager's every whim. We suggest that the attitude of empowerment that leads to assistance and facilitation will not be limited to the formal organization. Indeed, the manager will benefit from the empowerment of certain informal organizations (i.e., those consistent with organizational goals). This is accomplished by creating alliances with and obtaining feedback from the informal organizations, and then responding with actions beneficial to organizational goals.

15.4.8 Reversal of Motivation Mechanisms

> " . . . drive out fear . . . "

This is clearly a major reorientation from Theory X to Theory Y, and it forms the center point of management's obligations. Because it is so central and so paradoxical, we give this simple edict extended attention.

Clearly we are dealing with an application of the theory of rewards, which is one of the most absolute of the principles that we have discussed. It is also the one most abused. This is difficult to understand, because it is so simple, yet this may be the reason that it is so neglected and misapplied. Although we discussed this thoroughly in Chapter 2, the issue of fear was not considered directly. The dilemma we have here is that you cannot have positive rewards without negative rewards, since by definition, *the absence of a positive reward is a negative reward*. Thus mere concern about the absence of a reward can be considered as fear; not a fear of loss but a fear of not receiving a positive gain. We could probably make a very strong case that the elimination of this type of fear is well beyond the realm of management capability. Further, much of the stimulation that many obtain from their work could be classified as a controlled fear. We could go on with this argumentation, but instead, let's just state the obvious: It is necessary to apply different approaches to "drive out" different types of fears.

Few would overtly question the wisdom of implementing this obligation. The question is: How can this be accomplished? Managers have two perceptions that stand in the way of accomplishing this obligation: (1) denial—"what fear?" and (2) their perceived value of fear—"there's not enough fear." These two are very closely related. Managers who attempt to practice Theory Y will see no reason for fear. They internalize (correctly) that they pose no threat to anyone who performs in a reasonable way. As a result, little effort is made to drive out fear, because the perception is that *if it should not exist, it does not exist*. At the other extreme is the Theory X manager, who believes that more fear is in order. Obviously, this person will have no motivation to drive out fear. So who will?

Some who are enlightened will cease their denial and recognize that there are concerns, if not outright fear, which all people have about their jobs. We might expect them to attempt to drive it out. However, they should recognize that it is impossible for management alone to drive out all fear. Figure 15.2 lists several types of fear—and their consequences—in a priority order: from those we would definitely need to drive out, to the types that definitely serve a useful purpose. This provides an example spectrum of fear/rewards on which we can base our discussion.

Type of Fear	Consequences/Benefits
Fear of management retribution for errors	Cuts off communication, originality, innovation.
Fear of change	Prevents continuous improvement.
Fear of rejection by one's peers	Neutral within itself; if the peer group is collectively motivated to accomplish organization goals, this is positive; otherwise, it is negative.
Fear of lost income due to a failure to keep up with technological change	A positive personal motivator that should lead to continuous self-improvement, although low on the hierarchy of needs.
Fear of lost self-esteem due to a failure to produce a quality product	Highest on the hierarchy of needs, this fear can hardly be considered negative.

Figure 15.2 Types of Fear and Consequences

The first of these is clearly within management's control, and it is probably the object of Deming's directive. Everyone needs some motivation for their actions, but fear causes a stressful and resentful workplace as opposed to one that fosters cooperation, which is one of the most essential elements in creating synergism. To drive out this type of fear, it is essential that management create improved metrics and corrections within the control system. Traditionally, employees are measured by how well they follow orders and subject themselves to management. Thus the more fear that they had of getting out of line, the more likely they were of being rewarded. Managers who did not keep their people in line were considered by their managers and peers to be weak.

The improved control system measures against objectives that are developed directly from the goals of the organization. Although we would not expect these to vary from those of management, it is the latitude that is given in their realization which spells the difference. Under traditional control systems the goals are interpreted and measurements are made by the next-higher level of management. When managers miscommunicate the goals (i.e., improperly translate them into objectives) or fail to apply proper measurements, this infects the entire system. (This is the reason that so much resentment occurs when a fellow subordinate is perceived to be "kissing up" to the boss.) The evolution of this change starts with a direct statement of goals to the subordinates and their involvement in turning these goals into objectives from which measurements can be derived. At the completion of this evolution, we would expect the subordinates to be equal partners even in the definition of goals.

The change in the correctional component of the control system is as important as the change in the measurement elements. While subordinates can be integrated into the goal and objective-definition process rather easily, the transformation of the measurement to one of optimally shared peer review is quite problematic. Clearly, the two go hand in hand, since as long as the assessment is purely the responsibility of management, fear will continue to stifle the organization.

The interrelationship between the elimination of the fear of management reprisals and the other four "fears" given above is quite straightforward. Starting with the second, they are considered in Figure 15.3. Major changes must occur throughout the fear spectrum over time. Fears must evolve from those that were quite counterproductive to organizational goals to those that will produce both personal and organizational benefits.

It is recognized that the goal of driving out fear, like the other goals of TQM, is never fully realized. Indeed, it is questionable whether it can be, since the introduction of a more positive and cooperative workplace

- *Fear of change.* This is closely related to the fear of management reprisals, and it is the primary deterrent to continuous improvement. By providing everyone with a knowledge of, and equal participation in, the development of organizational goals (or, at least to start, objectives), the fear of change will be greatly reduced and replaced with an enthusiasm for positive change. Clearly, this change must be perceived to be in the interests of everyone within the organization, a perception that can only be established if it is reality.

- *Fear of rejection by one's peers.* In the Theory X organization peer pressure is generally against management. The goal integration process is intended to reverse this by bringing the informal organizations into the goal- and objective-definition process before the fact. Once goal integration is achieved, peer pressure becomes the dominant force in motivating the cohesive effort to attain organizational goals. Thus this ubiquitous and indestructible commodity has been turned around to serve rather than oppose organizational goals. In the ultimate matured organization the fear of losing synergism becomes the greatest of positive motivators.

- *Fear of failure to keep up with technological change.* The on-the-job training discussed in Section 15.4.6 tends to drive out this fear, recognizing that without it as a motivation, the training programs will not be as effective. However, this is not the threat of being fired, which is a fear of management, and tends to be blamed on management as opposed to the true cause (lack of productivity). Rather, by participating in and understanding organizational goals, all members understand that the inability to keep up technically spells the disaster not only of the organization but also of every person who fails to engage in continual, life-long improvement.

- *Fear of lost self-esteem due to a failure to produce a high-quality product.* The combination of the reward of increased self-esteem and the fear of losing it provides the ideal motivation for quality improvement. With the introduction of effective integration, the source of motivation shifts from one of positive fear (of failure) to one of positive reinforcement (for realizing success).

Figure 15.3 Modified Views of Fear

brings with it the fear that this very environment might be lost. However, this is certainly not the same type of fear that has been driven out. The fear of harming a friend is certainly quite different from the fear of harm from an enemy. Thus it is not the elimination of fear per se that is the obligation of management, but the elimination of the type of fear which characterizes Theory X management practices.

Driving out fear is closely linked with a proper assessment of blame. It is totally impossible for fear to be driven out as long as managers are in a state of denial, blaming their subordinates for all organizational problems. Further, it must be recognized that the specific managers may be only a secondary cause of the problem. As indicated above, a flawed control system within most organizations creates practices that propagate fear. These practices must be replaced by others that evolve an organization which is friendly and which inspires cooperation by its very nature. This theme is continued as we discuss the next management obligation.

(We realize that we have strayed quite far in this subsection from what Deming may have had in mind, but we found no better opportunity to elaborate on the subject of fear.)

15.4.9 Elimination of Departmental Barriers

> "Break down barriers between departments."

In most cases the problem of departmental barriers is caused by the current reward system within the organization which creates incentives for competition as opposed to cooperation (another case of countercontrol). This implies that it would be difficult to improve the organization in this regard even if every manager were totally competent. This situation is bound to continue as long as the primary means of administering rewards is through the department manager, and the department managers are rewarded solely for their respective departments' productivity. The Theory X approach of taking this to a higher level only compounds the problem.

Although this answers many questions with regard to observed deficiencies within traditional organizations, alternative approaches are very difficult for most middle managers to accept. Most often, suggestions for better relationships between departments are received quite defensively by department heads. This is evidence that they accept responsibility for any incompatibility among their departments. However, as friendly as they may be toward each other, and as much as they encourage their subordinates to be the same, it is virtually impossible to integrate their efforts totally under the current reward and resource allocation structure.

The problem is compounded further by two types of competitions that result, both for organizational resources. The first is the most obvious —salaries of the employees within the department. This is generally a zero-sum game, in that the pot of company or divisional resources allocated to this is usually budgeted for a given time period. Thus when individuals within a given department receive more due to their superior performance, those within another must receive less. It seems clear that across-the-board increases are totally counter to the theory of rewards and once firmly entrenched within an organization will spell its rapid demise. So the solution lies in an alternative mechanism that creates a control system which assures continual improvement.

A second, closely related competition is for company resources to perform the department's tasks more efficiently. Since the rewards of individuals within the department are (at least perceived to be) determined by their productivity, each department seeks resources whereby that productivity can be improved. This is especially true of the departmental managers, who have every incentive against allowing their people to support other departments if this does not accrue to the benefit of their respective departments. Again, the zero-sum game makes this problematic. Every time that one department gets something it will cause a negative reaction in the others, since they are being put at a disadvantage.

The first step toward eliminating this problem is the use of horizontal integration in the formation of group activity. However, there might still be a stigma attached to conspiring with the enemy, and the result is a rather dysfunctional team. Until rewards are no longer perceived to be based on departmental competition, no proactive cooperation will be forthcoming from the formal organization.

So Deming has, in this obligation of management, laid on the organization an extremely difficult task. However, since it is the organizational structure that is the problem, only a change in this structure can provide the solution. We might suggest a number of potential remedies, including (1) interdepartmental reviews involving peer group members, and (2) vertical integration on groups coupled with management personnel assessments. Combinations of these might be set to evolve within any organization. However, this is not a simple problem. Indeed, the restructuring of the reward system is by far the most difficult aspect of fully integrating the organization. But it must be changed so that management can meet this obligation and get all of the various components of the organization in tune with its goals. An optimal solution to this problem will evolve from the mechanisms within each organization that will ultimately be established to address these issues. The next three obligations of management also affect the reward system.

15.4.10 Honest Maturation

> "Eliminate slogans, exhortations, and targets for the work force..."

These are symptoms of a management by objectives approach, which, you will recall, Deming opposed so strenuously. They are ineffective as goal elements of the control systems, which must be in place if there is any hope of control. Because they are ineffective as standards against which employees are measured by their respective mangers, no process improvements can take place, and the employees tend to get blamed when these artificial goals are not met. The same rationale applies to the next management obligation.

We call this *honest maturation*, recognizing that a maturing process is required as part of the evolutionary transition away from the traditional Theory X–based management structures. Actually, these particular gimmicks of manipulation are just examples of the types of things that need to be eliminated to evolve the open management style that is essential to the type of participatory management that this obligation requires. Maturity is required by both management and their respective subordinates. We are not trying to denigrate those who are mature in their knowledge and practice within the current management structure, and we are not speaking in terms of absolute maturity to a superior state of morality. Rather, we are using this term in the sense of a gradual change in attitude between managers and their subordinates (and vice versa) which must continue to grow over time.

15.4.11 More Effective Measurement

> "eliminate... numerical quotas..."

Numerical quotas are just the formalization of slogans, exhortations, and targets. This might be viewed as a step in the right direction toward getting departmental objectives fully specified and out in the open. However, they tend to focus and evolve control systems that are based on quantity as opposed to quality. Thus they sacrifice the process itself for the sake of an arbitrary goal.

This obligation might appear to be against all forms of measurement. This would conflict heavily with sound management principles, since without measurement there cannot be control. Those familiar with Deming's approaches realize that, consistent with his strong statistical quality control background, he strongly advocated measurement to determine the extent to which quality was improving. Thus this obligation is not to eliminate measurement per se but to formulate metrics that were effective for meaningful control. The realization of this obligation will find groups establishing their own control systems and determining those measurements that are essential to them.

15.4.12 Pride of Workmanship

> "Remove barriers . . . to pride of workmanship . . . "

Self-realization is at the very top of the hierarchy of needs, and it is that which motivates all good managers. Theory X assumes that those managed are not motivated by such higher-order needs. Clearly, Deming did not agree. This is especially true in the technical organization, where almost all nonmanagerial personnel are highly technically trained.

We could list a number of barriers to pride of workmanship, but most of them have been listed above. The removal of these barriers has also been considered in empowering all persons within the organization so that they are more completely in control of their own destinies.

15.4.13 Long-Term Education and Retraining

> "Institute a vigorous program of education and retraining."

As contrasted with the management obligation given in Section 15.4.6, this one places an emphasis on a systematically developed *program*. This goes far beyond on-the-job training to get everyone up to speed with the job at hand and the evolving nature of the management structure. It represents a proactive effort by management to (1) anticipate the changing nature of technology and markets, (2) determine the education and retraining needs that will be required, (3) translate these needs into specific programs (internal or external) that will satisfy these needs, and (4) motivate everyone in the organization to take advantage of the programs that are established. Like all other management obligations, this will be supported by group efforts that contain both the vertical and horizontal integration necessary to assure success, which brings us to Deming's final obligation of management.

15.4.14 Teams

> "Put everybody in the company to work in teams to accomplish the transformation."

This recognizes work as a social activity and the necessity to take advantage of the synergism that accrues when people work together cooperatively. This has been a recurring theme in the first 14 chapters, especially those on informal organizations and group dynamics.

15.4.15 Closure on Deming's Fourteen Obligations

Although these are called the fourteen management obligations, Deming clearly recognized that management could not do it all: "Top management *and everyone else* must feel pain and dissatisfaction with past performance, and must have the courage to change" [emphasis added] (DEMIN85,11). The "and everyone else" includes all levels of management and those who are being managed as well. Most members of technical organizations play both roles to varying degrees, especially as they become empowered to make those decisions which only they can make best. We could easily imagine fourteen obligations of subordinates that would complement Deming's fourteen management obligations. Thus the process of maturing applies equally to everyone in the organization.

Also inherent in the quotation above is the implication of blame. While management should blame the management structure of the organization, it is certainly not healthy for anyone to blame their managers. The fact is that the management structure is not any given manager's fault. True, managers could apply sound management principles within their own organizational components to assist in evolving an improved structure. And true, they could certainly improve their technique even within the context of the traditional Theory X–oriented organization. But the problem is something that has grown up over time and existed long before the current managers were in place. The solution requires a systems approach that involves everyone in the organization, and we suggest some methods for that in Chapter 24.

At this point we wish to consider some of the approaches that seem to have evolved within organizations as they began taking Deming's fourteen obligations of management seriously.

15.5 Quality Management Approaches

In this section three approaches associated with the implementation of TQM are presented: quality function deployment, statistical process control, and total employee involvement. These are covered from a conceptual as opposed to a procedural point of view. We reserve a fourth approach, customer orientation, for the next section, due to its importance as a major driving force of TQM. In the final section we compare TQM with technimanagement.

15.5.1 Quality Function Deployment

In the traditional organization the quality function is concentrated in a certain group of people who usually have special training in the area of quality control. The major function is one of inspection and providing feedback to the processes to reduce defects in the future. This staff-group model has been adopted in many technical organizations. For example, in major software design and development projects, testing is often performed by a uniquely qualified staff group that does this on a regular basis. This verification and validation (V&V) looks not only at the function of the code but also determines if that function was that specified by the original requirements.

The problem with this arrangement is that it tends to pit tester against developer, in a sort of competition. Indeed, cases have been documented in which this competition has gotten fierce enough to cause severe conflict. The author is directly familiar with one major software development organization for the U.S. Army in which the V&V people were treated as outcasts from the rest of the organization. (The atmosphere changed completely when they entered the room.) A second problem is caused by the separation of the measurement from the correction component of control. By the time the fault is detected, the cause might have disappeared or the tester might not be able to trace the cause of an identified fault. In either case, the correction made is purely on the surface and does not get back to the improvement of the process that created the fault.

As its name implies, quality function deployment has the primary goal of distributing the quality function throughout the organization. This requires that everyone acquire some training in quality control procedure and methodology (i.e., measurement and testing). However, everyone does not need the detailed training afforded previously to the quality control staff group. They need only that which will be adequate to integrate the quality function to their own personal tasks. In software development this might include a knowledge of test case development and the use of automated testing tools.

Quality function deployment does not eliminate the need for specialists in the quality control and statistical areas. Instead these individuals become the facilitators of the groups charged with continuous improvement of specific subsets of the product or operations.

There are specific techniques employed in implementing quality function deployment, the details of which we refer the reader to the literature. Essentially, they take the form of charts that map customer needs across the organization, indicating the specific component of the organization which is currently addressing that need. This establishes a benchmark from which three alternatives may be considered: (1) increase, possibly from zero, the level of involvement of the particular component with regard to the customer need; (2) decrease, possibly to zero, the level of involvement; or (3) allow it to remain the same. As these alternatives are proposed, so is horizontal integration to enable the various components to complement the efforts made elsewhere. Thus, while initially quality function deployment techniques are analytical in nature, their ultimate goals are to synthesize a holistic approach by harmonizing all components of the organization around the common goal of quality improvement and customer satisfaction.

15.5.2 Statistical Process Control

We make no secret that we are avoiding many of the quantitative aspects that have accompanied many TQM implementations. This is done for three reasons: (1) we are convinced that unless the basic philosophy and principles are first understood and adopted by management, these techniques will do little good, because, in fact (2) most of these techniques are directly obtained from, and in many cases, simplifications of, quality control and teaming techniques which are very well documented in the literature (some quite dated), and (3) these techniques can get in the way of acceptance and understanding of the underlying principles. It is the third point that is of greatest concern. The mistaken belief that Deming's principles can be implemented top-down by forcing a set of quantitative procedures on the organization will do more to kill their potential for productivity than anything else.

The theory for developing control metrics and applying them to the control of processes can be obtained from literally thousands of textbooks on the subject. However, these presentations invariably omit consideration of the human element. The key difference is in the orientation in which they are applied. Instead of their being a mystical black box understood only by statistical gurus, now metrics are developed and obtained at the project group level. Instead of being a device to prove how bad certain components of the operation were performing, they are now used internally to sustain constant improvement. Staff specialists become facilitators to the teams as opposed to alternative communication links to upper management. These (example) differences characterize TQM as we interpret it in the context of technimanagement.

From this one should not infer that these measurements and their proper analyses and application are not important. Although difficult for most of the uneducated labor force, the basis for statistical analyses can be understood by virtually all technical personnel. The concepts of inherent variability of measurements, the need for a sufficient sample size, and the methods for comparing sample measurements obtained at random can be conveyed quite readily. Although it is not expected that everyone will become an expert in statistics, neither is this necessary, since everyone does not have to take the measurements, process the numbers, and draw the conclusions. However, everyone can gain an appreciation and an understanding of the critical role that measurement plays in the control process.

It is this understanding that results in employees who create and implement their own metrics. By-products also include the elimination of fear and improved customer orientation. Thus, as opposed to an inspector, the quality control expert provides the expertise necessary to enable the spontaneous development of control measurements, proper methods for accumulating data on these measurements, and effective translation of these data into information that is useful for modifying the processes (i.e., correction). But this must emanate from and be totally integrated into the engineering, design, and development process itself. It cannot be an add-on function.

We are obviously speaking of a mature process, one in which both management and subordinates are united in their view toward improving customer satisfaction. Because of the inherent maturity of most technical organizations, the ability to bring about this climate within a few years is quite feasible. This cannot be done, however, without total employee involvement.

15.5.3 Total Employee Involvement

Total means two distinct things in this context: (1) all employees from the

CEO to the lowest-paid hourly employee must be involved, and (2) they must all be involved completely. If the first of these is problematic (and it usually is), the second becomes unattainable. Most organizations practice some form of distributed management at some level. There is some horizontal integration through committees (or whatever they might be called), some employee feedback, some involvement in the decision-making process, some direct regard for the customer, and so on. Unfortunately, these efforts can be counterproductive even though they might be bringing some immediate productivity gains. For example: (1) the management belief that they are "already there" provides no incentive to evolve further, (2) those who are already empowered by the organization might see further evolution as a threat, and (3) employees who are not empowered might be quite skeptical about attempts to change things (they suspect it to be a move to take further advantage of them). Although it is clear that some precepts which are consistent with TQM are probably essential to any modern organization, especially a technical one, the implications of a static partial implementation might be worse than none at all.

Unfortunately, the situation above is a reality that must be faced. We have emphasized how impossible it is to make major paradigm shifts quickly. The key word above, however, is static. Of necessity, the implementation will never be totally complete. It will never include everyone to the ideal degree. So the title of this section is a goal that admittedly is not totally attainable. It is the recognized direction of the organization, however, which is most important. Recognition is the key to growing involvement. The direction toward distributed management must be recognized as being positive to everyone in the organization, and it must be recognized that the organization is moving at a significant pace in this direction.

15.6 TQM Infrastructure: Customer Orientation

A common subtheme of TQM it that "the voice of the customer be heard throughout the organization." It is customer orientation that unifies management and their subordinates in what we have been calling goal integration. To establish this organizational infrastructure, a procedure emerges that includes (1) defining the customers, (2) prioritizing the customers, (3) involving the customers, and (4) orienting the organization to the customer. After defining these activities, some consideration will be given to the transition.

15.6.1 Defining the Customer

Because the process of defining the customer is taken so lightly, it is often neglected or even omitted. There should be a formal definition of customers by each component of the organization. This should include both *internal* and *external* customers. External customers are fairly easy to define, although they are generally defined in categories rather than individually. Simply put, the definition is in the answer to the question: Who purchases our products or services? Internal customers consist of the other components within the organization which a given component supports so that they can satisfy their customers (this is a recursive definition).

Generally, each particular component of the organization will have a different set of customers. Components that deliver product or services directly to external customers will have one set, whereas those who support primarily other components will have a completely different set.

Managers should not get their personal customers mixed up with those of their component. In the matured organization, managers recognize that their only customers are their subordinates. However, this requires that the subordinates have their customers defined and are acting within the prioritization that we discuss below. If this is not the case, as is true in the traditional organization, the manager will be responsible for "looking over" his or her subordinates to their customers. Similarly, a service component within the organization might be required to "look over" the components which they serve if, in fact, those components have not defined their own customers adequately. This is a complicating factor when the organization does not move uniformly and consistently toward customer definition, a situation that is characteristic of a lack of upper-management commitment.

15.6.2 Prioritizing the Customers

If a component has a unique customer, its definition of goals and objectives will be quite simple. This is very rarely the case. Usually, organizational components have both internal and external customers competing for its resources. Actual external customers might range into the hundreds. The actual number is not as much of a problem, however, as the type of customer. As a simple example, a department that produces two types of products, software and consulting services, would have more problems in prioritizing customers than would one that was only in the consulting business. When there are several types of products/services, it usually helps to prioritize the type first, and then prioritize the customer within the product type. Similarly, the customers can be grouped into types (e.g., government and private industry) and these groups prioritized. It may not be necessary to prioritize specific customers if the same activities that will satisfy one will satisfy all.

The process for accomplishing this can get extremely complex. Usually, it is not until the prioritization process is initiated that the true ramifications of customer definition become apparent, and a redefinition and enlargement of the customer list should be expected. The customer definition process should involve as wide a representation of the organizational component as is possible. Brainstorming and consensus-building techniques should be applied. The prioritizing of customers is necessary since all customer needs cannot be addressed simultaneously.

15.6.3 Involving the Customers

The next step in constructing the customer infrastructure is to create a list of customer *needs*, beginning with the highest-priority customers. It is preferable that all customers be addressed, but usually this is not possible and concentration will be given to the highest-priority customers. The term *needs* is used rather loosely to refer to anything from problems to desires, depending largely on the product or service being marketed. It is critical to involve the customer at this point, since rarely do the members of the organizational component have a full customer perspective. The number of customers that should be involved depends on the number of different product and customer types. One or more customers might be included on the team to determine and prioritize customer needs.

The process of involving the customer is facilitated by creating groups that address the customer needs defined and prioritized above. Generally, customers will be included and given preeminence on these teams to give the team a complete customer perspective. Thus the activities described above are all focused on the creation of a customer-oriented infrastructure which augments the traditional organization to communicate customer needs *directly to the lowest levels of the organization*.

15.6.4 Orienting the Organization to the Customer

We dare not leave the impression that customer orientation can be attained merely by the establishment of groups, even if their purpose is almost entirely to serve the customer. In the traditional organization each employee's main customer is her or his supervisor. As long as the reward system of promotions and raises depends primarily on satisfying the boss, the orientation of subordinates to their management (as customers) is inevitable. The problem with this arrangement is that (1) it requires the person at the top of the hierarchy to have a complete understanding of customer needs, and (2) it requires that these needs be communicated adequately down the chain of command. Customer orientation turns the organization on its head. The lowest levels become subservient to the customers, as defined by their organizational components, and each level up the hierarchy serves the next lower level of management as its primary customer. In this way the entire organization becomes totally supportive of the customer.

Clearly, this cannot be accomplished overnight, and there are severe problems associated with the alteration of the reward system. This requires a transition to take place over time which will enable employees at all levels to change the paradigm of thought that locks them into the traditional structure of management.

15.7 TQM and Technimanagement

Since this chapter is concerned with TQM, and since virtually all of the precepts of TQM are consistent with technimanagement, we would expect the reader to anticipate a clarification as to the difference. Are we pushing technimanagement to replace TQM as the next management *fad of the year*? We hope to dispel that suspicion in this section. First, we hope that the major principles of TQM are not seen as a fad of the year, but rather as a major, 180-degree change in management philosophy. As we have defined it above, TQM is to be credited with the most significant major advancement toward Theory Y since its formulation by McGregor. This being the case, why technimanagement?

To answer this question, Figure 15.4 contrasts TQM and technimanagement with regard to the scope of their applications. Our discussions above might have been misleading in this regard, for we intentionally stressed the application to technical organizations. In fact, there are many aspects of TQM in the literature that have little application to organizations composed primarily of technical professionals. While some of the underlying principles have yet to be totally proven in production organizations, we have no doubt of their application to technical organizations.

The importance of the transition cannot be underestimated. It is fairly easy to visualize the ideal organization five to ten years out. If we could just start from scratch with a set of operational rules based on sound principles, perhaps we could do away with the need for a transition. But we do not have the luxury of starting from scratch. We are all in organizations which, while having some Theory Y aspects, are still essentially Theory X based. The problem is the mindset of all members of the organization at all levels (paradigm problem). They have never before seen Theory Y activated on a

Technimanagement	**TQM**
Technimanagement is based on a wide range of established management principles (see Chapters 1 to 14 and 16 to 23).	While it is clear that TQM contains the essence of many of these same principles, it does not consider them all.
Technimanagement is concerned only with the technical organization.	TQM finds its roots in the study of production-oriented organizations.
Technimanagement can be much more definitive in making applications of established principles since technical organizations have a head start as far as training and maturity of the workforce is concerned.	TQM of necessity must rely on directives ("obligations") and techniques for its initiation.
Technimanagement recognizes as essential, and therefore provides a model approach to the transition mechanisms.	TQM does not address or provide methods for the necessary transition for its effective implementation.
Technimanagement is extensible; thus it subsumes those aspects of TQM that are applicable and discards those that are not, just as it does with all the management theories of the past. As the transition continues and optimal levels of distributed management are sought for the various degrees of maturity of the organization, it is expected that other principles will also be added as needed.	While TQM is evolving, it is doing so under such a wide variety of names and auspices that its evolution is difficult to track.

Figure 15.4 Technimanagement versus TQM

large scale. A transition plan is necessary to provide the bridge from the contemporary model to an organization that is learning inherently to better itself without outside intervention.

This is an opportune time to recap where we are and where we are going. The first fourteen chapters presented most of the underlying precepts of technimanagement in terms of established principles. In this chapter we summarized the application of these principles to support TQM, as we interpret Deming. The remaining chapters lay additional groundwork for the transition to technimanagement. We discuss the largest obstacle to change—managers themselves—in the next chapter. This will be followed by two chapters on conflict, the conflict that always accompanies change. This will prepare us for an additional consideration of personal power. Topics of negative feedback, leadership, expert limitations, and communication lay additional groundwork for Chapter 24, which completes the discussion on transition begun in the preceding section.

16

The Clinical Approach

16.1 Definition

Although direct quotations are not used, many of the thoughts of this chapter were documented by Schoen (SCHOE57) and the reader is encouraged to review his work further for perspectives on this important subject. We use the term *clinical approach* to form an analogy between management science and the art of medicine. (It would be as accurate to say: management art and the science of medicine.) In treating a patient, a medical doctor has incomplete knowledge of the body condition and chemistry that is obtained from direct observation, direct measurements, past history, and testimony from the patient. Based on this, a diagnosis is made to move the patient from the current deficient state to an improved or healed state. Along the way, observations are made to assure that the remedial measures are having their intended effect, and, if not, adjustments are made. Essential to this process is the complete *objectivity* of the doctors. For example, if they determine that a mistake in diagnosis has been made based on further evidence that has come to light, it is essential that they make the necessary adjustment, as opposed to taking offense that their original judgment is called into question.

This is an excellent model for the transformation of a sick organization (or component) to one that is functioning with the full enthusiasm and support of its entire membership. According to Schoen, effective organizational and personnel decisions could only be made if an objective, clinical manner were adopted. He enumerated the characteristics of the clinical approach that are summarized in Figure 16.1. These are detailed in separate sections, as indicated in the figure.

Characteristic	Subject (Section)
Understanding and accepting the members of the organization as they are (the present)	Perception of current reality (16.2)
Recognizing, respecting, and not being threatened by differences in opinion	Differences in opinion (16.3)
Understanding the parallel nature of the logical and emotional content of words	What is said and what is meant (16.4)
Understanding the impact that managerial decisions will make on the behavior of other organizational members and components (the future)	Perception of future reality (16.5)
Understanding the informal organization	Informal organizations (16.6)
Understanding the inherent effects on behavior of moving up the hierarchy	Ego growth and ego protection (16.7)
Objectivity in obtaining feedback and adjusting to changing individuals and organizational structures	Objectivity and the control system (16.8)

Figure 16.1 Characteristics of the Clinical Approach

Much like each human being, each organization has its own personality, strengths, weaknesses, and potential diseases. As an organization matures there is tremendous potential value to experience within a given organization if it can be applied objectively to that organization. We call this phenomenon *organizational experience*. As in the human organism, however, age can work equally against progress when it is reflected in insurmountable bad habits. Three factors can frustrate the potential power of organizational experience: (1) distorted perceptions of reality acquired over time; (2) the consequences of all (mostly wrong) decisions accruing from these distortions, further reinforcing these false perceptions, leading to even further misperceptions and even worse decisions (countercontrol); and (3) attempts made to apply successful experience obtained in one organization (or component) to another in which the underlying cause–effect mechanisms are quite different. Thus both success and failure can result in a misuse of organizational experience. Although these two extremes might seem to be essentially unrelated, an unsuspected common thread joins them at their sources: the unrelenting ego of the manager. But let us go back and start from the beginning to lay a firm foundation for the clinical approach.

16.2 Perception of Current Reality

It might seem that all we need to do is open our eyes and look at the organization in order to see what is going on and the underlying causes for these individual and collective organizational behaviors. *Wrong*! It is so easy to sit on the sidelines and critique, whether it be the Sunday afternoon football game or the Friday morning staff meeting. However, there are very few managers who do not feel totally overwhelmed with their first management assignment. And, we dare say, if you had the job of getting into the head of a 300-pound lineman, *overwhelmed* would be the quintessence of understatement. This is simply because what seems so easy from a lower-level perspective (i.e., making decisions that satisfy me), become extremely difficult when viewed from the next-higher level (making decisions that satisfy . . . ?).

A proper substitution for the question mark above tends to crystallize the clinical approach. Four alternatives come to mind: (1) me, the manager; (2) my boss, a middle manager; (3) my subordinates; and (4) everyone. The initial and natural response is generally 1, especially in organizations where subordinates believe that their management has all power and is making decisions primarily for management's own benefit. Thus, when arriving at that position, it is only normal to want to "join the club." Hire an extra personal assistant, clobber the subordinate who did you in, take your long three-martini lunches, and so on; after all, you have earned it! Hopefully, you recognize the short-term thinking involved here, so after dabbling with 1 for a little while, you quickly mature to 2. But then you soon find out that you have a rebellion on your hands. Indeed, you cannot please your boss without the full support of your subordinates, and they are clamoring for more of the organization's resources to accomplish their work. Your boss is accusing you of being a wimp and so are your subordinates. So you have quickly bypassed 3 and gone to 4. It used to be that you only had one boss; now everyone is putting the pressure on you.

What does this have to do with a perception of current reality? *Everything*. We hope that after reading the first 15 chapters of this book you recognize that all four of the alternatives posed above are dead wrong. This affects your perception of current reality, for as long as you are making decisions to try to satisfy any of these constituencies, you will not be able to formulate a clear perception of current reality. So first let us accept this simple principle:

> Decisions based on a false sense of current reality can be correct only in the wildest of outside chances; the closer the manager's perception to reality, the more chance that the decision will be correct.

The clinical approach demands that managers get outside themselves and gain this knowledge, mainly by their interaction with other members of the organization.

In our analogy, the patient is the organization. Thus the correct answer is to make decisions that satisfy *organizational goals*. If properly defined, these will be customer based. To the first-line manager (e.g., the project leader) this is the project itself. This is the first and most basic and essential element of grasping current reality: knowing the objectives of your organizational component. In our medical analogy, the goal is to get the patient well as expeditiously as possible. Objectives are defined to reach that goal most expeditiously. Once defined, all other decisions are focused purely and clearly on accomplishing those objectives, not on satisfying individuals' transient desires.

Although this element is totally necessary, it is not sufficient; rather, it is merely a prerequisite to a series of other clarifications of current reality that must be made, such as: (1) How is the organization currently structured to accomplish these objectives? (2) What improvements in this structure are in order? (3) How will these improvements affect other organizational components? and (4) What flexibility do you have to make unilateral changes in the structure? These, and a whole host of other questions, establish your current impact on and flexibility within the organization.

We appeal to the novice manager to look around at those who have been in their managerial positions for relatively long periods of time. While these people have much to provide you in terms of experience and history, be wary of their conclusions. Their stationary positions indicate that they might not have an accurate perception of reality, *especially if they are dissatisfied* in that position. Be wary of those who complain all the time but seem impotent in making changes. In many cases they have failed to take into account people's changing natures, which ultimately will be reflected in both individual and organizational behavior. Thus their reality may be one of the past organization, not its current status. A primary contributor to this misperception is the blocking that takes place when differences of opinion are encountered.

16.3 Differences in Opinion

This section and the next are fundamentally in support of that given above: to provide the means for acquiring the more accurate perception of reality so critical to the clinical approach. In the mature, fully converted organization, differences in opinion are considered as major assets. However, in the confrontational situations that have been fostered by Theory X, they are often perceived to be threats. Much can be learned by exploring your own personal emotions in the face of criticism. Ideally, a manager should feel more threatened by the absence of criticism than by it presence. This is especially true during group decision making.

Basic to understanding the nature of management decision making is the realization that only in very rare circumstances will there be total agreement as to a given course of action. Thus the absence of the presentation of alternative opinions is a sure sign of intimidated or complacent subordinates. It is not that they agree, it is just that they do not see it as being to their personal benefit to voice these opinions. They see it as a loss of their personal power. The result has a twofold downside: (1) potentially improved alternatives are not brought to the table, thus hurting the "project," which is the first interest of the manager; and (2) the bottled-up opinions gain potential power of their own, waiting to be released at the most inopportune and unexpected time.

To foster openness, criticism should never be received defensively. Instead, it should be analyzed for the information it conveys, which is a critical component of the manager's perception of reality. It is not that the criticism is necessarily valid, although that possibility must certainly be entertained. As important, however, is the information conveyed with regard to the perception of other members of the organization. This leads to our next subject.

16.4 What Is Said and What Is Meant

A primary reason not to get defensive over criticism is that the words spoken rarely address the real issue. Criticism and other forms of complaining need to be addressed at their root causes. For example, it is not unusual for individuals who complain to be (at least implicitly) disciplined by management. However, the problem is not the fact that the complaint has been lodged; in fact, this may be the beginning of the solution. Every statement made conveys some information that is useful to the manager in constructing a perception of reality. The literal meaning of the complaint itself might also be quite misleading in that it tends to skirt the real issue. The most important reality is reflected by the honest opinions of the subordinates. If managers are open as they should be, they would hope that the perceptions in the minds of their subordinates accurately reflect reality. This is true for the following reason:

> A major prerequisite for goal integration is for all persons in the organization to share an identical perception of reality.

Deceptive and manipulative managers attempt to instill a perception of reality within subordinates which is quite different from their own, and then wonder why goal integration is impossible.

A typical scenario arises when a subordinate perceives that she or he is being treated unfairly by the manager (for whatever reason: let us suppose that it is not even valid). Fear prevents the subordinate from bringing this to the manager, perhaps because it cannot be proven. However, the suspicions linger, and subsequent actions of the manager are interpreted in this light. At some point a controversial issue arises that is aired for group discussion. The person with the misperception reacts negatively to the direction suggested by the manager, not because of real reservations, but because of the previous perceived unfairness.

Variations of the foregoing scenario arise continuously, not only between the manager and subordinates, but between managers and between subordinates as well. We hasten to add that this is a common human characteristic: that of displacing aggression until what seems to be an opportune time and then venting it on an object completely different from the original cause. We do not believe that it is caused by dishonesty as much as it is caused by fear (the type of fear that Deming strove to drive out). Thus the subordinate should not be blamed for this behavior—few of us fail to engage in it. Further, few of us are truly cognizant of it when we are doing it. The subjective reasoning process (i.e., emotions) are so intertwined with our rational senses that it is difficult for most of us to determine when our emotions are overriding our reasonable senses. When this happens, however, our sense of reality suffers tremendously, as we reason that our own behavior is caused by A when in reality it is being caused by B.

Now here is the question: If we cannot tell when our own statements are emotion driven as opposed to being logically objective, how can we tell this about others? It may not be totally impossible, but the uncertainty involved requires great caution. Again, we appeal to the clinical analogy. Medical doctors are often warned not to treat themselves. Their objectivity is considerably different when treating others. They must diagnose with considerable skepticism, recognizing that they must pay very close attention to short-term reactions and adjust the treatment accordingly. This approach cannot work without inherent self-criticism.

The proactive countermeasure to unsure communications is to drive out the fear that causes them, thus enabling subordinates to discuss problems directly as they arise. We recognize that this is strictly an ideal, and one that might only be implemented in very mature implementations of technimanagement. In the meantime it will be necessary to deal with the symptoms of the problem. The worst thing that a manager can do is to try to convince subordinates that their motives and reasoning are flawed (i.e., read them the paragraph above). This is true because you have absolutely no proof that this is the case—only your own suspicions. And even though you might have a 95 percent certainty of being correct, that is not sufficient for you to have the audacity of trying to psychoanalyze someone else's mind. Even a 1 in 20 error rate in this regard could prove fatal.

The answer is to apply the clinical approach to conflict resolution. Stay perfectly objective and do not get yourself into the same situation that you might suspect in others. Do not act on suspicion or emotion, but continue to collect the facts even as the resolution process unfolds. (This subject is so important that we devote the next two chapters to it under the headings of functional and personal conflicts.)

To this point our improved perceptions have concentrated on past and existing reality. However, this is of value only if we are able to develop the ability to project this into an accurate perception of future reality, which we discuss next.

16.5 Perception of Future Reality

If any of us had a crystal ball, all of our problems could be solved easily and readily. All decision making depends on imperfect assumptions about the future. It therefore stands to reason that the more intelligent we are about this subject, the closer our decisions will be toward being correct. The most important component of building an accurate perception of the future is that of having an accurate perception of current reality. If we do not know the current reality, we cannot hope to estimate the impact of our actions on the future.

It is important for each person to grapple with his or her own future reality. Often, we spend so much time worrying about the actions of others that we lose sight of that which we control. There is only one thing that we do control and that is our own actions. Each of us has an *alternative action space*, a set of alternatives that we are free to exercise at any given point in time. To formulate our perception of future reality, it is essential first that we define our alternative action space with regard to any given issue that confronts us. Infeasible and obviously counterproductive and self-destructive alternatives can summarily be eliminated.

Given the remaining alternatives, two questions should be asked: (1) What is going to be the personal impact of each of these alternatives on all other members of the organization (i.e., how will they react, and what are the short- and long-term ramifications)? and (2) What will be the impact on the organization itself? The difference between the experienced manager and the novice is that a sufficient number of alternatives have already been "tried out" on both individuals and the organization that data from a fairly high probability of predicting outcomes have been accumulated. The difference between wise experienced managers and bitter experienced managers is that the former have converted these data into information and have modified their behavior so that the alternatives they now select have a very high probability of success. This is the essence of the clinical approach. It involves the construction of an internal control system that modifies personal behavior to accommodate both the weakness of the manager and the changing nature of the organization. Most important, it sets a direction of continual improvement, and therefore it requires a certain degree of humility to actuate it.

The single greatest problem is a failure to recognize the consequences of the default (doing nothing). It is essential that your component of the organization not remain static with regard to the transition toward technimanagement. This is the reason that we stressed the necessity for taking a proactive stance with regard to issues. Unless they are anticipated and dealt with in the present, they will become the firefights of the future. Ultimately, you will be wiped out fighting the alligators and putting out the fires.

16.6 Informal Organizations

We considered informal organizations in detail in Chapter 12. The fact that Schoen's list included its consideration reemphasizes its importance. A major part of the behavior of the organization results from the reaction of the informal organizations as opposed to the formal organization. Clearly, to make the clinical approach work, managers must have an understanding of the reactions their actions will have throughout the entire organization. It can hardly be said that they have realistic perceptions of current reality, much less that of the future, if they do not understand the principles that determine the behaviors of the informal organizations.

16.7 Ego Growth and Ego Protection

In our discussion of the Peter principle (Chapter 10) we made it quite clear that the organizational positions that managers hold serve to modify their behaviors. This is obvious from a functional point of view, since their responsibilities are usually changed significantly with the change in position. Not so obvious, however, are the internal changes—those of attitude toward others and self-esteem. We package these concepts within the simple word *ego*.

Any person who just got a promotion and has moved into a new office with all the privileges and perquisites of rank and then denies that this is an ego trip is just plain lying. Perhaps this is one of those things that must be experienced to be understood. Although most people will freely admit to the elation they felt at such times, few will admit that this creates a change of behavior on their part. That is, they will generally claim that the decisions which they make as a manager will deviate from those which they would have made (if that were possible) were they still in their previous position.

We submit that the perspective of the new manager can be changed significantly due to the ego factor, and, under ordinary conditions, increased experience has just as much chance of increasing this change as it has in reducing it. Thus, once promoted, some never return to their prior perspective. Although they might have perceived of their management as being closed and dictatorial, they cannot imagine that others would perceive this of them. They feel quite comfortable having their organizational component totally under their control, and they cannot see why others would object to that. In fact, it is only when they are threatened with a loss of control that they get uneasy. The symptoms vary, but only in degree.

This is relevant to the clinical approach, since the organization is inherently doing a behavioral-modification job on most of the managers who populate it, thus robbing them of their objectivity. The following summarizes the problem:

> The greatest barrier to objective decision making is the manager's own ego.

Since the organization uses ego as a motivator to attract more talented people to its higher ranks, the prognosis is not good. This supports the organizational entropy theories discussed in Chapter 13 as well as the Peter principle.

If we trace the emotions of the new manager (or one recently promoted), we might find ample reason for the reversal of this process. Many new managers tend to get overwhelmed with the magnitude and uncertainties of their new jobs. If anything, this should promote ego deflation. However, human nature is such that new ego growth is generally protected. That is, it is quite difficult for new managers to admit that the job is too much for them. Denial.

So what is the solution to this problem? We stated above that the clinical approach sets the direction for continual improvement, and therefore it requires a certain degree of humility to actuate it. The process of "turning the organization on its head" requires an admission that management does not have the capacity to make all the decisions, which might also help. But in the final analysis it is up to the individual manager to exercise control over his or her own ego, as we discuss in our final subsection.

16.8 Objectivity and the Control System

The primary role of management is to establish control systems. Managers who cannot establish a control system within their own personal lives will have considerable difficulty in understanding how to establish them in their organizations. An example is the control of the ego. There is nothing wrong with the elation and celebration that comes with promotion to a higher level. The problem arises when the ego gets out of control and blinds the manager to current and future reality. Clearly, an overinflated ego interrupts the self-improvement process, by giving the impression that it is no longer necessary. This is controlled by most good managers, organizational pressures notwithstanding.

Beyond the personal control system, objectivity must be maintained in obtaining the feedback and adjusting to changing individuals and organizational structures. Feedback (the result of measurements) and adjustment (correction) are essential elements of the control process. The most effective feedback is negative feedback, since it does not just reinforce established directions. This is so important that we devote a complete chapter to it (Chapter 20). At this point it behooves us to delve much deeper into the areas of conflict and conflict resolution, since this extends consideration of the clinical approach.

17

Principles of Functional Conflict

17.1 Definitions and Causes

This chapter and the next deal with conflicts. Although these are relatively short chapters, we felt compelled to separate them due to their differences and the importance of the respective subjects. The implementation of technimanagement requires major changes in the way that we view management, and for that matter, work activity in general. This cannot be accomplished by edict, it must evolve over time. One of the greatest deterrents to this transition is the conflicts between individuals that will inevitably occur once this transition is set in process. It is not that technimanagement will cause these conflicts—as we shall see, they are inherently present within all human organizations. In fact, the principles of technimanagement will avoid and deal with conflicts before they become problems. (Implication: Functional conflicts are not problems.)

We anticipate that the conflicts which will arise in the course of the transition will be used as an excuse to reverse the progress that has been made. The typical bureaucratic response to all conflicts is to view them as problems and impose top-down rules that are intended to solve all such problems forever. Instead, this burdens the entire organization with rules that divert it from its original goals. (Recall that this was a major reason that Theory Y did not make much progress during the 1960s despite its large number of proponents.)

Conflicts within an organization transitioning toward technimanagement fall into two distinct classes: (1) those which are normally expected in the course of organizational activity, and (2) those which are caused by resistance to the conversion process. Both of these should be anticipated so that they can be handled in a proactive manner before their underlying causes become serious problems. Appropriate countermeasures will prevent many conflicts. However, the changing and unpredictable nature of human interactions makes it virtually impossible to anticipate every possible conflict that might arise, so we will also consider conflict resolution.

We have subdivided the subject of conflicts into those which are positive and constructive (functional) and those which tend to be destructive (personal), which are handled in the next chapter. We define a *functional conflict* to be:

> A significant difference of opinion that arises between individuals or organizational components when behavior that is perceived to be rewarded by one component of the organization is perceived to be frustrated by demands made by another component or person.

Significant in this context means that it results in negative, counterproductive behavioral changes on the part of either or both of the parties in conflict. At the lower levels, functional conflicts usually start with a collision between two people (or components) in a competition for organizational resources, but rarely is it limited to these people because of the actions of the formal and informal organizations. At the higher levels these conflicts can easily mushroom into other issues, such as relate to organizational vision and customer priority.

There are many human interactions within the organization that resolve resource sharing and other issues which never become conflicts. These are effectively resolved by the informal organization before they become *significant*, and thus they should not be identified as functional conflicts. Management should welcome all such informal arrangements to further synergism. We shall see that it is the further cultivation of these informal accords that is the ultimate resolution of all functional conflicts.

The most important aspect of the definition of functional conflicts is that the causes of these collisions are *people trying either to define or do their jobs more productively*. This leads to the first and most basic principle of functional conflict:

> All healthy organizations will inherently produce functional conflicts.

This necessarily implies that an organization without functional conflicts is not healthy. We explore this further when we discuss the benefits of functional conflicts in Section 17.2.

Because functional conflicts are a sign of health, it is essential that functional conflicts be differentiated and handled separately from personal conflicts. Management usually learns about conflicts well after the informal organizations have had one or more shots at them. However, when they surface, they are quite obvious. The first thing that an astute manager will do is to assess a conflict to determine whether it is functional or personal. In this regard there are two types of errors that can be made: (1) mistaking a personal conflict for a functional conflict, and (2) mistaking a functional conflict for a personal conflict. Of these the second is more costly, since it can have the effect of suppressing, and possibly trivializing, the functional conflict. *Nothing can be worse for morale than for people who are truly concerned about their productivity to be treated as though they are personally the source of functional conflicts.* Thus we break out a second principle for approaching conflicts in general:

> All conflicts should be approached as functional conflicts until sufficient evidence demonstrates that the conflict is largely personal.

This admits to the fact that most conflicts are a mixture of both functional and personal. Rarely is this a 50—50 mixture. The assessment as to whether a conflict is largely personal or largely functional can be determined by using a clinical approach. If a functional change (e.g., an addition or reallocation of resources) will solve the problem, it is functional. However, if one of the conflicting parties will inevitably continue the conflict in any event, it is personal. For example, if the attitude is "I don't care what Joe does, I am going to cause him trouble," and then Joe is faulted for "hogging the workstation," this is largely a personal conflict since it will not be solved merely by adding a workstation. Personal conflicts are almost always brought forward under the guise of being in the interests of the organization in order to manipulate the formal organization into doing someone's dirty work. A failure to understand this can be very costly to a manager, since by this time the informal organizations are usually quite involved.

The opposite, putting forth a functional conflict as personal, rarely occurs intentionally. For example, the attitude expressed, "Joe would be a great guy if he just didn't hog the workstation," indicates a functional conflict that is beginning to get personal (as most of them do when left unresolved). However, resolving it as a functional conflict leaves both parties with a positive attitude toward each other and their jobs. Attempting to trivialize it as a personal conflict can be extremely counterproductive (e.g., "can't you guys just get along?")

To differentiate effectively between the two types of conflicts, begin by assuming that all conflicts are functional and attempt to resolve them as discussed in this chapter. This places the burden of proof on the conflict being personal, which will lead to errors on the safe side. When it is clear that there is no work objective that has led to a given conflict, it must be concluded to be personal and handled by the methods given in Chapter 18.

In this chapter we continue by illustrating the benefits of functional conflicts and then go on to prescribe methods for their resolution.

17.2 Benefits of Functional Conflicts

Managers should not view functional conflicts as a threat, but should take full advantage of the benefits that can be derived from them. While some functional conflicts can and should be avoided, their complete absence is a sign of complacency. Complacency, in turn, is a symptom of the following possible underlying problems: (1) suppression of functional conflicts, (2) failure to effectively address functional conflicts, or (3) forced settlements that are perceived to be unfair or unjust. In all three cases the result is the same: Employees stop bringing functional conflicts to management for formal resolution. In the absence of formal resolution the informal organizations handle the problems but not necessarily in the direction of resolution. In fact, components of the organization might conduct informal feuds that expand over time. By the time that the formal organization discovers the problem it may well have progressed to the point that effective resolution is impossible.

To appreciate the benefits of functional conflicts, it is important to recognize the downside of organizations in which they do not occur. Most very large bureaucracies are virtually devoid of functional conflicts. At one time the organization was young, vital, and filled with them; but upper and middle management "worked them out" over time with laborious procedures and protocols. Now they surface only with new people; but they are quickly indoctrinated into the way that the organization handles these types of problems. So if purchasing takes two months to get a computer that you could go down to the corner and purchase in less than an hour, who cares? You still get your paycheck. And the absence of the new computer is a good excuse not to have your final report in on time.

Functional conflicts should be recognized as arising out of people's desire to be productive in performing their respective functions. Thus a functional conflict is an organizational problem, not an individual employee problem. Disciplinary action against people involved in functional conflicts results in that behavior not being repeated. This is one of the greatest causes of complacency and adherence to existing paradigms. Figure 17.1 summarizes several major benefits that can be attained from functional conflicts.

- *Improved productivity.* Since they are caused by people striving to perform their tasks more effectively, effective mutual resolution of the conflict will almost assuredly result in greater individual effectiveness.

- *Areas for paradigm changes.* Each point at which a functional conflict arises is a potential area where a major paradigm change might produce a breakthrough in productivity and/or quality.

- *Stimulate alternatives.* One of the rarest commodities in an organization is original thinking. Conflict when properly directed can stimulate the creation of alternative paradigms which can be extremely beneficial to the organization.

- *Relieve organizational tensions.* Properly handled, the resolution of functional conflicts brings various people closer together in establishing a common purpose.

Figure 17.1 Benefits of Functional Conflict

It is important to observe that the benefits of functional conflicts can *only be attained when functional conflicts are effectively resolved.* Thus conflict suppression has a tremendous downside. Equally damaging is the failure to identify functional conflicts when they occur, which prevents management from dealing with them at all.

The question arises: If functional conflicts are this beneficial, should they be cultivated? Conflict should certainly not be encouraged for its own sake. Functional conflicts should be seen as a symptom of a combination of good employee motivation and failure of an informal organization to resolve the conflict. It is an effect, not a cause. Thus the idea is not to foster functional conflicts but to identify them and transform them into benefits to the organization.

Functional conflicts are much like tornados. They are the result in beneficial changes in the atmosphere which usually bring needed rain. The short-term effects of a tornado can be personally devastating. However, the rebuilding brings people together, it produces jobs, and ultimately, the restoration of the damaged buildings is superior to that which existed before. Although all this is well and good, we do not create jobs and rebuild our cities by arbitrarily bombing them out.

It should also be understood that in a healthy organization, functional conflicts will occur whether management knows about them or not. Thus the question is not how to cultivate functional conflicts but given that functional conflicts are a normal occurrence in a healthy organization, how they can best be identified and resolved by management. Resolution is handled in Section 17.3; we deal with identification immediately below.

The process of functional conflict identification (or more aptly called *discovery*) can range from the straightforward to the very oblique, depending primarily on the confidence and trust that the affected individuals have in their management. If functional conflicts have been handled effectively by management in the past, there is a good chance that they will be trusted in this regard again. Thus the effective addressing of functional conflicts is probably the most important contributing factor toward their being discovered in the future.

If functional conflicts have been viewed in the past as more of a nuisance than a benefit, chances are that the process of discovery will be more involved than a direct subordinate appeal. Inspiring subordinate confidence in this regard will have to evolve as part of the transition to technimanagement. It cannot be mandated. However, the process can begin by two proactive steps, one to eliminate the punishment and the other to provide a positive reward. These are outlined in Figure 17.2.

- Assure subordinates that no blame will be assigned when functional conflicts arise.
    - Conduct logical reviews that apply the procedures discussed in Section 17.3.
    - Educate all members of the organization in the nature of functional conflicts.
- Communicate the necessity for a participative resolution to functional conflicts and your belief that this is an integral part of the self-improvement process.
- Follow through. When a functional conflict arises, be sure that the reaction is not the typical Theory X pattern of blame, rule making, and top-down resolution.

Figure 17.2 Proactive Steps toward Functional Conflict Discovery

It is clear that there is a very strong interplay between the resolution of past functional conflicts and the potential for reaping the benefits of functional conflicts in the future. Indeed, we can visualize a control system in which the organization moves toward more and more effective resolution of functional conflicts based on the past experience of resolving them effectively. Managers should continually learn from their past successes and failures in this regard to improve their approaches continually—a philosophy that is certainly consistent with the clinical approach discussed in Chapter 16.

17.3 Resolution of Functional Conflicts

Most functional conflicts can be resolved quite easily. Get all the parties to sit down and discuss the problem and a consensus will usually develop quite quickly. This seems overly simplistic, but it is the basis for all conflict resolution. Everything that we will say in this section is merely an elaboration of this basic theme. The mistake that is usually made is that of blame attribution and blame avoidance. The ground rules for this meeting should be set: *Management already accepts all the blame. Management should have had this meeting several months ago and worked it out, but did not. The conflicting parties are to be commended for their desire to increase their productivity and work together toward a solution.* With these ground rules, most functional conflicts will be resolved within the hour. Management must also recognize the possible need for increased resources, or decreased demands, which might be required.

Let us crystallize this basic principle as follows:

> The resolution to all functional conflicts resides within the parties themselves, since if the managers were smart enough to resolve them, they would not have arisen in the first place.

Thus, when functional conflicts arise, it is the fault of the management. But recognition of this is just the beginning of the solution. Since management is not omniscient, functional conflicts will arise despite the most brilliant and concerted efforts. The solution lies in utilizing the group dynamics mechanisms described in Chapter 14 for allowing the organization to generate an optimal solution. The alternative can be illustrated by considering how functional conflicts are often mishandled by management, which is covered in the display below.

> **Mishandling of Functional Conflicts**
>
> - *Trivializing as personal conflicts.* An attempt is made to assign blame to one or more of the employees so that their "improper behavior" will not be repeated. Sanctions might even be imposed. This is a real morale killer inasmuch as there was no improper behavior to begin with. This is the natural reaction of egocentric managers, who are so possessed with avoiding blame that they must assign it to someone else as quickly as possible.
>
> - *Forced settlements.* This is very close to the one given above, although there is an admission that the conflict is functional (i.e., caused by the organization). However, rather than getting the people who fully understand the problem to resolve it, managers apply their own solutions, usually by creating a rule and imposing it top down. Again, their egos will not allow them to recognize that the collective wisdom of the organization is superior to their own.
>
> - *Denial.* This is an outright failure on the part of management to understand and address functional conflicts. If the formal organization does not address a functional conflict, it will become a major topic of the informal organizations. Although this is not always bad, it is unpredictable. If there is a good informal relationship between the affected parties, their chance of working it out informally is quite good. On the other hand, if this does not exist, the informal organizations might further ameliorate the conflict into a feuding situation.

Formal action will never totally circumvent informal action, and this bears additional discussion. The resolution of functional conflicts by getting the parties to sit down and negotiate under the loose oversight of management is a temporary formalization of the informal organization, and this is the key to its success. In other words, it formalizes what would occur in the ideal situation. Ideally, when competitions for organizational resources arise, the parties involved will work out a solution apart from the formal organization which is totally consistent with organizational goals. In fact, this difference resolution happens all the time in most organizations, but as we discussed above, the differences never obtain the significance where they fit the definition of functional conflict.

To illustrate, let us revisit our example of the competition for the new workstation. Suppose that Joe and Joan both need access to the same workstation and it seems that every time Joe needs it Joan is on it, and vice versa. Joe and Joan get together and decide that since most of his requirements are in the morning and hers are in the afternoon, each will get priority at those respective times. Although this fits our definition of a collision, it did not result in a difference of opinion, and certainly not one that negatively altered behavior. However, this is a simplistic example, and if several other people got involved, the arrangement between Joe and Joan might become irrelevant.

This process of allowing the parties to resolve their own functional conflicts will work in most cases, since the differences are caused by the motivation to satisfy organizational goals and not by personal animosities. However, there might still be difficult problems, especially in the areas of resolving the organizational (or component) vision or the specific methods for accomplishing objectives. Recognize that we already have a paradigm for resolving this type of problem, since functional conflicts are clearly problems in meeting (internal or external) customer requirements. As such, the recommended approach for their resolution (as described in Chapters 14 and 15) is the establishment of a group (task force or work team) involving the parties that would meet as needed to address the problem. Managers would serve as facilitators to this process, not trying to impose a solution or find a villain, but in a supportive role providing the following: (1) additional resources where justified, (2) an organizational representation when there is a conflict between two organizational components, and (3) mediation to try to bring the conflicting parties together. This support role is essential both for effective resolution of the current functional conflict and for purposes of enabling discovery of future conflicts so that they can be addressed proactively.

The discussion above indicates that there are proactive ways of avoiding functional conflicts through the informal organizations. These preemptive measures should be encouraged by management (as superior to conflict resolution) in that they cover a multitude of issues that would otherwise swamp the formal organization. Effective organizations cannot exist without these continuous cooperative efforts outside the formal chain of command. The maturing process that unites individuals and components of the organization within a common customer orientation will greatly reduce the number of functional conflicts that will arise. However, human nature being what it is, we would never anticipate their total elimination, nor do we believe that this would necessarily be desirable.

In conclusion, functional conflicts are relatively easy to resolve because they are caused by the very thing that is essential to synergism: high motivation and goal integration. The solution essentially takes advantage and stays out of the way of the positive momentum that originally brought it about. However, the same cannot be said for personal conflicts, as we shall see in the next chapter.

18
Resolution of Personal Conflicts

18.1 Statement of the Problem

Seeing (in Chapter 17) that properly handled functional conflicts can be constructive, we now turn our attention to those that are almost always destructive. Personal conflicts are inevitable in any organization, and they cannot be eliminated totally. However, they can be controlled. If not controlled, they can feed on each other, multiply, and ultimately tear the organization apart. Unlike the natural entropy increases, however, which act over a period of time, these have a disrupting effect analogous to the energy release of a bomb. Even if organizational devastation is prevented, the presence of personal conflicts almost certainly prevents the attainment of the synergism that will enable the organization to attain its full potential in both quality and productivity. This is because the degree to which personal conflicts are avoided or resolved largely determines the affinity of the individuals to each other within the organization.

The resolution of personal conflicts can be one of the greatest challenges for the manager. As with functional conflicts, the first step in their resolution is their discovery, and in this regard, there is little to add to our discussion in Chapter 17. In summary, the greater the trust and confidence that subordinates have in their management, the more freely they will communicate their needs for all conflict resolution. If they are not communicated, the informal organizations will be their only resort; and while the informal organizations can be extremely useful for many organizational purposes, the resolution of personal conflicts is rarely one of them. This is because informal organizations thrive on internal harmony, so we would expect them to reorganize themselves spontaneously to isolate the parties of any conflict within separate informal organizations.

To a large extent the degree of confidence that subordinates have in their managers depends on their past history in effectively dealing with both personal and functional conflicts. The purpose of this chapter is to furnish advice in this regard. We begin with the proactive approach toward avoiding personal conflicts. Recognizing that this is not always successful, we will move on to the resolution of conflicts involving subordinates and then those in which the manager is personally involved. Finally, on a related subject, we discuss internal conflicts, since these are often the root causes of personal conflicts.

18.2 Preventing Personal Conflicts

Clearly, the best situation within an organization occurs when none of its members are at conflict. Most management philosophies assume that personal conflicts are not normal, and for this reason the manager does not take a proactive stance toward them. This is a major error. Human nature dictates that personal conflicts are inevitable when people are together for extended periods of time on a continuous basis. Thus for managers to prepare for them and to treat them as natural events would seem to be a much better approach. It also establishes a completely different perspective on the part of managers, since their preparation will avoid the rush to blame that characterizes most novice managers.

Obviously, managers cannot prevent their subordinates from getting into personal conflicts. However, they can do many things to establish an environment that minimizes their occurrence. The principles of technimanagement that we have covered thus far suggest several recommendations, which are summarized in the display on page 264. These proactive steps will not guarantee the elimination of all personal conflicts (and hence the next two sections). However, they will go a long way toward establishing a climate of cooperation as opposed to confrontation in which all conflicts can better be resolved.

Proactive Steps to Minimize Personal Conflicts

- *Establish an attitude of cooperation with your own peers.* Subordinates will pick up on this attitude quite quickly, and it will be reflected especially in their relationships with members of these organizational components

- *Do not speak in a militant manner about any others in the organization.* It is quite common for insecure managers to run down their counterparts in other components of the organization. This is contagious, and it will ultimately reveal itself in personal conflicts.

- *Discourage all criticism of others.* This is difficult, since often such criticism is valid, and its discouragement itself can be criticism. However, usually there is a mitigating circumstance or a positive aspect that can be cited. Obviously, criticism cannot be forbidden, but it can often be diffused. The most effective way to discourage criticism is leadership: be liberal in statements that are positive, truthful, and sincere about others. When criticism is due, deliver it face to face and in private, which brings us to our next item.

- *Deal with your own personal conflicts in a straightforward manner.* The resolution of your own personal conflicts with the least possible involvement and blame will set the example, since the presence and resolution of your personal conflicts will be known via the informal organizations. More important, failure to do this can result in the spawning of unresolved personal conflicts between your subordinates and others.

- *Encourage informal organizations that promote amiable relationships between all members of the organization.* Let the informal organizations grow: In a healthy formal organization, they will solve problems that you do not even know about.

- *Establish the customer orientation.* When the steps are taken to define and prioritize customers and then define and respond to their needs, a unity of purpose emerges that can often transcend personal differences.

18.3 Personal Conflicts Involving Your Subordinates

Before starting we must reemphasize that the ultimate solution to personal conflicts is their prevention. Prevention involves the anticipation of, and dealing with, an *issue* before it even becomes a conflict. This can be practiced by individuals, but managers cannot practice this for their subordinates. We stated above that personal diligence in this regard on the part of the manager was one of the keys to establishing an environment of cooperation. (Also, if the conflict does not involve one of your subordinates, do everyone a favor and stay as far away from it as you possibly can—it's not your business.)

Given that everyone will not be practicing proactive conflict avoidance, and given that even those who are will occasionally stumble into a conflict situation, we will move forward to address managerial actions to mitigate the damage when the inevitable personal conflicts arise. As with all other aspects of technimanagement implementation, a direction must be set and the organization must be allowed to mature in that direction before anything approaching the ideal situation will occur. In this case, the ideal situation is where all members of the organization are dealing proactively with divisive issues before they are allowed to become conflicts. In situations where this level of maturity has been attained, personal conflicts will be self-resolving by the parties involved. Again, we realize that this is the ideal, but unless we have this ideal vision in mind, it is impossible to set a direction for it.

We are defining a control system with regard to personal conflicts in which success is defined by the organization incurring fewer and fewer of these as time goes by. A failure to meet the ultimate goal is identified whenever a new personal conflict surfaces. The correction should address not only the current problem but also the failure of the control system to have prevented the conflict in the first place. This is where typical Theory X–based methods go totally counter to the interests of the organization. For although they might address the specific problem with a short-term procedure- or rule-based solution, they do not address the underlying cause of the problem.

Effective preparation requires some fundamental understanding of causes before getting into the procedure for addressing personal conflicts involving your subordinates. We saw that functional conflicts are caused by people within the organization who are trying to meet what they perceive are organizational objectives. The causes of personal conflicts are not nearly as simple to define. Sometimes, functional conflicts get personal, and if this is the case, the first step is to resolve the functional conflict, as discussed in Chapter 17. On the other hand, if a functional conflict is merely being used as a cover for a personal conflict in order to manipulate the formal organization, solution of the functional conflict will solve only part of the problem. Unfortunately, rarely is there a clear distinction, nor is there any assurance that a manager can assess the motives of a subordinate. All we can do is to use the clinical approach based on the verified historical facts (i.e., what has been observed and reported by reliable witnesses). In this regard, objectivity is most critical to an accurate perception of reality.

The adjacent display presents some of the potential causes of personal conflicts. To be effective, managers must address the basic cause, not just the symptom of the conflict. This list is certainly not complete; it is given more to illustrate the complexity of the personal conflict than to formulate solutions.

> **Representative Causes of Personal Conflicts**
>
> - *Personal jealousies.* These are caused by factors outside the work project itself (e.g., romantic or other "friendship" status issues). They might be tied to relationships in the formal organization, but usually aspects of the informal organizations exert much more influence.
> - *Professional jealousies.* These will most often be disguised as functional conflicts, and they might well have started out as such. They clearly involve status within (or recognition by) the formal organization. However, the influence of the informal organizations on this underlying motivation cannot be ignored.
> - *Perceived unfairness.* The difference between this and the first two items is that the perceived cause of the conflict (i.e., management) is not one of the parties involved in the conflict. Often, conflicts arise between those who allegedly are being shown favoritism and those who believe it is being shown. This differs from jealously in that the person who is motivating the conflict does not want anything possessed by the second party of the conflict—the main gripe is with management. However, not desiring a conflict with management, the conflict is initiated with the (supposedly) favored person.
> - *Personal problems.* This is not independent of the other three causes (nor are the others mutually exclusive). In all cases the failure to address an issue prior to its becoming a conflict brings to light a personal problem on someone's part. Unfortunately, managers are rarely in a position to tell exactly who that someone is. However, there are some conflicts that arise for rather strange and seemingly disconnected reasons. We know, for example, that sexual dysfunction can on occasions occur in a perfectly beautiful marriage due to stress in the workplace. It should take no leap of faith to recognize that the broad range of family and marriage problems might cause personal conflicts in the workplace. Even more significant problems might be caused by substance abuse and other personal obsessions and addictions. Although not all personal problems cause personal conflicts within the work environment, if only 5 percent of them do, the resulting number would be significant.

This foundation enables us to move toward the formulation of a first principle in the resolution of personal conflicts within the organization. Usually, when the personal conflict surfaces it does so with a host of smokescreens. The rhetoric usually has little or nothing to do with the cause of the problem. Further, the chances of the manager determining the cause might be very low. This is just as well, if the following principle is observed:

> As a manager it is your responsibility to assist in the resolution of personal conflicts, not to assign cause or blame.

This must be accepted, because you may never know the primary cause. It might be anything from personal jealously to a suppressed personal problem. As a manager it is not your job, nor within your expertise, to isolate the causes of problems. When you attempt to do this, you are treading on very perilous ground. Your direct involvement might alter the cause (you might even become part of it). Neither is it your job to solve the problem, which is a common misperception of Theory X managers, who believe that a solution can be mandated (e.g., "now shake hands and apologize to each other . . . "). This, too, is generally well beyond the capabilities of anyone other than the parties themselves. The steps described below will enable you to assist the parties of the conflict to solve their own problems—not just this time, but also when issues arise in the future.

We hasten to add a very important qualifier here: We are discussing the conflicts that arise between what should be considered normal people. It is recognized that no one is perfectly normal, so this discussion is one of degree. Certain people have clearly identifiable abnormalities which, if not addressed directly, will continue to cause personal conflicts. These cannot be handled by the conflict resolution method presented below (and indeed if this method does not work, perhaps this is a symptom of a larger problem). However, in most cases the problem, such as mental illness or alcoholism, has already been very clearly identified by peers and management alike for some time. For these cases, let us state is as emphatic a way as we possibly can:

> If a root cause of a personal conflict is *known*, and that cause has the possibility of resulting in physical or psychological harm to any of the people involved in the conflict, the respective person or persons should be referred to the appropriate professional counseling.

All organizations have access to professional counselors, whether in-house or within the general community. Find out who they are, and keep the list available. Do not wait for an incident to occur before making this preparation. If your company does not have this staff on call, see that provisions are made in this direction.

The word *known* was accentuated in the principle given above, recognizing that we may never know anything with perfect certainty. Thus we are speaking about when the overwhelming balance of evidence indicates something to the degree that it would be accepted by most reasonable people. Actually, the burden of proof is highly dependent on the consequences of the behavior exhibited. The consequences of dealing with a compulsive liar might not be grave if everyone knows, understands, and accepts the person (i.e., they are willing to adjust) and the person performs admirably despite this defect. On the other hand, if suicide or violence is threatened, action might be warranted even though there not be as much proof of a long history that clearly establishes a problem.

With that important detail clarified, let us address the issue of conflict resolution. We repeat for emphasis: it is not the manager's job to solve the problem directly; imposed solutions will only have a short-term effect. However, if your objective is to facilitate a solution and support both (all) individuals as they show good faith in moving toward a solution, the chances of long-term success improve dramatically. Consider the following guidelines in proceeding:

1. *Keep the conflict as confined in scope as possible.* This is the overriding principle, which will be manifested further in other guidelines given below. This is done to protect the people involved, not for damage control (although it does serve this secondary purpose as well). It must be recognized that few conflicts escape knowledge by some of the informal organizations, so total privacy in the resolution is never attainable. However, that is the goal, and it will be appreciated and respected by the people involved. In pursuit of this goal, the manager should not participate in spreading information about the conflict except that which might be essential to facilitate the solution.

2. *Conduct separate private interviews.* This should occur separately with each party of the conflict to get an objective assessment of perceptions. It might be reasoned that since managers are to avoid being judgmental, ignorance of the facts would be an asset. Further, presentation of the facts is going to be quite biased and probably quite inaccurate. However, to refuse to give the participants a "fair hearing" will inevitably create a perception on both their parts that you are on the other party's side. The purpose of the initial interview is not only to dispel this misperception, but also to loosen up the communication process. The following guidelines should be observed:

a. *Refrain totally from being judgmental.* This is difficult, and exceptions might be allowed for obvious errors of perception that will work out clearly to the person's discredit should they not be immediately corrected. Your job is to collect facts at this point, not to begin "solving" the problem. It is particularly difficult to be objective and totally impartial in listening to the second party after hearing from the first. However, fairness demands that this information not be exchanged.

b. *Use a nondirective interview approach.* This is recommended specifically to avoid being judgmental. In this approach virtually all responses are reflected back for more elaboration. Thus the response "Joe is aggressive" would be reflected back with a question such as "What about Joe do you find aggressive?" In no case would you respond: "I never found Joe to be aggressive" or anything else that might be argumentative. It is critically important that this be done in private with the utmost discretion and that adequate time be provided to give each of the parties a fair hearing.

c. *Separate perceptions from reality.* Typically, you will be inundated by facts. This is healthy; the worse problem is if a person does not want to talk about it at all, which is rare. The nondirective approach usually serves quite well to get both people to respond, which is the main purpose of this exercise. Recognize however, that emotions, not facts, cause behavior. What you are hearing are not necessarily the facts, but they are indicative of the perceptions within each person. As long as this is recognized, they provide you with invaluable information. Because you are not responsible to determine ultimate truth, use the data presented to gain a knowledge of opinions and emotions. This separation is to improve your decision-making ability with respect to the process. You cannot force others to accept your opinions—they will have to come to improved realizations for themselves.

d. *Involve only the parties who are directly affected.* This goes without saying from item 1, but we repeat it for emphasis because of its overriding importance. The only major objective of the interviews other than to apprise you of the parties' perceptions is to move both parties to the following step.

(The recommended steps that follow assume that you discover that the issue is one of a difference in opinion, not a clear case of ethical or moral wrongdoing. *Clearly, cases of ethical or legal impropriety require remedial action with the appropriate judicial authority, either within or outside the organization, as appropriate.* The guidelines here are for resolution of personality differences, not of legal or ethical problems. Hopefully, these actions will preempt personal conflicts before they get to the point of requiring legal or other disciplinary remedies.)

3. *Initiate direct discussions between the two people, excluding you.* For reasonable people, communication is the key to conflict resolution; and most people are reasonable. Getting the two parties together without you is a no-lose situation, and it should be a minimum perceived threat to the parties. After unloading on you, they should be willing to deal more effectively with one another. There is an implied guilt on the part of anyone who would refuse to participate in this step. An additional motivation for getting together involves understanding the following: *If agreement is reached within the subject meeting, there will be no further negative ramifications for either of the parties, and the matter will be communicated no further than that.*

Assuming that the meeting takes place and is successful, the assumption should be made that the conflict has been resolved. At this point it is essential that you have a firm concept of what constitutes *resolution* on your part. We recommend the following definition:

> A manager should feel that a personal conflict is totally resolved when every indication is given by the offending person that sincere efforts will be forthcoming to establish a better relationship. Apologies and win–lose situations should be avoided entirely.

It is not important that you understand all the ramifications of the solution. However, it is important that both parties get back to you (perhaps together) and assure that they have a common understanding as to the resolution of the current problems as well as their efforts to avoid similar conflicts in the future.

Sometimes personalities are such that one party could not find it within herself or himself to agree to do this. In this case, or in the case that the discussions between the two people do not come to a common resolution, the next step will be required.

4. *Meet with both parties together.* Prior to this meeting a decision will be required on your part: Attempt to resolve the issue or submit it to further mediation. You should be able to select between these alternatives at this point, since you will have the information provided by your initial interviews augmented by feedback from their meeting together. If you feel that your efforts at resolving the problem will have a good chance of success without causing negative residual effects, you might

elect this option. However, it is imperative that you not assign blame. (If one or both parties accept it, leave well enough alone.) The more important question is not "who has the false sense of reality?" but "why do they have this perception?" By eliminating the cause of the misperception, hopefully the two will be able to assume a normal relationship based on a common understanding.

If this fails, or if you do not believe that you can bring about a solution simply by correcting misperceptions, then the next step is to move to an arbitration agreement. Given that you have made this decision, *the discussion should no longer be on the problem itself.* Rather, it should concentrate on reaching an agreement as to the arbitration process. For example, one process is to allow each party to select one arbitrator to represent them and then allow the two arbitrators to appoint a third "tie-breaker." Once an arbitration mechanism is resolved, both parties should agree that they will present their respective cases to the arbitration panel, and whatever that group agrees will form the basis for the resolution of the conflict. It is amazing how often just the mention of this leads to a recycling back to a "step 3" resolution, which is preferable. However, if not, you should be ready to implement the arbitration method.

5. *Implement the arbitration method.* It is very beneficial that you remain impartial and not interfere in the arbitration process. This is the solution of the two parties involved, and they have agreed to live with it. Certain management actions might be necessary in the interim, such as temporary reassignment if the subordinates are having difficulty working together.

6. *Avoid the last resort.* Termination/resignation is indicative of failure. People who are not matched to their job requirements and therefore are incompetent and nonproductive are candidates for permanent relocation. However, these issues have not arisen in any of our discussions of personal or functional conflicts. Admittedly, it might be resolved by management that a given person just does not "get along" in the organization, due to the long history of conflicts with fellow employees for no legitimate cause. On the other hand, some of the most brilliant, creative people just do not have the human skills that other people seem to have naturally. While expelling the "troublemaker" might seem to be an easy solution, it might in the long run create an organization that is mediocre and intolerant of anything but compliance.

These guidelines are intended to move toward resolution of personal conflicts with a minimum of external involvement and embarrassment. It is recognized that every conflict is unique, and liberal modifications will be required in certain circumstances. These are guidelines to illustrate the general principles, as opposed to a formula solution.

Certain people will become virtually indispensable to technical organizations over time because of the detailed knowledge they have of the particular projects. Major problems arise when these people get into personal conflicts with their colleagues. One of the greatest challenges for the technical manager is to handle this in a fair and equitable way. For, while the technical characteristics of your subordinates enable them to contribute, it is up to the manager to combine their talents such that they remain productive. This is your problem, not theirs. They have little control over the way that the educational systems and past work environments have shaped their personalities, at least not at this time. The idea that the manager is going to change these personalities to the extent that they suddenly assume some magic compatibility with one another is wishful thinking. The personality differences and their incompatibilities are a reality that must be faced. However, as we saw in Chapter 14, these very same differences, when appreciated, can be quite beneficial in bringing about group synergism.

Proactive creation of a workplace where friendliness is rewarded, coupled with initiatives to resolve conflicts that do occur, will go a long way toward mitigating these potential problems. In addition, a balance must be maintained in promoting a mutual appreciation for all members of the organizational component without patronizing any of its members. This is an ongoing effort that requires attention during virtually every interaction that a manager has with members of the organization. The minute that appreciation for this is lost, major conflicts will not be far behind.

18.4 Personal Complaints Against You

It seems reasonable that when we have no animosity against someone else, we would expect that they would have none against us. Nothing seems to come to managers so much as a surprise as someone who has a complaint against them, especially when they do not believe that they are involved in anything controversial. Their immediate response is to believe the offender to be irrational, although with experience they realize that unjustified negative feelings come with the managerial turf. Generally, the more successful you are, the greater the unwarranted animosity toward you. This is another example of success bringing with it the seeds of failure.

We used the words *unjustified* and *unwarranted* above from the manager's (your) point of view. If these were in fact unwarranted or unjustified from the other person's point of view, that person would be irrational. But we are generally not dealing with irrational people in the workplace (again, exceptions should be referred to the appropriate assistance). If not irrational, the person with the complaint feels that she or he has a legitimate reason for this. There are three possibilities: (1) it could have no basis in fact, in which case the person has a false sense of reality; (2) it could be valid, in which case you have a false sense of reality; or (3) perhaps you both have misperceptions. Regardless, these problems all share a common remedy: establishing a common understanding of one another.

It is essential that managers anticipate that they will be involved in conflicts despite their best efforts to the contrary, and often this will occur without their knowledge. We are not advocating that the manager be paranoid and suspect everyone. However, the very acts of information gathering for decision making often cause issues to arise spontaneously. When these issues turn into conflicts, recognition is essential to preventing major damage to yourself or the organization. In brief: Don't deny it, deal with it!

Many of the same principles can be applied that were discussed above in the resolution of personal conflicts involving subordinates. To review them as they might apply here, consider the following:

1. *Hold an initial private meeting.* Get together with the "offended" person and discuss the problem in as nonconfrontational a way as possible. It should be understood by both of you that if agreement is reached, the dispute will absolutely go no further, and no one else will be involved. This will pose the least threat (of lost face) and provide the maximum incentive to settle. Most important, if settlement is forthcoming at this point, follow through and keep the conflict as limited in scope as possible. Most disputes are resolved at this point through a demonstration of good faith on your part and a willingness to admit mistakes. The following guidelines are suggested:

   a. Be objective and avoid getting defensive (which will now be much more difficult since you are directly involved). Use nondirective interview techniques, but do not try to set any traps. Synthesize the facts objectively and determine the true cause of the problem. Recognize that subordinates will generally "hold back" and give explanations that you might want to hear rather than those which are totally open. Make an allowance for this—it might be a matter of courtesy and kindness as opposed to dishonesty or fear.

b. If it was something that you did which was improper or even questionable, apologize and take full blame for the misunderstanding. Follow up by doing whatever is needed to rectify the problem (e.g., be sure that all negative effects are eliminated, talk to others who might have similar misunderstandings, etc.). Do not leave the issue up in the air by formulating a compromise or deal. If you are at fault, admitting guilt is by far the easiest way out, and it can be extremely power building, as we will see in the next chapter. *Never be afraid to admit that you made a mistake.*

c. If the problem cannot be construed to be your fault, it will be contingent on you to communicate effectively with the person on your side of the issue. If you have made some mistakes in handling the problem (such as not addressing it soon enough), set an example by admitting your shortcomings before presenting your side. In making your presentation, be as concise and objective as possible. Do not be judgmental, but assume that the problem is a lack of information and/or communication as opposed to the other person being at fault. Put the best possible construct on his or her words and actions at all times. This is necessary to facilitate communication, since any overstatement or excessive defensiveness on your part will destroy your credibility. Emotional responses are a sign of weakness; they should not be necessary if the facts are on your side (which they must be or else you should have taken the option in item b).

(Once again, it is the direction, not the absolute position, that is important. If the problem cannot be resolved at this point, and the conversation seems to be cycling continually into a repetitive argument, move away from the issues to the subject of a subsequent meeting in which you can bring one or two other mediators into the conversation.)

2. *Hold follow-up meeting with mediators.* The purpose of a mediator is to observe and possibly comment, not to arbitrate (i.e., impose a solution), at least not at this point. (If used at all, arbitration should be postponed as long as possible, since at that point you lose control of the solution. If you are perceived to be right, arbitration will be seen by the other party as more of a threat, which sets up a win–lose situation.) The choice of mediators should be controlled as well, and you should give thought to their identity before the initial meeting (step 1) takes place. Additional considerations are:

a. Maximum effectiveness will be attained by choosing people who are as close to and respected by the other party as possible. Selecting mediators who are clearly on "your side" will be totally ineffectual, and it might be necessary to allow the other person to enter into the selection process. Ideally, they will not be at a higher managerial level, again to minimize the possible threat. Select people who have no prior knowledge of the issues of the conflict, and do not "prime" them for the meeting in any way. Set up the meeting for the earliest possible time that is convenient for all. Keep the process as informal as possible. Again, just the suggestion of this might precipitate a solution.

b. The same guidelines apply to the meeting with mediators as applied to the first meeting. Allow the observers the opportunity to contribute to improving communication whenever they feel it is appropriate. They should recognize from the outset, however, that they are facilitators of communication, not judges in the matter. Ideally, there will be no winners and losers. If a resolution follows, all participants will be pledged to keep the resolution of the issues in as limited a scope as possible. However, it should be made clear that failure to come to agreement at this point will lead to a management review of the entire set of issues involved with the conflict. This "threat," coupled with the reasoning of the offending party's colleagues, should lead to resolution.

3. *Submit to higher management.* At this point, a failure to resolve should be submitted to the next-highest managerial level for resolution. Most personal conflicts are readily resolved at the first meeting, and the few remaining are almost always resolved at the second. However, some people will not listen to reason, so other remedial action is required. We believe that if the first two meetings are ineffective in bringing about a resolution, no further informal efforts of a peer-review basis will help.

(We agree that this sounds awfully Theory Xish, and therefore we feel compelled to comment. We would hope that the organization would be mature enough that your management would use the approach given in Section 18.3. If this is not the case, you might seek a solution yourself by means of arbitration. In any case, we are dealing with only a very small subset of conflicts at this point, which hopefully you can avoid altogether during your career.)

Regardless of the step above at which resolution was forthcoming, assure that you have truly settled and not just imposed a verbal glossing over of the underlying issues. In this regard the *restoration of a normal rela-*

tionship, or periodic reinforcement meetings, is quite important. Offer to have lunch together, or otherwise establish an informal relationship which will assure that any further issues which might arise are controlled before they are allowed to blossom into conflicts. There is a tremendous value in not claiming victory or doing anything else that might create incentives in the direction of future conflicts.

Again, these guidelines are presented to illustrate the underlying principles of conflict resolution as opposed to a strict procedure to follow in all cases. Flexibility is essential to assure that the highly personal aspects of each person are considered in developing the approach to any given conflict resolution.

18.5 Role Conflict

We mentioned above that people who cause conflicts often have problems with their personal lives. That is, the real source of these problems is not their colleagues, as they perceive. The psychological causes of these problems are complex and well beyond the scope of this book. Nevertheless, we must apply whatever measures are at our disposal to keep counterproductive conflicts to a minimum, even though all of the ramifications of their causes may not be understood. This is a major key to obtaining and maintaining a healthy level of morale.

Those internal conflicts that most typically affect the interpersonal relationships in the workplace are most often attributed to role conflicts. A *role conflict* is an internal problem that can ultimately surface in conflicts between individuals. Kazmier states that a role conflict "results when an individual is faced with two or more role expectations that are incompatible. The individual in question cannot satisfy both role expectations simultaneously" (KAZMI69,158). There are a number of roles that most people play within the formal organization, informal organizations, and the person's private life.

An example that we already discussed with regard to the informal organization takes place when informal leaders are promoted over their former peers in the formal organization. Attempting to function as both a formal manager and an informal leader over the same group almost always leads to a role conflict. Another obvious example occurs when company projects conflict with personal family demands. Probably the worst-case scenario is when a manager asks an employee to perform an act that is contrary to the employee's personal value system (e.g., "tell him I'm not here").

The analogy between role conflict and personal conflict is quite interesting when considering methods of role conflict resolution. We merely

Sec. 18.5   Role Conflict

"split" the person into two separate personalities and then let those two personalities resolve their differences by the methods described above. Of course, this cannot be done physically. However, if this can be internalized by the person, most simple role conflicts can easily be resolved.

Role conflicts are relevant to managers first in their personal and corporate lives. Few mangers serve only one "master," both in terms of their managers/subordinates and the various projects to which they are assigned. This is a very frustrating problem, and a failure to deal with it invariably leads to the squandering of time on low-priority tasks. As suggested above, one procedure for resolving these conflicts is to visualize yourself as several people of whom you allocate to each a role. Then allow these imaginary people to resolve their conflicts just as you would were they your subordinates. This role-playing exercise is best handled when written notes are made to assist the visualization process. This might include the name of the role being played, the main person(s) who are satisfied by this role, the activities that this role requires, and the time demands that result.

Essential to this exercise is a unifying goal from which a resolution of priorities among the various roles can emerge. It is not necessary for one role to eliminate another if there is sufficient time to allocate to both and they are not incompatible. However, in situations in which it is virtually impossible to play both roles (even at different times) a decision will be necessary to eliminate one of the conflicting masters. Once this decision is made along with the necessary time allocations, the conflict should be eliminated. (If the problem is mainly one of time management, there are many techniques available in the literature, as well as computerized devices to assist in this regard.)

A second reason that role conflicts are relevant to managers is that they might reveal themselves by personal conflicts between subordinates. One of the most common complaints of subordinates is that there are multiple conflicting demands being made on them. Few technical employees have only one assignment. But even more demanding than the time conflicts are the psychological conflicts that can result when a person is forced by the organization to split loyalties, and these often manifest themselves as personal conflicts. The key to being able to advise subordinates on the resolution of role conflicts is to go through the exercise of getting in order the various inevitable role conflicts within your own life. The same formal procedure that you used to resolve your own role conflicts can easily be transferred to your subordinates.

To close our consideration of role conflict, consider the role conflicts that all managers must face between their individuality and the uniformity that is imposed by the bureaucracy of their organization. All managers are plagued constantly with the necessity to determine just how far to "push" the other individuals and components of the organization to accomplish the purposes of their particular components. There is a built-in role conflict inherent in all organizations, which was defined quite well by Kast: "The general societal value system emphasized imagination and independence as dominant in business success or life in general. The concept of bureaucracy has been a countertrend. It de-emphasized the human element, particularly capriciousness, and stresses the formal structure of positions and established procedures" (KAST70,269). In short, to what extent do you follow established procedure and act as the organizational "good guy" for the boss, and to what extent do you "do what is right" for the benefit of your organizational unit?

From the organization's point of view, the objective is to bring these two options together so that there is no conflict (i.e., the satisfaction of organizational requirements will inevitably accomplish the goals of each organizational component). But this objective is accomplished rarely. One of the primary roles of middle management is to establish the management structures and mechanisms that eliminate conflicts between the bureaucracy and the effective functioning of the organization. But this is a major challenge, especially in large organizations where many rules have attained a life and justification of their own.

Once the problem has surfaced it would be nice if we could develop a precise solution. However, we can think of no advice or guideline that can improve on that given by Kast: "The most appropriate behavior pattern for progress toward the top of modern, large-scale organizations seems to be forceful, creative individualism, which is *not too far out of line*" [emphasis added] (KAST70,271). Unfortunately, you must decide what is too far out of line for your own organization.

The necessary implication of this statement, however, is that you are not going to progress unless you do get somewhat out of line. The philosophy of technimanagement demands it, and if your manager is any good, she or he will, as well. In turn, this implies that you will demand that your subordinates "get out of line" occasionally, which means that they will not comply with your every wish. But if we inspire it from the outset, we can hardly call it conflict when it results. Instead, we chalk it up to a healthy organization.

The resolution of personal conflicts strongly affects your personal power as a manager. This is the subject of the next chapter.

19

Individual Entropy and Personal Power

19.1 Definitions

As functional conflicts often mask personal conflicts, so personal conflicts often mask power struggles. These fall into two classifications: individual and organizational. We define an *individual power struggle* to be a conflict between individuals in that the power differential between the parties is inadequate to predetermine the outcome. Although there are usually several parties involved in an individual power struggle, usually only two are central to the conflict—those who have the most to lose. Individual power struggles that become further generalized to involve entire organizational components will be referenced as *organizational power struggles*. Individual and organizational power struggles are quite relevant to technimanagement in that its implementation requires that the structure of the organization undergo changes caused by empowerment. An understanding of the nature of personal power will determine the degree of success of the individual manager in the transition to technimanagement.

To enable us to discuss this further, consider the terminology introduced by Katz:

> *Influence* includes virtually any interpersonal transaction which has psychological or behavioral effects. *Control* includes those influence attempts which are successful, that is, which have the effects intended by the influencing agency. *Power* is the potential for influence characteristically backed by the means to coerce compliance. Finally, *authority* is legitimate power; it is power which accrues to a person by virtue of his role, his position in an organized social structure. (KATZ66,220)

Clearly, these four terms are degrees of the same thing, which we are just calling *power* in its general sense. It is important to recognize Katz's distinctions, if not his exact terminology. Power begins with influence, which is as likely to emanate from an informal organization as from the formal organizational structure. The fact that a greater degree of power is necessary to exercise control fits the definition of control given in Chapter 2, especially with regard to the ability to bring about correction when deviations from the standard are detected. While Katz restricts the use of the word *power* to just one of its degrees, we concur that this is the central meaning (or degree) that it generally assumes. The relationship between authority and power is consistent with our discussion of authority in Chapter 5. Finally, we use the term *organizational power* to refer to the organization's potential (or that of a component) to accomplish its goals; similarly, an *individual's power* refers to a person's potential to reach her or his personal goals.

These definitions demonstrate how the concepts of power are integrated into all aspects of organizational operations. It is quite easy to reorganize by making name and personnel changes that do not affect the basic implied power structure. Considerably more difficult is *alteration of the power structure with only nominal changes in the formal organization*. This is the key to evolving technimanagement. As we shall see, the problem is not the actual loss of power; it is the perceived loss on the part of some who mistakenly view empowerment as a zero-sum game.

In this chapter we continue with the application of the second law of thermodynamics to personal power. We then consider the sources of power, defining the mechanisms by which individual entropy can be reduced. Finally, the paradoxical nature of power is discussed.

19.2 Second Law of Thermodynamics

We discussed the application of the second law of thermodynamics to organizations in general in Chapter 13. There we saw the general pattern of an organization to muster a tremendous potential for accomplishment, which is correlated with low entropy, tight cohesion, unity of purpose, and most important, synergism. As this organization matures, we find the natural tendency toward increased entropy, just as in the physical realm the natural trend is toward disorganization, randomization, and decay. Protection, such as paint, can stave off the physical process of decay, and the addition of energy can once again reduce entropy within an isolated physical system. Similarly, we saw the countermeasures that management must take if the organization is to continue to function effectively.

The relevance of this to the study of individual power is quite straightforward. We might state the following as a corollary to the principle of organizational entropy:

> The second law of thermodynamics also applies to both individual and organizational power.

Given our discussion in Chapter 13, it is interesting to begin by contrasting these two applications of the second-law analogy. It would seem that the application to the total organization is merely a logical extension of the application to individuals and organizational components. Although this is generally true, there are some significant qualifiers, as outlined in Figure 19.1.

- Control
    - People have more control over their own power entropy than they do over the organization's.
        - Individual entropy is determined by the person's behavior. The control over it is practically absolute.
        - *Example:* A person is free to move between organizations, although there might be a temporary sacrifice of power in doing so.
    - Failure to realize the source will result in false blame of others for virtually all one's ills, which results in a dispensation of power in nonproductive directions, further increasing personal entropy.

- Source and Exercise of Power
    - The person does not acquire power from the organization, the organization acquires its power from the people who compose it.
    - The power of managers must be recognized to be indirect and limited to the influence they can exert over others.
    - People may have only limited ability to exercise power due to the many other people who might provide counterforces to those initiatives, increasing individual entropy.
    - Individual exercise and loss of power does not necessarily increase organizational entropy. It might decrease it significantly, depending on whether individual power is being used to increase or decrease organizational power.

Figure 19.1 Contrast between Organizational and Individual Entropy

Despite these qualifiers, from a holistic point of view, the organization's ability to accomplish its goals is very much dependent on: (1) the individual power of its members, and (2) this power being directed consistently toward accomplishing organizational goals. This leads to the following principle:

> An organization's power is determined primarily by the interactions of the power potentials exerted by its members.

Recognize that even the most powerful of companies have gone bankrupt and the most powerful of nations have fallen because they have failed to recognize this simple principle.

To illustrate the validity of this principle, consider the two extreme scenarios described in Figure 19.2. Most organizations are a mixture of these two extreme scenarios. These provide a further contrast between organizational power and individual power, illustrating the relationship between the two. Obviously, the ideal situation is where all members of the organization have high power potentials and they are exercising them consistently for the accomplishment of organizational goals.

- *Each member has very high power, but this power is exercised primarily against others within the organization.* We say *against* in the sense that it is being used in an attempt to diminish their power as opposed to accomplishing the goals of the organization. In this scenario we have every member with very high potential power derived from the variety of sources that we discuss below (e.g., knowledge, associations, abilities, etc.). However, since these power sources neutralize each other, the organization can no longer accomplish its goals. It is doomed, although the people within it might well survive to exercise their power within other organizations at other times.

- *Each member has relatively low power but it is exercised together toward common organizational goals.* This could result in a very strong and powerful organization, which would certainly have a much higher probability of success than that described in the scenario above.

Figure 19.2 Two Extreme Scenarios of Individual/Organizational Power Mix

The discussion above has used the terms *potential power* and *exercised power*, thereby defining them implicitly. It helps to define these terms more precisely for this context. Going back to Katz's definition of power, which we agreed was about at the center of the relative power spectrum: "*Power* is the potential for influence characteristically backed by the means to

Sec. 19.2   Second Law of Thermodynamics   283

coerce compliance." This definition includes the characteristics both of potential and exercised power. It talks about the "potential for influence" but also the "means to coerce." *Understanding the difference between these two aspects of power is the key to its effective use and the prevention of individual entropy increase.*

Although most managers probably perceive a difference between these two aspects of power, they are probably considered only rarely, and then implicitly, in the decision-making processes. Yet if the last sentence of the preceding paragraph is true, this should be an explicit consideration in every decision made. To what extent is this decision a perceived exercise of power? To what extent is it diminishing your potential to exercise this power in the future? If these questions are not articulated mentally, they cannot be addressed intelligently. Explicit consideration of these issues enables a restatement of the second law of thermodynamics as applied to individual power:

> A person's potential power diminishes to a much greater extent than the extent to which this power is exercised.

This is true even when it is used effectively to accomplish organizational goals. However, its importance goes up dramatically when personal goals are not consistent with those of the organization and the power is exerted against others within the organization. This leads to a simplified closely related corollary:

> The most effective individual power is that which never has to be exercised.

It seems obvious that this principle is counterintuitive, with the persistent exercise of power in the world today (in a broad variety of ways: military, legal, pressure group, etc.), to say nothing of the brainwashing effect of our media in support of this mayhem. We do not expect many people to accept it, even with the proof that we present. Therefore, we expect power to continue to be misused and abused.

Before defending and showing the validity of these principles as they apply to the technical organization, let us deal with the arguments that might be launched against them. Figure 19.3 presents some expected examples. Most other counterarguments can be handled similarly; but we will go on to provide positive proof of these principles.

Power used is no longer available. This holds in the physical realm. To think that the opposite is true in the organizational realm is folly at best. A manager might have the potential power to fire a subordinate. However, exercise of this power without the concurrence of the formal and informal organizations usually disempowers the manager, making any

- *If you do not demonstrate your power, no one will know that you possess it.* This shows a complete misunderstanding of the nature of true power. A person who has it need only show it in the rarest of cases. As usual, the maximum gain is attained by everyone having an *accurate* perception of reality. Unfortunately, raw demonstrations of power rarely portray this. They are accompanied by all kinds of false rumors and attributions of motives. They beg for explanations that are not believed. If I must exercise power, I would rather do it in private.

- *If you do not exercise your power, people will think you are a weak manager.* This is a clear appeal to the ego, and thus goes counter to the clinical approach discussed in Chapter 15. If people think you are a wimp but the job is getting done, the project is a success, and everyone is working together, they will soon be corrected. The temporary misperception is a small price to pay for success.

- *If everyone would play by these rules, fine, but that is just not the way life is.* This, we feel, is probably the most compelling of the arguments, and we cannot deny that it applies outside the realm of technical management. The very nature of our world is such that certain elements can be controlled only by force. But this is the management of technical organizations, not the state prison system! We are dealing with rational, educated, and intelligent members of the organization, who should take great offense at being classified with the reprobates of this world. We have consistently maintained that the cultures out of which all of our organizations have evolved have basically had to bow to this argument; but a primary goal of technimanagement is to evolve new organizational structures that are not based on old assumptions that no longer hold.

Figure 19.3 Counters to Arguments Supporting Free Exercise of Power

subsequent exercise of power (especially the power to fire someone else) virtually nonexistent for the foreseeable future. We can argue further that if it is done with the concurrence of the formal and informal organizations, it is hardly an exercise of personal power; indeed, to refuse to fire someone under those conditions would be the exercise of power.

We went one step further in asserting that even when effectively exercised, more power tends to be dissipated that which is actually used. This implies that there are times when power must be used, and that it can bring about organizational and individual goals. Effective use is analogous to the efficiency of the conversion of potential to kinetic energy in an engine. It can range anywhere from a highly efficient transition to one that totally wastes the fuel. This can be measured with excellent precision in the realm of mechanical engineering. Although there is an analogy, this efficiency cannot be measured in the organizational realm, since individual power cannot be quantified. However, the idea that power increases with use is as ludicrous as the concept of a perpetual-motion machine. No machine is ever able to translate potential into useful power at 100 percent efficiency. Automobile engines have efficiencies well below 30 percent, due to friction, incomplete burning, and just the general structure of the mechanism itself. The organization's ability to translate potential into usable power is probably much worse, because of internal stresses, wasted motion, infighting, and the like. In the worst case of raw power exercises, all of the energy is dissipated without accomplishing any organizational purposes. In this analogy the greatest cause of inefficiency is destroyed synergism, the oil of the organizational engine.

Even in the military realm we can see the wisdom of not consuming power. Indeed, most civilized peoples realize that the exercise of power is an admission of failure on the diplomatic front. Hitler's apparent successes in this regard gave rise to necessary and justifiable reactions from the West. In this example, was the expended power not justified? Justified yes, but optimal? No! No one would argue that a better course of action would have resulted had someone had the wisdom in the 1930s to prevent this situation from ever coming about. So it is in exercising personal power. There will be times when managers are forced into it by the militant actions of others. But by no means does this mean that there might not have been mechanisms that could have prevented these counterproductive acts in the first place. The time to begin installing such mechanisms is now, in anticipation of the inevitable. Unfortunately, instead of these proactive steps, most tend to think more in terms of preemptive actions to stave off the inevitable. As a result, they become part of the problem. Suppose that everyone had this philosophy. (This is a fair test in all civilized societies.)

This leaves us with a big question: How do we exercise influence over the organization and our subordinates without dissipating our power? The answer is to set a course in which personal power continues to be built by the influence that is exerted (not the power dissipated). This is the essence of leadership. This can be understood if we understand the true source of individual power as it is acquired within the technical organization.

19.3 Sources of Power

Personal power is one of the most misunderstood aspects of any organization. Consider President Bush, who had the highest approval ratings of any president in history (of such ratings) after the successful Desert

Storm campaign. Within the space of one year it was dissipated primarily by false perceptions of reality on the part of his advisors. Unless we understand the nature of this beast, it will invariably destroy us. This is because what poses as personal power is no such thing at all; it is but an illusion. We suggest the following to formalize this fact:

> For power to be real it must be internalized—it cannot depend on the perceptions of others.

By *internalized* we mean that its existence must be an accurate perception of reality on your part. This can only be true if you understand the true source of power, if you know that you have acquired it, and if you know that you are willing to use it. The *perceptions of others* change so dramatically over time that any such power base is temporary at best. Thus the personal power of most of our athletes, performers, and politicians is of high intensity but very short duration. We say most, because there are some who are willing to recognize and make the necessary preparations for this very thing (i.e., they have internalized it and it no longer depends on their popularity). However, their numbers are low.

Consider many of the contemporary and popular writings on power, which pitch power as perception. They give detailed instructions as to the rituals, embellishments, and entrapments of these perceptions. But when finished they leave the reader with the impression that power accumulates from every other source but strength of character. We appeal to history and beg the reader not to get caught up with the mere appearance of power. We are not proposing a quick fix that will land you on top overnight only to find your Waterloo (or Watergate) at the height of your career. On the contrary, these principles of power will sustain you for the duration of your career. However, they can be applied only with a patience that comes from the confidence of knowing that you have an accurate reflection of the reality of true (not just perceived) power creation and growth.

19.3.1 Time Value of Power

This heading is derived from an analogy to what economists and bankers call the time value of money. Recall from your engineering economy courses that something costing $10,000 ten years from now will not cost nearly that much today. In fact, anyone who understands the time value of money would gladly agree to defer paying this price for 10 years, since the amount of money that would have to be invested to obtain this amount 10 years in the future is considerably less than it is today. The application of this principle to power is nearly perfect: *Power builds with time.* To see the validity of

Sec. 19.3 Sources of Power

this relationship, consider the following from Cleland:

> One of the project manager's greatest sources of authority involves the manner in which he builds alliances in his environment—with his peers, associates, superiors, subordinates, and other interested parties. The building of alliances supplements his legal authority; it is the process through which the project manager can translate disagreement and conflict into authority (or influence power) to make his decisions stand. (CLELA68,239)

This relates to time in that the building of these alliances requires time. This might seem to be an obvious disadvantage of the new manager, the hire from the outside, or the person transferred. However, this depends totally on the effectiveness of the existing members of the organization in building such alliances. If, in fact, these are perceived but not reliable alliances, we can see where the novice might have a tremendous advantage on entering a new organizational environment. However, it is essential that patience be practiced, even in exercising influence, much less power. Hastily constructed opportunistic alliances are not the power sources to which Cleland refers.

Our section title might be misleading in that it implies that time alone is sufficient to build potential power for the individual. The analogy to investment is complete: As money must be invested properly, so alliances must be built properly. The extension of the analogy is informative in further defining what is proper. In investing we can go for the very secure—the bank savings account or certificate of deposit (CD). Although insured by the government, the downside is the relatively low interest rate. All other investments require considerably more risk, usually putting most of the principal at risk. Similarly, managers do not enter into alliances without some risk. The degree of risk is determined largely by the number of conflicting alliances and the amount of time spent in alliance building as opposed to minding the home front (i.e., building alliances with the informal organizations within their organizational components). As an example of these optimization considerations, it might be of far greater benefit to have a very strong alliance with someone who in turn has strong alliances with many others than it would be to attempt to have relatively weak alliances with all of them yourself.

These are trade-offs between alternatives that are largely opportunistic in nature. Which committees do you volunteer to join? Who do you seek to assist outside your organizational component? Each additional commitment will eat into the time that you can spend with your subordinates, making sure that things are not getting out of control due to your pursuit of influence elsewhere. This is a major power balance, since if your own operation goes down the tubes there is little that your influence elsewhere can buy you. (Exceptions are legion, but we are convinced that they justify only a very short-term view. Remember, we are viewing this from a total career perspective, not just a few years of high-riding.)

To assist in answering these questions, consider the following principle, which summarizes the discussion in terms of power growth:

> Managers' power accrues with time as (1) their organizational components help achieve organizational goals, and (2) the managers and/or their organizational components contribute to other organizational components meeting their goals.

The counterintuitive nature of this principle is that it does not consider power growth to be a zero-sum game, as is the case in many popular management philosophies. A manager's ultimate power is reflected in the ultimate success of the organization as a whole. Let's face it, who has less power than the CEO of an organization that has just filed for Chapter 11? The ultimate loss of power is exclusion from the organization (i.e., getting fired or having the organization disappear), and it does not matter how much power was enjoyed prior to that point. However, even the lowliest of managers has some power (a job, if nothing else) when the overall organization is prospering.

The primary measure of managers' power is in the behavior of their assigned organizational components. If it is accomplishing organizational goals by surpassing those objectives specifically assigned to it, the manager assumes greater value to the organization. This requires two factors to be in place: (1) goal integration—the manager and the organizational component must be consistent with the goals of the organization in general, and (2) performance. Of these the second is the most important, and group synergism is the most critical ingredient to this. We say this merely because if the capability of the component to be productive is apparent, its power will be evident to everyone, and modification in direction should not be difficult.

The organization can least afford to offend those managers who are contributing most to organizational goals. In addition, there is a tendency to pile more and more responsibility on them because they have shown a propensity for success. Thus it is essential that the organizational component have a high morale, be highly motivated, and have goals consistent with the total organization. *No strength or number of external alliances will overcome a lack of synergism within your organizational component.* Unfortunately, the temporary success in depending on external alliances often leads to grave errors in placing these external alliances above those with the informal organizations which ultimately drive the manager's organizational component.

Although we place the external alliances second to internal alliances, they are by no means unimportant. The organizational component cannot function in a vacuum—generally it needs the cooperation of many other organizational components. But similarly, these other components need the cooperation of the manager's home component. So there is a natural interaction that will automatically define those alliances that should be given greatest proactive consideration. This is essentially the definition of internal customers prescribed by TQM. Once this customer definition takes place, the manager, and the organizational component as a whole, should take those steps to establish the alliances that will keep those organizational units happy, *even if it is to the detriment of some short-term metrics used to evaluate the home component.* Attaining the optimal balance between satisfying external and internal customers will maximize the empowerment of both the manager and the organizational component.

Again, we notice that true empowerment is not a zero-sum game. By making the other "customer" organizational components successful, and thus empowered, power accrues to both the home component and its manager. But this works with subordinate customers as well! The empowerment of a manager's subordinates further empowers the manager via the power attained by the organizational component. This leads to an extension of Cleland's principle, which is stated below:

> The alliances formed by a manager's subordinates multiplies the power of the organizational component, and hence that of the manager.

The jealously that motivates many managers to feel they have to be involved in all such alliance building attests to the counterintuitive nature of this principle. Cleland's quotation above implies that it was the manager who must build all the alliances, but enlightened managers recognize their own inability to create and control alliances, especially those within the informal organizations. Thus they view alliances of their subordinates as a major power builder. In Chapter 12 we saw that these alliances are going to occur in any event. The worst managers bury their heads in the sand and fail even to recognize their existence. Better managers just tolerate them. The best managers encourage them and thereby establish their own alliances with them, both directly and through the influence of their subordinates.

Given an understanding of the time value of power invested in alliance building, it becomes the managers' task to create personal alliances that will further organizational goals. But much more than this (as an end in itself) is the leadership example set for subordinates to do the same. This leadership should go beyond the reward and support system; it needs to include establishment of the mechanisms by which subordinates will build alliances by default (e.g., interdepartmental work teams). We would almost go so far as to say that if these mechanisms are properly established, they will force subordinates to do this. However, alliances cannot be forced. They can be forged only when both parties recognize the mutual benefit that will be derived therefrom. It is contingent on the manager to assure that an environment exists where these alliances can be initiated and grow. In this regard, the effect of a timely nod of the head and smile on the face cannot be underestimated. However, the true test comes with the actual sacrifice of the organizational component's resources to support these efforts.

As a qualifier of the major premise of this section (that power builds with time), recognize that this does not contradict application of the second law of thermodynamics, which essentially has it deteriorating with time. The second law applies when things are left to themselves, which is not what we are discussing in this section. On the contrary, we have emphasized that the manager must be proactive toward (1) the operations of the home organizational component, (2) the building of alliances with the informal organizations at all levels, and (3) the leadership that inspires and rewards subordinates for doing the same. If these are applied over the long term, power will accrue to the manager; if they are neglected, expect entropy to increase as the organizational component moves in the opposite direction from technimanagement.

19.3.2 Influence Spectrum

At this point it serves us to revisit the terminology introduced by Katz above in which the degrees of power potentials were neatly packaged by the words *influence, control, power,* and *authority*. Kast (KAST70,309) suggests a different organization of these degrees in what he calls the *influence spectrum*, which consists of the following:

- Emulation
- Suggestion
- Persuasion
- Coercion

Hopefully, the reader will not get bogged down in the inconsistent semantics, but rather, attempt to understand and consolidate the valuable principles presented by each.

The most powerful method for affecting organizational behavior is by inspiring *emulation*. We have been calling this leadership. This occurs when the example set is so compelling of success that others will emulate it without it even being suggested. Human nature assures that this occurs constantly within organizations. Leadership will be exerted, whether by you or by someone else. However, there is no assurance that the behavior emulated will be beneficial to the direction of the organization. In any event, individual or organizational behavior modification by emulation is the only pure power builder. It requires no risk of power loss, and no power is exercised in its accomplishment. It requires a consistency of purpose and behavior, however, which few possess and which can be developed over time only with concerted effort. (Note the difference between this first method and the others: This one is purely voluntary, whereas those that follow require initiating action on the manager's part.)

Suggestion is the next most subtle form of influence. The implication follows that if the item suggested is of such overwhelming logical veracity, reasonable people will not be able to resist it. The primary problem with this approach is the bias that the presenter of the suggestion invariably brings to the table. We have discussed this elsewhere (i.e., the fact that certain people will not accept a suggestion just because it comes from a given manager). Although this is unfortunate, it is a reality of human nature that cannot be ignored without risk. However, the suggester risks very little in terms of power dispensation, and if one or two more influential people should pick up and run with the suggestion, it might readily be accepted. Flexibility is important as the suggestion gets modified and hopefully improved. Remember, you merely suggested it, there should be no pride of authorship. Otherwise, we enter the next degree in the spectrum of influence, which is persuasion.

Why so many feel compelled to start with *persuasion* and not give the more powerful behavior modification techniques a chance is an intriguing aspect of human nature. (It probably has much to do with our culture of winning and losing, which we discuss in the next section. Or perhaps it can be attributed to impatience.) We say that this is less powerful because it consumes more personal power than does emulation or suggestion. Persuasion implies arguments that must be launched. Success depends on them first falling on open ears (i.e., a mind not stuck on a current paradigm) and then being understood. They can easily be deflected by intentional misunderstanding or simply by taking a reasonable position to an unreasonable extreme and arguing from the point of exception. Although we might suspect such to be dishonest, this behavior is all too ingrained in our culture. Thus we see little hope in most cases for pure persuasion, and submit that when it does appear to occur, it is usually a disguised form of coercion.

Of course, *coercion* is a raw discharge of potential power, and like shorting out a battery, it can be quite costly. It represents a hemorrhaging of the organization, an internal bleeding which, unless addressed by surgery or the like, will probably result in loss of consciousness and ultimate death.

The influence spectrum is not discrete, and the continuum between the four nominal degrees should be understood. Also important is the exponential nature of the power loss as we move in the direction of coercion. Suggestion is only slightly more energy consumptive than emulation, which may, in fact, have a net energy gain. Although persuasion is significantly more power consumptive than suggestion, coercion is an order of magnitude greater.

Again, we have defined but not solved the problem. Clearly, to influence by inspiring emulation is the ideal, but how is this accomplished? Let us proceed to get more insight into this interesting subject by once again dealing with the ego of the manager.

19.3.3 Winning and Losing

A major part of the problem is tied up with our culture and the athletic paradigm: (1) winning is everything, or perhaps the only thing, and (2) for every winner there must be one or more losers. Our media are unwilling to allow this paradigm to be contradicted, and society not only follows faithfully along, they also pay the bill. Unfortunately, everything from warfare to politics to business is viewed using this same paradigm. On the contrary, organizational reality dictates that often to win requires us to *unwin*, since the mere perception that we have won will cause us to lose. (We will see that unwinning is not the same as losing.) Certainly, overcoming by coercion causes a tremendous loss of potential power, which, until restored, cannot be used to attain personal or organizational goals. Further, when one person loses, everyone loses. Thus competitive athletic-type competitions within the workplace tend to kill the cooperative spirit necessary to accomplish organizational goals effectively. If the athletic paradigm is to be useful, it should be played out between entire competitive organizations, not within the organization itself.

Nevertheless, there will be battles for the hearts and minds of the membership. It is essential that you rise above these by adopting the clinical approach (i.e., by keeping your ego out of it in the realization that these battles do not conform to the commonly accepted athletic model). This recognition will enable a pattern of leadership to evolve that has as its center the realization of organizational goals. Let us consider the following principle in this regard:

> Those who would shape organizational behavior by inspiring emulation must do so by adhering to principle rather than responding to personal attacks.

This is an extension of the clinical approach, but it also touches on the issue of individual power. Nothing creates an increase in individual entropy, with its accompanying confusion and power loss, like a turf battle. The old adage "don't take yourself too seriously" certainly applies. Your own personal interests are not necessarily those of the organization, or more important, those of your subordinates (the people who make you what you are). It is very easy for these to be confused, and many managers seem incapable of even conceiving the difference. Reality dictates that the failure to make this distinction can be fatal with regard to your career within the organization.

Robert Townsend put it this way: "There is a time for engagement and a time for withdrawal. A time to walk around the job. A time to contemplate it—and a time just to laugh at it" (TOWNS70,66). When are these times, and how are they determined? The clinical approach dictates that we get outside ourselves and look at the situation objectively from the point of view of *what is best for the "project," the organizational component, or the organization as a whole*. Consistency of purpose is essential, but consistency of action can be fatal. The concept of *unwinning* is simply to accept what is best without claiming victory. Unwinning is equivalent to unlosing. It is the satisfaction of the victory without the glory. [It is that feeling that you get when you take that lunker you caught (that would look great on your wall) and release him back to the lake to live some more.]

We also admire Townsend's advice with regard to winning, losing, and compromising. In essence, he said that it was necessary to know when and how to win, lose, or compromise. He favored the clean win and the clean loss over the compromise, however. This was because of the recognized downside of the win. Most organizational "wins" are costly in terms of power losses. ("You got your pound of flesh on that one, don't expect another for a good long while.") The same, however, carries over into most compromises. Since no one has clearly "won," everyone is looking for compensation in the future. Never, ever confuse unwinning with compromise!

It might seem that we are espousing that you should constantly take the loss, since this consistently builds your potential power. If so, we might ask, for what? Why should we build all of this power if we are never going to use it? (Good questions, but they miss the point.) We are not implying that the successful manager throws in the towel from the outset. First and foremost, enough foresight should be available to avoid the situation of ever getting to a win–lose situation. However, our fallible minds are much too weak to be able to circumvent every confrontation proactively. Thus when a battle does erupt (and it most assuredly will), our behavior should be one that objectively defends the organization in general, and our component of the organization in particular (since we have assured that we are consistent with organizational goals). As J. R. Walker put it, "We're not looking for a fight, but if one comes our way, we're going to enjoy every minute of it."

Although "winning" under these circumstances requires a power loss, it should not result in any sense of loss on your part. It is a fair sacrifice. The key, however, is knowing when to engage and what is worth the energy discharge. The old admonishment: *Don't die on a small cross* holds true. There is a time to make the sacrifice of power. There is a time to bloody and get bloodied. But unless this is really going to provide a net gain for your view of the organization, lay back, take your unloss, and wait for the day when winning will provide some major gain. If it is not worth laying your very job on the line for it, chances are that it is too trivial to take up organizational time fighting about it. If you can anticipate a sense of satisfaction for defending what is right (even if you lose!), chances are it is time to take up the cause.

In this regard, we hate to be trite, but it really is *how you play the game*. Are your actions based on principle or ego? Are you willing to make short-term sacrifices for the cause of the organization? Are you willing to put those mechanisms into effect that will avoid win–lose situations? These are questions of how to play. This is not a matter of winning or losing; it is one of unwinning and unlosing.

It should feel good to lose once in a while, but not *too* good. Winning by coercion should never feel good. With this in mind, let us take another cut at the true sources of power within the organization.

19.3.4 Informal Sources and Uses of Power

In Section 19.1 we saw that power built with the time that was required to establish relationships to accomplish organizational goals. These relationships tend to overcome the necessity for using the formal authority that is bestowed by the organization. In establishing these alliances, the manager is creating and strengthening informal organizations that will serve organizational purposes in the future. At the same time, the manager promotes the same thing at lower levels to maintain and promote harmony and cooperation within the organization. Thus the informal organization is a major power source.

We now turn our attention to other factors that intrinsically bestow power to individuals throughout the organization. As a basis, consider Kazmier's statement: "In addition to the formal source of power a number of informal sources exist, such as *knowledge, ability to reward, place in the communication system, and uniqueness of skill*. . . . A group with formal service authority may have additional organizational power because it can choose to give priority to some requests" (KAZMI69,166–167). Each of these sources of power is discussed individually in the below display. They are highly interrelated, and those with the highest potential are closely related to the establishment of alliances discussed in Section 19.1.

Informal Sources of Personal Power

- *Knowledge.* The ultimate as a portable power source, it can take various forms, which we prioritize from most to least potent:

    - *Market knowledge.* This begins with knowledge of the customers, who they are, what their needs are, what their needs will be in the future, and how they can be satisfied. This provides the basis for organizational goals and objectives. However, to maintain this as a power base, the person must be able to sell in both directions: to the customer as well as to the organization.

    - *Technical knowledge.* Since ever-increasing degrees of technical knowledge can be purchased within the job market, it is imperative that those who depend on this for their power base continue to stay abreast, maybe even get ahead, of the field.

    - *Organizational knowledge.* Knowing who the key movers and shakers are in the organization is important, but this knowledge alone does not assure success or even power. It is also essential to know how these people can be influenced.

- *Ability to reward* (*and, implicitly, to punish*). This takes two forms: the formal organization (perquisites, raises, promotions, etc.) and the informal organizations (acceptance of one's peers).

- *Place in the communication system.* Due to position and not inherent skill, this can be a very unstable and vulnerable power source. This is especially true if this power source has been used to the disadvantage of others over time. If it has been used to the advantage of the organization and to build constructive alliances, it becomes more reliable. This is due to the use to which this power has been put, as opposed to the mere presence of the power itself.

Informal Sources of Personal Power (Continued)

- *Uniqueness of skill.* This might be such a unique skill to the particular organization that no outside training or experience could create it. Quite potent in the short run, it can be quite vulnerable, since increasing technology can make any unique skill obsolete. Any attempt to wield this power would be extremely counterproductive should the skill no longer be needed by the organization.

- *Formal service authority.* These are those organizational components that formally provide support to others as their internal customers. The theory is that they can leverage this into power for their organizational component, their members, and ultimately, their manager. This is the most vulnerable, especially if it deteriorates into organizational blackmail.

Additional consideration on *formal service authority* as a source of power is in order. Often there is a false perception of the value of this power source, especially in technical organizations. All managers of service units, whether it be the network administrators or the secretarial pool, should be aware of the natural tendency for these organizational components to acquire power well beyond that which is justified by the service they provide. In fact, there is a very strong tendency for these types of service components to withhold their unique skills in order to increase their power within the organization. This is obviously counter to organizational goals, but because they often report to a manager who is a level or two above the managers of the components they serve, it is difficult for them to be held directly accountable for this shortcoming. Neither is it their fault that their power grows in this way. They merely make the natural assumption that since their technical ability is so crucial to the organization, they must exercise power over the production components of the organization. (This is another example of viewing power only in the short term and not understanding the ultimate downside of its misuse.)

Starting as a demoralizing aggravation, this can grow to a point where it becomes totally intolerable to the production components, at which time some reorganization or other Theory X solution is usually imposed to try to solve the problem. Obviously, this is not an effective power-building strategy in that the power accumulated by these service groups comes right out of the hide of the production components.

These types of problems can easily be foreseen (since they are inevitable), and any manager responsible for such service components should be proactive in resolving them before they become problems. The solution is to assure that a customer orientation is assumed and to obtain feedback directly from the production components for purposes of assessment. More important, however, mechanisms should be established such that alliances are established between key lower-level members of the service component and their counterparts in the production components. The worst thing that can evolve is the necessity for going through the chain of command to obtain necessary support, a sure symptom of Theory Xism.

In concluding this section, recognize that power sources are important for the restoration of power to counter individual entropy increases, which are the natural consequences of individual activities within the organization. Losses of potential power result from the exercise of influence (at various levels), resolution of conflicts (especially those "won" or compromised), and most dramatically, when the use of coercion cannot be avoided. Ideally, managers will take proactive steps to circumvent these power-loss pitfalls by forging alliances that empower all their associates. However, when an inevitable power loss occurs, damage control is in order. It will require the restoration of destroyed alliances, and it might call for the development of a capability within your organizational component involving improved knowledge, ability to reward, and positioning within the communication system. All of these factors take time, but the important aspect of personal power growth is quite analogous to organizational maturity: It is the direction that is important, not the absolute status at any given point in time.

19.4 Power Problems

Technimanagement calls for major power enhancements, which only on rare occasions will disempower anyone. However, as much as this might be communicated throughout the organization, we anticipate that certain persons will never believe it. This is because of the existing Theory X paradigms, which essentially measure power as the ability to control the actions of others as opposed to the ability to accomplish organizational goals. The very methods for implementing technimanagement will bring about conflicts. According to Kazmier, these follow from the sociological or political evolution essential to its implementation: "Almost inevitably, patterns of alliances in the organization develop in conjunction with the use of political methods, with certain groups finding themselves in a position of 'natural conflict'" (KAZMI69,168).

While these upheavals accompany any change, the radical differences between traditional and technimanagement will create new patterns of alliances within the organization that will not be accepted readily by everyone. Managers who are dedicated to the transition to technimanagement must take proactive steps to anticipate and deal with this resistance before it has a chance to establish a countertrend. This requires including those who are anticipated to cause conflicts, not only within the new mechanisms that will be established, but also within the mechanism-design process. They need to be included because some of them are the best technical and managerial people in the organization.

This brings us to another aspect of human nature that we ignore at our peril. No one is threatened by someone without power. Thus the novice manager will be well received by new peers who have little to risk in befriending and helping the new kid on the block. However, once you start practicing technimanagement and liberate your people to their full potential of productivity, this acceptance will change. This is especially true if your upper management is not oriented toward and motivating a transition on its own. But even so, the more success that you have and the more alliances that you form, the greater are the chances of your being isolated (can we say sabotaged?) by your other peers and their constituencies in middle and upper management.

We stated that the ideal personal strategy was one of individual power growth through mutual empowerment of your colleagues at all levels of the organization. We prescribed decision-making strategies that preserved and built personal potential power by principle-based decisions that favor leadership over coercion. We advised a long-term approach that would anticipate and accommodate setbacks and set the direction for a continuous increase in individual power. Climbing this mountain will be easy at first, and the primary concern will be your home component. However, as with mountain climbing, the higher you get, the more perilous and steep is the terrain.

This is another example of where success brings the seeds of its own downfall. In other contexts we discussed the personal ego problems with which the successful manager must cope. At the same time there will be pressures from the outside to conform to the bureaucracy and not "make waves." We have coined a saying to keep things in their proper perspective which applies here:

> That which makes you good makes you bad.

(Note that this has many nonpower applications, and we will revisit it in Chapter 22 when we talk about limitations of experts.) The analogy of the mountain climber is an excellent one, for there are always steeper and more challenging slopes to conquer. What makes the mountain climber good is the motivation to prepare for the challenge. The greater the motivation, the greater the resolve to train to be the best. But this same motivation will not be satisfied with anything but the highest mountain under the most challenging of conditions, and each success just further hones the skills in this direction.

The knowledge of this particular inherently self-destructive aspect of human nature is not at all of recent origin but is deeply rooted in Greek mythology: "The fabled Icarus of Greek mythology is said to have flown so high, so close to the sun, that his artificial wax wings melted and he plunged to his death in the Aegean Sea. The power of Icarus' wings gave rise to the abandon that so doomed him. The paradox, of course, is that his greatest asset led to his demise" (MILLE90,3). Power creates its own conflict, and the temptation to use it cannot be understood until it is possessed. Between this ego peril and the pressures that accrue from the natural enemies of the powerful, the pathway of the successful manager is indeed perilous. It is no wonder that most decide to merge with the bureaucracy rather than embarking on it.

The only countermeasure to the Icarus paradox is an accurate perception of reality. Attaining and retaining this is our major concern in the next chapter.

20

The Negative Feedback Principle

20.1 Introduction

We use the term *negative feedback* as most would the word *criticism*, both constructive and destructive. Feedback emphasizes that this commodity is a critical part of the measurement element of control. It is certainly not the only measurement available, but it is the most easily blocked or ignored. Rarely are mechanisms established to assure its generation. Thus ranking among the most counterintuitive notions presented in this book is the *negative feedback principle*:

> Your greatest asset is negative feedback about yourself, regardless of who it comes from.

Even if this principle can be accepted mentally, we expect extreme difficulty for most people in attempting to practice it. The primary purpose of this chapter is to prove its validity and then provide some personal methods to assure its proper application.

If you believe the principle behind the Icarus paradox, if you believe that success brings its own seeds of failure, if you recognize that your ego is your greatest enemy, you have to believe that your greatest asset is negative feedback—greater than all your alliances, all your friends, all your power, and all your money. For all of these things can be lost if you do not have the proper perception of reality. Wake up, open your eyes, and look around at those who are riding high one day and crash and burn the next. What was it that killed them? Why were their senses producing countercontrol? Was it not that their perception of reality was distorted because they surrounded themselves with "yes persons"?

The second point that needs elaboration is the last clause: *regardless of who it comes from*. Most people can accept criticism and advice from people for whom they have particular respect. Others can "accept" it only from a superior—actually being forced to act accepting while remaining in a subconsciousness state of denial. Some can accept it from no one. Instead, they manifest their state of denial by blaming every possible person around them for their problems, especially the person who has the gall to actually criticize them. We begin by addressing this problem specifically and then go on to discuss the appropriate reaction to criticism. In two additional sections we extend these basic concepts and demonstrate their importance in the context of continual personal improvement.

20.2 Killing the Messenger

The author once had an administrative assistant who was very well wired into a number of the informal organizations. Time and again she would dump her negative feedback at my feet, and inevitably I would respond defensively. (This was a mistake that I now try to control.) Fortunately, her concern was more with the department than with my favor, and I can still hear her rejoinder: "Don't kill the messenger!" When you think about it, the ancient practice of killing the bearer of bad news is one of the stupidest things imaginable. In fact, it is so stupid that it is hard to imagine it even being practiced or, for that matter, even imagined. Here is someone risking life and limb to warn you of an impending problem, the knowledge of which is your only hope for countering its negative effects. And you kill that person!

Question: Why would anyone be a messenger under these conditions? Certainly, if they knew their fate, they would hardly run to their death. There are many other places to run. Answer: They will learn their lesson, if they survive.

The fact is that most managers "kill" their messengers off to the point of extinction. Every time that we defend our actions and get upset at criticism, we are killing our messengers. They might still be alive, but they will no longer be the bearer of bad news. Kill them all and you can sit in your office with the perception that everything is going great. If you ever (ever, ever!) get this perception, it should be accompanied by a hollow feeling of impending doom. The worst problem is the one of which you have no knowledge.

To show that we are in good company, we appeal to Townsend: "Every chief executive should find someone to perform this function and then make sure that he can be fired only for being too polite" (TOWNS70,68). This is the very opposite to killing the messenger, and it reflects an attitude that is critical to continuous improvement forever. We will now explore the creation of this attitude, which begins with the proper reaction to criticism.

20.3 Reaction to Criticism

There are two aspects to a person's reaction to criticism: internal and external. Internal reaction should be relatively constant, and we discuss this first. External reaction depends on the validity of the criticism and the environment in which it is given.

20.3.1 Internal Reaction to Criticism

Given the negative feedback principle, the ideal internal personal reaction to criticism should clearly be one of acceptance and appreciation. Human nature being what it is, however, this ideal can be realized only with great discipline. There are several reasons for this. The natural reaction to criticism is one of denial. This is not a logical reaction. In reality, all criticism has some essence of validity. Assuming basic honesty, it is either an accurate perception of reality per se, or it is an accurate statement of the perception of the one doing the criticizing, even if this perception is not accurate. In either case, this is extremely valuable information without which control might well be lost. (We recognize the third alternative: the critic intentionally trying to manipulate you by your belief in the negative feedback principle. However, such deception should become readily apparent.)

Why, then, denial? We can only surmise that it has something to do with personal insecurity. Those who are secure will not fear criticism. They will see it as essential to complement their efforts and make up for their own inabilities. It will not be a threat to the secure because they are not espousing themselves to be perfect. They recognize that flexibility and continuous improvement are essential to their success.

The internal reaction to criticism is quite critical for two reasons: (1) it will reveal itself in your external reactions, and (2) it will affect your ultimate response to the criticism. Both of these are critical. The external reaction will determine the quality of feedback in the future. However, the response (behavioral change) that results from criticism is probably even more important. If the reaction is defensive and there is a tendency to blame others, there will obviously be no improvement resulting from the information that is forthcoming.

This gets back to the clinical approach, especially with regard to objectivity and maintaining the "project" or organizational goal orientation. Recall that this approach placed the manager's feelings secondary to the accomplishment of these objectives. The practice of this approach leads to a depersonalization of the criticism. It is not a jab at your ego, it is a suggestion for better accomplishing organizational goals, and that is first and foremost in your mind. With this continued discipline, it becomes almost impossible to take criticism personally because of its recognized value to the organization.

As we consider reactions to criticism, recognize that the difference between criticism (negative feedback) and conflict includes degree and possibly, intent. Criticism might be the first sign of a potential conflict. However, our consideration of both functional and personal conflicts in previous chapters assumed that these were not mere differences in opinion or recommendations. The intent of criticism is often quite positive, which is not characteristic of conflict. By the time that issues have progressed to the point of conflict, they are beyond the considerations given in this chapter. On the other hand, since criticism is an early warning for potential conflicts, application of the principles given here might serve to stem issues before they mature into conflicts, which is consistent with the recommended proactive approaches toward preventing conflict.

20.3.2 External Reaction to Criticism

We have probably all seen managers react defensively to advice, which we perceived to be no criticism to them at all. This reflects the fact that *all advice is implied criticism*, even though it might not be intended as such. Nevertheless, when this happens it causes a number of questions to form with regard to the cause of this external reaction; among them: Why does this person feel threatened by these apparent attempts to help? The external denial becomes a major deterrent to resolving the real problem. While the manager was not even perceived to be a part of the problem, now clearly she or he is. A meta-problem forms that is related to determining if the first problem actually exists and possibly assigning blame. Neither of these issues address the first problem. Clearly, meta-problems are characteristic of the organization in a state of confusion (increasing entropy).

Your external reaction to criticism is one of the most influential aspects of your leadership ability. All leaders, formal or informal, are subjected to criticism, and to imagine otherwise is a perversion of reality. How they handle it determines whether they use it to build or diminish their power.

The first step toward having a proper external reaction to criticism is to have a proper internal reaction to it. This establishes objectivity and places you in a state of control over yourself. The most important aspect of using negative feedback effectively is to understand it thoroughly. Emotional reactions often prevent understanding by (1) cutting off the people who are criticizing before they can explain the problem thoroughly, and (2) preventing an objective assessment of the true meaning of the statements made. The first is quite annoying to the person who might be at great risk in doing you this favor. The second is even more damaging in that often the logic of the complaint disguises its real essence. That is, the particular thing complained about is only a symptom of the underlying problem.

To probe this deeper, consider external reactions in light of four possible scenarios created by two factors: the validity of the criticism and the forum (public or private) in which it is launched. This leads to the situations discussed in the paragraphs that follow, from the most to the least desirable.

Valid criticism in a private forum. This is the most desirable forum and form of criticism, from the points of view of both advising others and receiving useful information. Valid criticism indicates that an accurate perception of reality exists on the part of the person doing the criticizing, and the private forum assures the greatest objectivity on everyone's part. The external reaction to this type of criticism should be totally positive, appreciative, and further, encouraging of both the validity and the forum. Excuses and blame should be avoided, although explanations of mitigating circumstances that might not be known to the other party are certainly in order. Rarely is any criticism totally valid (i.e., the person doing the criticizing will generally not have all the facts). This imperfection should not render the criticism invalid in your mind. Work with the other party and attempt to fine-tune your mutual understanding. This mutual understanding, and the assurance on your part to do everything that you can to overcome the fault, will promote the positive direction of your organizational component.

Invalid criticism in a private forum. This is an order of magnitude more complex than the situation discussed above because the major issue becomes the cause of the "invalidity." Several possibilities exist: (1) your perception might be flawed, and the criticism might indeed be shown to be valid, (2) the other party might be addressing a valid problem but using invalid facts, (3) the other party might be honestly mistaken as to the facts and the existence of any problem, or (4) the other party might be trying to deceive for a wide variety of reasons. The private forum enables a great

Sec. 20.3   Reaction to Criticism   305

deal of latitude in dealing with these possible situations, and probably makes the last alternative unlikely. Figure 20.1 addresses these possibilities. In summary, all criticism received in a private forum should be appreciated for the value that it serves in communicating an improved knowledge of the environment in which you must function.

- Thoroughly understand what is being presented; repeat back your understanding (nondirectively) so that there is absolutely no lack of understanding.
- If there is any doubt of the validity of the criticism:
  - Call "time out," suspend your discussions temporarily, and get the facts from independent sources.
  - Allow things to cool down for a day or two while you are accumulating the facts.
  - Get back to the other party; a failure to do this will almost assuredly create additional problems.
- If the facts being presented are only the symptoms (or perhaps a smokescreen) for some other underlying problem:
  - Identify and address the true cause.
  - Continue to show sensitivity to the other party, since the fact that it is being handled in private is indicative that there are no ulterior motives.
- If the other party is honestly mistaken, the reaction should be a positive exchange of information, not a defensive display.
- If the other party demonstrates ulterior motives:
  - The reaction should still not be defensive.
  - The information obtained enables you to acquire better knowledge of the perception and intentions of the other party (and possibly an informal organization).
  - Most problems of this nature can be resolved if there is at least one person who remains reasonable and makes an honest attempt to communicate effectively.

Figure 20.1   Reaction to "Private" Invalid Criticism

Valid criticism in a public forum. The problem of maintaining objectivity is greatly increased by the criticism being aired in public. If nothing else, this should emphasize the need for not criticizing anyone publicly.

But how do you handle personal public criticism? If you recognize it to be valid, the best approach is to admit the shortcoming and pledge any remedial actions that will be necessary. As opposed to this being a sign of weakness, as is generally perceived, this is a demonstration of strength. Essentially, you are saying that you stand corrected (i.e., even though you admit that you were mistaken, you will make any adjustments that might be required and continue to do your best). In this regard, you strengthen your position considerably by avoiding denial and not making excuses. Rather than trying to mitigate the situation, which illustrates that you might not take full responsibility, maximum power is attained by assuming full responsibility. Things are not right, you recognize this fact, and you are going to fix it. This also tends to obligate others to you who are not as secure (e.g., your subordinates), since they do not need to defend themselves. Above all, do not take this criticism (or yourself) too seriously: Chances are that in a few weeks everyone will have forgotten about it. (Absolutely never attempt to transfer criticism to your subordinates. If they are at fault, this is your responsibility, so accept it. Handle criticism of your own subordinates privately.)

Invalid criticism in a public forum. It might seem that if power is accumulated by accepting blame for valid criticism, the same would hold true for that which is invalid. Since few would naturally assent to accepting full responsibility even when they are totally guilty, even fewer would accept this premise, and neither do we. The advice given above was only for those cases where the criticism is largely valid. In situations where you are being attacked without due cause, the effective handling becomes much more problematic. The steps recommended are given in Figure 20.2. Invalid criticism in the public forum is the most difficult to address effectively. Too violent a reaction tends to indicate guilt, However, those who utter the trite "I refuse to dignify that criticism with a response" already have. Some response is in order since every member of the organization deserves your dignity, and your power probably depends more on effective resolution of invalid criticism than it does the valid criticism (which is relatively easy to handle). If handled properly, some of your critics will become your greatest supporters.

The subject of personal power emerged in several aspects of the discussions above. It goes without saying that the handling (or mishandling) of criticism rarely occurs without a power shift. *The very purpose of issuing criticism is to create an increase in power on the part of the person doing the criticizing.* This is the reason that it is sometimes done in an underhanded and public way. It is surmised that by bringing you down, their power might be increased. We recognize this as the *zero-sum fallacy.* Responding in kind is exactly the effect that is desired, and when you give way to this, you

> - Find out whatever you can that is valid about the criticism and agree with that before continuing.
>   - This demonstrates strength.
>   - It provides an area of common ground within which agreement can be attained.
> - Recognize: The objective is to attain consensus, not win an argument.
>   - Picking at valid points demonstrates propensity toward opposing personalities as opposed to facts.
>   - The most effective resolution results when no one loses status.
> - Proceed to deal with the invalid aspects by:
>   - Offering potential reasons for the misunderstanding while refuting the facts that were invalid.
>   - This provides the other party with an escape from the dilemma that has been created.
>   - *Example:* "You had no way of knowing this, and I can understand why you arrived at that conclusion, but ..."
>   - Avoid issues that cannot be definitively refuted, since this will only cause a perception of confusion to observers.
> - Once the more definitive misperceived facts are effectively refuted:
>   - Offer publicly to get together with the person to resolve your differences further; this will be seen as a position of strength, since:
>   - It gives the adversary the opportunity to respond without losing face.
>   - It indicates your willingness to negotiate, thus taking away the initiative.
> - Follow up with a meeting in which you can apply the criticism-handling and conflict-resolution methods presented in Figure 20.1.

Figure 20.2 Reactions to "Public" Invalid Criticism

have effectively allowed someone else to control your behavior. This causes a major waste of personal power.

One of the major problems with implementing the principles given above is our own perception of reality when we are the focal point of criticism. Can we tell what is valid and what is not valid? Can we tell when we are at fault? Most important, can we muster the security to admit what we suspect? One of the major difficulties in this regard is vacillation. Generally, this occurs when we overreact to criticism only to recognize later that it was valid. Or we might be in a hurry to admit our fault, only to find out later that the actions which you took were justified and should have been defended. In this regard, recognize that a primary problem is timing. This is caused by the tendency to surrender timing to the critics, causing the trigger to be pulled before aim is taken. A major initiative in timing can be secured by inspiring criticism.

20.4   Inspiring Criticism

Negative feedback is of such value that its generation requires much more than an effective reaction to it. Indeed, one reason that we advise against forceful reactions to criticism is that they lead to suppression of the generation of this valuable information in the future. It is virtually impossible to convince anyone that you want constructive criticism when your response to criticism is defensive. Thus, effective reactions to negative feedback comprise the first step toward inspiring it in the future. However, it is not proactive, and additional steps for inspiring criticism are necessary if healthy mechanisms of continuous improvement are to evolve.

Before going on, we must relate some disastrous attempts to inspire criticism. Recognize that criticism is something that you are going to get when you ask for it. The most disastrous failures come when the manager is really fishing for approval and then gets defensive of all the nit-picking incidentals that are forthcoming. In this regard we urge you to refrain from inspiring criticism until you possess adequate maturity to handle it. Handling includes not only the ability to respond positively to all criticism, but also to keep from taking yourself too seriously by distinguishing between the significant and the trivial.

Those who are supportive of the manager usually are the most reluctant to criticize, mainly because they do not want to "get personal." Yet these are the people who could provide the most useful negative feedback, due to their objectivity. Leadership in this regard demands that criticism be depersonalized. Consider the following recommendation in this direction:

> Blame the organizational mechanisms as they have evolved, not the people in the organization.

Do this because it is valid, not for purposes of manipulation. However, its effect is to depersonalize criticism against everyone, including the manager, and direct it toward the existing mechanisms as they have evolved. This means that the solution to the problem (subject of the criticism) is the reworking of those mechanisms, *not the discipline of individuals*. It is not a matter of your having the strength of character to "handle it." It is that you recognize that things can be better and that you recognize that you do not have all the answers. So you invite criticism to get these potential solutions onto the table.

The inspiration of criticism is not a difficult process given the will of the manager to accomplish it. Most subordinates are more than willing to give their opinions if they know that these opinions will be appreciated and used. The problem is one of fear of reprisal from the manager. This is the reason that the manager must depersonalize the process and communicate to subordinates that they will be rewarded, not punished, for providing feedback to the manager, regardless of how negative it might be.

20.5 Inspiring Acceptance in Others

In addition to generating essential information, a second major benefit of inspiring negative feedback is the effect that it has on the subordinates' ability to accept criticism. The manager who cannot accept criticism generates subordinates who have difficulty in handling it as well. In this regard we propose the following leadership position:

> Assume and announce your intention to assume full responsibility for all problems within your organizational component, and take total full blame for everything that goes wrong therein.

Again, you do this because it is valid, not for purposes of manipulation. As manager, it *is* your fault, because if you were smart enough and/or had the perfect leadership characteristics, you would be able to anticipate and prevent every problem from occurring.

This rather radical assumption of responsibility has the effect of providing leadership to your subordinates to do the same in their respective organizational components. This attitude of blame might appear to contradict that given in the Section 20.4 ("blame the organizational mechanisms"). However, as manager you are responsible for the organizational mechanisms within the organization. Although this might seem to let your subordinates "off the hook," it should be clear that the evolution of technimanagement will distribute these management responsibilities. Thus, by admitting your shortcomings, you establish the basis for evolving better mechanisms that distribute management responsibility. This provides the leadership for subordinates spontaneously to assume blame (as opposed to being blamed) for their individual operations. They will ultimately come to recognize that their maturity depends on their accepting the total blame for what goes on in their units of the organization. (Notice that *all* and *full* preclude you from the attitude that: "Now that I have accepted my share of the blame, you should accept yours." Leadership defies coercion.)

There is an inverse relationship between synergism and the way that criticism is administered and handled by the manager. If it is administered or received in a spirit of blame, morale will suffer. On the other hand, if there is a mutual attitude of self-improvement, criticism can actually build morale. If the one criticized perceives that the criticism is being administered to further her or his career, it will be received with the proper spirit. In fact, criticism will be sought by subordinates, just as it is sought by the manager. However, it will be resisted if it is perceived to be administered to "get something on" the subordinate, possibly to gain some future advantage.

In a society where 50 percent of marriages end in divorce, we appeal to the family unit as a model of ideal critical communication at our own peril. However, we have all observed the frankness within family units. Why is it that some seem to work despite constant criticism, while others break down over the most trivial of disagreements? It seems clear that it is the intent of the criticism and not the content. When there is a clear perception that the criticism is being lodged to help, the immediate negative reaction is easily forgiven for the longer-term good. On the other hand, when the criticism is merely for the purpose of punishing the other party, the result is disastrous. This serves to formulate a guideline for both the giving and the receiving of criticism:

> Criticism should be received with the assumption that it has been given in the best interests of the organization; criticism should be given only when the motivation is consistent with the best interests of the persons being criticized.

Will this work? Actually, if you want a guarantee that it will work, you are resorting to manipulation. The appropriate question is not: *Will it work*? but: *Is it right*? This is the major difference between manipulation and leadership. Assume the blame (responsibility) for your organizational component and others will accept their responsibilities. Blame others and you can expect your subordinates to do the same.

Since we have introduced some concepts of leadership in this chapter, it is appropriate that we elaborate on them in the next.

21 Leadership

21.1 Counterintuitive Notion

In the first 13 years of its existence, Avis Rent-a-Car had never turned a profit. Robert Townsend took over, and within three years it had earned $9 million. A review of Townsend's explanation of this success reveals his belief that very little of this turnaround was due to mandates that he issued. On the contrary, he attributed much of it to a counterintuitive philosophy of leadership that he attributes to the ancient philosopher Lao-tzu. We excerpt the following quotations from Townsend both to demonstrate how consistent these ancient thoughts are with the precepts of technimanagement, and to demonstrate their applicability in the modern world: "To lead the people, walk behind them." . . . "As for the best leaders, the people do not notice their existence. The next best, the people honor and praise. The next, the people fear; and the next, the people hate. . . . When the best leader's work is done the people say, 'We did it ourselves!'" [from the writings of Lao-tzu, as quoted by Townsend (TOWNS70,80–81)]. The first two sections of this chapter are based directly on these excerpts. In the remaining sections we elaborate on some aspects of successful leaders and managers as seen through the eyes of other authors.

21.2 "Walk Behind Them"

[The fully counterintuitive nature of this saying came home to this author the first time that I presented the sayings of Lao-tzu to a class. On an essay test, about 90 percent of the students misquoted Lao-tzu's saying as: "To lead the people, you must follow them." Of course, as their instructor, this error was my fault. I should have been able to foresee that technically minded students (or possibly all students) have a difficult time perceiving the difference between "walking behind" and "following." Of course, I cannot interview Lao-tzu, and I do not claim to have performed major acts of scholarship over his writings. However, it should be clear that the word *follow* is the diametrical opposite of the word *lead*. Thus, in my ignorance, I had led my students to accept a nonsense statement. Clearly, I was not walking behind them.]

It should be clear that *walking behind* entails a large number of factors absolutely none of which involve following those led. This is understandably difficult to comprehend in a society so devoid of true leadership. Indeed, most of our (successful) politicians are virtually driven by the public opinion polls. (Successful at getting elected, that is.) Thus they are not leading as much as they are attempting to follow the majority to assure reelection. This might seem to be the ideal functioning of democracy if it were not the radically shifting nature of these polls, which demonstrates the volatile influence that informal organizations are exerting over the masses. Such inconsistency is symptomatic of populist leaders.

This surfaces a seeming paradox. Is it possible to lead someone to participate in following without yourself *being* a follower? The answer is an emphatic no. Leadership requires the leader to establish the way, to set the example, and to demonstrate the willingness to make the sacrifices. If leaders are not willing to follow, how can they expect their followers to be willing to do the same? However, the paradox is resolved by recognizing that the people led are not the same as the people followed, at least not within a given circumstance. Further, *the best leaders do not follow people, they follow principles*, and they have no objection to their followers doing the same.

A most definitive demonstration of leadership is recorded of one of the greatest leaders of the Old Testament, Joshua:

> Now therefore fear Jehovah, and serve him in sincerity and in truth; and put away the gods which your fathers served beyond the River, and in Egypt; and serve ye Jehovah. And if it seem evil unto you to serve Jehovah, choose you this day whom ye will serve; whether the gods which your fathers served that were beyond the River, or the gods of the Amorites, in whose land ye dwell: but as for me and my house, we will serve Jehovah. (Joshua 24:14–15, American Standard Version)

Notice that Joshua was following principle, not the crowd. In fact, he made his statement before even knowing their opinions. He was definitive and left absolutely no room for misinterpretation. No trial balloons, no vacillating, no room for the spin doctors. No coercion and no manipulation. Just pure leadership.

But how, you might ask, does this comply with the title of this section? Quite simply: Joshua was not dominating, he was serving. In fact, there was a clear choice that he allowed them to make, and he was willing to accept their decision in this regard (for them, not him). His appeal was not for his own betterment; indeed, he had already sealed his own fate. His statement was to persuade them to make what he considered to be a decision in their best interests. So, he was *walking behind them*, serving them in whatever way that he could, but never compromising what he knew to be in their best interests.

The principles of technimanagement call for distributed decision making. When the mechanisms for this are established, decisions might be made that the manager is bound to *follow*. We submit, however, that in abiding by the decisions that can best be made at the lower levels, managers are not following the groups or individuals who made these decisions, they are following the principles on which those mechanisms were established in the first place. Thus they serve their subordinates, not by following them, but by consistently supporting the principles by which both the organization and their personal careers will be advanced. When a manager's subordinates have effectively defined their customers and matured to a full customer orientation, the only customers of the manager are the subordinates. This is a situation that can evolve only out of leadership, and when it is fully formed it provides the only environment in which leadership can reach its fullest potential.

21.3 Ego of the Leader

The effective leader never attempts to coerce, although there is very clearly an attempt to modify the behavior of those led. In technimanagement, this is accomplished simply by reinforcing motivations that already exist. The primary reinforcement, or reward, is provided by assuring that they possess a sense of accomplishment. *Effective leaders do not care who gets the credit, just so that the job gets done.* This is somewhat of an overstatement. They do care that the proper people get the credit, but they are not personally concerned with getting the credit themselves. In fact, they recognize the tremendous power loss that occurs should the credit accrue to them as opposed to their subordinates. Conversely, they will never publicly blame the shortcomings of their organizational component on their subordinates.

In Chapter 20 we stressed the need for managers to accept all the blame for the performance of their respective organizational components. This was the beginning of leadership to inspire subordinates to accept blame for their respective units. Similarly, effective managers will not attempt to seize the credit for that which is accomplished by their organizational components. Again, we stress that this is not a manipulative technique; if it is going to be effective, it must be believed as a matter of principle. Think about it: Managers accomplish nothing in their management roles. They might function as subordinates and set the example, but as managers they only realize their goals by the actions of their subordinates. Why then should not the subordinates be given all the credit for the component's accomplishments?

"*As for the best leaders, the people do not notice their existence.*" Does it bother you not to be noticed? Most aspiring managers want recognition. They want the power that comes from having dominion over others. This explains much of the organizational ills that we see round about us.

"*The next best, the people honor and praise.*" Is this bad? This question can be answered by an accurate response to the question: Is honor and praise bestowed in an effort to manipulate? The manager is usually not in a good position to judge, and for this reason all such ovations should be viewed with the gravest suspicion. In allowing yourself to be manipulated in this way, you surrender your principles for the enslavement of flattery. Did you notice how much funnier your jokes were after you got your promotion? How the criticism fell off? How much more polite everyone seemed to get? If you have a taste for honor and praise, there are plenty who are willing to accommodate, but they might not be the most competent or the most critical to your department's success. A taste of glory stimulates the appetite. So be careful, for it is an appetite that quickly grows to addiction.

"*The next, the people fear; and the next, the people hate.*" This is the destination of an ego trip that starts within the ready acceptance of honor and praise and ends with the use of manipulation and coercion. This is what happens when leadership is replaced with the formal authority that is assumed to be inherent within the organization. To avoid this pitfall, remember these final words of Lao-tzu: "*When the best leader's work is done the people say, 'We did it ourselves!'*"

Accomplishing this is not difficult; however, it might be difficult to swallow. If you have provided the leadership that is essential to success, you will have worked harder than everyone else on the team. You will come in earlier, leave later, and put up with more conflicts than anyone else—from above, from below, and from your fellow peer managers. But no one will notice. If you care about this, it will cost you dearly, for in your attempt to gain credit for yourself, you will accomplish just the opposite. Recognize that if you are doing your job as you should, they will say "we did it ourselves." The internal satisfaction that you get from this should be more than adequate, and the synergism that results will ultimately accrue to the increased success of your organizational component. And, that *is* what's important, isn't it?

21.4 Traits of Successful Leaders

Accomplishing the above is not simply a matter of will, although the discipline required can be attained only by a strong will. There was a time when it was argued that leaders were born, not made. This position is rarely defended today. However, there are many who seem to believe that only a select few are capable of acquiring leadership traits. There is no question that some have the advantage, due to past experience and natural demeanor. Given the modest beginnings of many who have attained successful leadership positions and the miserable failures of some who were clearly privileged, it should be clear that the traits of leadership are readily attainable.

Many traits could be proposed as supportive of leadership ability. This quotation by Kazmier summarizes this fact: "Although the discovered importance of specific personality traits has varied from study to study, three general trait areas—intelligence, communication skill, and the ability to assess group goals—have been found to be related to leadership success in a variety of situations" (KAZMI69,247). Figure 21.1 elaborates these three traits in more detail.

The area in which technically oriented persons are most vulnerable is that of general (as opposed to technical) intelligence. Being accurate and technically correct (while quite necessary) is not sufficient. It is also necessary to know those actions that will inspire confidence as opposed to doubt *at any given time*. This argues that the same actions that will produce success under one set of circumstances will fail miserably under another. This is particularly important for purposes of emulation: Just because a given tactic works for others does not mean that it will work for you, even under the same set of circumstances. One thing is certain: When the application of a given tactic produces a negative effect, the persistence in that behavioral pattern shows poor intelligence. The major problem in leadership is a failure to accept continuous improvement as the central personal goal of life. Having arrived, the successful upstart fails to adapt to both external and internal changes. Shutting off the reality switch, the VCR player is placed on hold, the picture never changes, and the leader becomes brain-dead. This is the antithesis of intelligence.

Our definition of effective leadership necessitates a positive, self-improving direction. Its negative counterpart, *manipulation*, while effective in producing short-term behavioral changes, ultimately fails. Some argue that Hitler was an effective leader. (Clearly, calling him a good or successful leader would be totally incorrect.) He controlled and manipulated a subset of people very effectively for a very short period of time. This does not fit our definition of effective leadership. He took the easy method of motivation (hatred), and both he and his followers were so blinded by it that they could see only what was immediately in front of their eyes. Indeed, had he fit our definition of being intelligent, he would have perceived the ultimate consequences of his actions. We cannot help but

> - *Intelligence.* Not being able to read Kazmier's mind, we might suspect that he was describing the characteristics that intelligence tests measure. For purposes here we accept the more militaristic definition of intelligence: *the ability to perceive reality in order to fully understand the consequences of one's actions.* This involves far more than that which can be measured by an intelligent test. For it is the ability to face and accept reality which is of so much greater value than the accurate technical assessment of it.
>
> - *Communication skills.* Many authors ascribe to communication the essence of both leadership and management. Put bluntly, the failure to communicate is fatal to effective leadership, independent of all other aspects. It relates very closely to our definition of intelligence since it is the leader who is responsible for being misunderstood. Vacillation, ambiguity, and all attempts at deceit are totally inexcusable. While they are quite tempting in that they appear to produce short-term gains, they should be recognized for what they are—the tools of manipulation.
>
> - *Ability to assess group goals.* We recall the words of Tom DeMarco which we quoted in Chapter 5, to the effect that the essence of effective leadership is the ability to inspire others to make short-term sacrifices for the long-term common benefit. This starts with the leader making the sacrifice first, which is the essence of this capability. Personal goals cannot get in the way. While almost everyone has the ability to assess short-term personal goals, few possess the ability to assess and balance short- and long-term personal goals. Even fewer are able to place both their long- and short-term personal goals aside to assess objectively their group counterparts. However, this is essential if leadership is to be enduring, which is a necessary characteristic to its being effective.

Figure 21.1 Elaboration of Kazmier's "Leadership Traits"

believe that his course of action would have been radically different. Manipulation never complements effective leadership; it stands totally opposed to it.

As we conclude this section it is important to recognize that no one has all three of these characteristics to their maximum extent. Further, a deficiency in any one cannot be overcome by a superiority in the others. Instead, there is a minimal level that is required in all three categories:

intelligence, communication skills, and the ability to assess group goals. This minimal level is above that which exists in the average person. Additional strength above this minimum level only adds to the person's ability to lead effectively.

To this point we have been talking about effective leadership in general. This might appear in either formal or informal organizations. In fact, it occurs naturally in informal organizations, while there is usually an attempt to induce it in formal organizations, which can be counterproductive. Since every member of an organization participates in both the formal and the informal organizations, these characteristics are relevant to all. Further, very few, if any, fail to play the role of a leader in some informal organization. It might just be with regard to directing a conversation between two people, but leadership is exhibited. Thus when we state that a certain above-average minimum level is required, we must qualify this by the circumstances. Thus, within a given set of circumstances, the person who possesses the three attributes to the optimal degree will generally rise to a position of leadership.

This leads to a final interesting principle of this section (which is a reiteration of the concepts introduced in Section 5.3):

> People who attempt to be leaders in all circumstances will soon find themselves exercising leadership under no circumstances.

While there are some who will assume leadership under many circumstances, effective leaders know when and where this will be beneficial (i.e., this is part of their intelligence base). This is particularly relevant to managers who might have tremendous leadership influence within their peer manager group but might not be able directly to influence their own subordinates. We refer to this characteristic of leadership scope as varying with the *circumstances of leadership*.

This leads us to a comparison of management and leadership, which we consider in the next two sections.

21.5 Traits of Successful Managers

For purposes of contrast, in this section we present some traits of successful managers as given by Yntema (YNTEM60). In the following section we make the distinction between these two abstract concepts. Yntema discussed the following traits of successful managers:

1. Ability to see and solve problems
2. Ability to deal with people effectively
3. Communication ability
4. Organizational ability
5. Persistent effort
6. A good memory

These are discussed in the paragraphs below in preparation for the next section, where we compare leadership and management concepts.

Ability to see and solve problems. These are two related but different characteristics. Most people intuitively perceive the symptoms of a problem, at least when they are directly affected by them. The identification of the cause of the symptoms (i.e., the problem itself) is usually not so intuitive. The misidentification of causes, or the identification of symptoms as causes, leads to the establishment of control systems that propagate rather than eliminate problems. Since the primary role of management is the establishment of control systems (see Chapter 2), failure to do so effectively is a sure ticket to failure. It is commonly stated that the effective identification of a problem is 95 percent of its solution. However, depending on the circumstances, this might be a major overstatement.

A second characteristic required for problem solving is effective countermeasure formulation. This is the most difficult intellectual component of problem solving, and it subsumes most of the other characteristics given below. In addition, the ability to solve problems takes more than problem identification and countermeasure formulation abilities. The final ingredient is what is commonly called the *self-starting* instinct. This is the characteristic to assume responsibility for implementing the countermeasure and to initiate whatever actions are required to get it implemented. Since initiation and assumption of responsibility both inherently include the potential for failure, courage is required to overcome the tendency to allow nature to just take its course. The effective manager recognizes from the concepts of organizational entropy that often the "do nothing" course is the most perilous.

Ability to deal with people effectively. The term *deal with* implies that there is an inherent problem that must be solved, which cycles us back to the first characteristic described above. In fact, very few problems that the manager must solve do not involve people. Functional and personal conflicts are normal in any organization. Unresolved, they get worse. As the fabric of organizational cohesion is slowly destroyed, organizational entropy increases. Slow oxidation builds heat, which can result in spontaneous combustion, which has its counterpart in the organizational firefight. These should be avoided if at all possible by proactive steps to establish mechanisms within the organization to relieve stresses on an ongoing basis. It is clear that the ability to understand and deal with people is a prerequisite to the effective establishment of these mechanisms.

Communication ability. The establishment of effective management mechanisms (control systems) also necessitates effective communication. This begins with the ability to listen, seek input from a variety of sources, synthesize the accumulated data, and formulate an accurate view of the organization. This is essential to determining the information that must be imparted in order to influence the behavior of the members of the organization. Before this information is conveyed, it must be packaged in messages that have the capacity to be understood.

Organizational ability. This ability goes beyond problem solving to include proactively addressing problems before they occur. Described as the establishment of management mechanisms above, it is the ability to get groups put together and then create an environment in which they can realize synergism. It also involves the ability to accumulate and organize relevant information about the organization to determine with high probability the impact of alternative managerial strategies and tactics. On a personal basis it involves the creation of filing systems that enable the identification and retrieval of information when needed.

Persistent effort. This is the proactive counterpart of patience. As alternative approaches of problem solving and control are tried, they will not bring about the desired effects as quickly as expected, and some may fail altogether. The ability to learn from failure without being defeated by it is essential. This process of maturing into a new position, and continuously improving an existing position, requires persistent, continuous effort.

A good memory. Although this might seem to be something that is largely a matter of natural ability, there are a number of things that can be done to improve this capability. To start, there are memory tricks that can assist the brain in connecting the right neural switches. Practice and persistent effort can make significant gains in this regard. However, no one has the capacity to retain all the information needed to function in modern organizations. Thus it is essential that personal organization systems be established to supplement memory and relieve it of unnecessary concerns. Appointment calendars, filing systems, and effective delegation are the tools of the trade, which in this day and age involves the assistance of a computer.

Our purpose in reviewing these characteristics of effective managers is to provide the background for a comparison of management and leadership properties. This is done in the next section.

21.6 Leadership or Management?

Are leadership and management identical? The terms are often used interchangeably, and we do not wish to argue semantics. It is significant that Yntema's list of attributes for an effective manager did not include leadership, as many other such lists do. However, it is clear that the characteristics given for the two terms in Sections 21.4 and 21.5 are quite comparable in most aspects.

To clarify the distinction and comparison somewhat, we offer this simple relationship:

> Management emanates from the formal organization's attempts to formalize leadership.

The first clause is clear; without the formal organization there would be no managerial positions. We say *attempts* to imply that this process is not always successful. The mere promotion of a person (or hiring one in at a managerial level) does not make this person the leader of that organizational component any more than placing a pair of glasses on a dog enables the dog to read.

This being the case, we might wonder why the characteristics of the manager and those of the leader are so closely aligned. The answer is quite simple: To obtain a managerial position, the successful candidate had to demonstrate leadership qualities to, and possibly even lead, the promoting (hiring) managers. Middle managers generally want their project leaders and other lower-level managers to be leaders and to produce the expected subordinate behavior by means of effective leadership. It is no surprise that the characteristics ascribed to managers are leadership properties. However, there are a number of problems that cause this seemingly ideal model to break down, some of which are detailed in Figure 21.2. The major point is that the circumstances of leadership tend to prevent the manager of a component from providing the primary leadership to that component.

This is not to say that managers cannot, do not, or should not try to lead the members of their organizational units. They can, should, and those who are successful do. This entire book has stressed the superiority of leadership over formal authority for accomplishing organizational goals. We are, however, trying to emphasize the fact (admittedly, once again) that the bestowal of a management position does not make a person a leader, even if all the attributes of leadership are present in the person. In fact, as we have shown, it might even disable the manager in this regard.

The fact that a manager is not the primary leader within the respective organizational component does not have to be detrimental to organizational goals. To keep them functioning harmoniously, it is only necessary that the manager establish and maintain excellent relationships with the informal leaders. By *relationships* we mean working arrangements in which neither

- *The selection is generally not made by the members of the organizational component.* The leadership characteristics that appear to be present to those who do the promoting might not be recognized by the members of the organizational component. This is a characteristic problem of traditional organizations, and it is manifested by the responsible middle manager hiring someone whom he or she "can be comfortable with." Rarely do the *circumstances of leadership* that affect middle management coincide with those of the members of the organizational component.

- *The circumstances of leadership as well as the characteristics of the manager change over time.* Someone who demonstrates the greatest of leadership potential can lose these characteristics. This change is usually highly correlated with major changes in power potential, either negative (loss of respect) or positive (egotism).

- *Problems persist even with heavy internal involvement.* Heavy subordinate input, or even allowing subordinates to choose their own manager, may not solve these problems. Subordinates will generally choose their most powerful informal leader, and the perils of this were detailed in Chapter 10. When the peer group reviews outside candidates, there is rarely the time to assess leadership characteristics to any great degree, so proxy measures of past performance (as documented on the résumé), communication skills, and general personality must suffice.

Figure 21.2 Complications in Expecting Managers to Be Leaders

formal nor informal leadership is exerted (although admittedly, it cannot cease to exist). This is required even when the manager is an influential leader to his or her subordinates, and major problems will arise for managers who cannot establish these relationships.

The lack of leadership within the organizational component need not be fatal; however, lack of leadership in general is. Managers who do not initially possess a constituency within their own organizational components must be able to build their power bases through other types of leadership. This might come through their marketing ability, which can be internal (influence within the organization) or external (the ability to create sales to or form alliances with other organizations). Of these, the second has the most potential for building power, since it is portable (i.e., it is not vulnerable to "who you know" within the organization).

Although the lack of direct leadership within the organizational component is not fatal over the short term, it should not be allowed to remain a static condition. Thus we spoke of power building above in terms of the leadership influence that is built up over time by sound decision making. This is done by persistent efforts that accrue to the best interests of the manager's subordinates. This rewarding behavior ultimately builds the manager's leadership capability within the organizational component. At the same time, relationships with the informal leadership should be strengthened continuously. This two-pronged approach toward evolving effective leadership is the key to the development of a successful manager.

We close this chapter by recognizing that there are a variety of leadership styles. Since technimanagement requires management to progress through a major transition process, this must be accompanied by an evolution in leadership style.

21.7 Leadership Styles

Traditionally, styles of leadership and management were so ingrained within each organization that it had its own look, feel, and personality. People within these organizations for their entire careers could not imagine that things could be done any other way. According to Kazmier: "The way that a manager leads is most importantly influenced by how his superior leads" (KAZMI69,251). This reflects a Theory X orientation, but it is probably largely true in most organizations today. However, as we demonstrate in Chapter 24, this does not have to be so. Today, most organizations are sufficiently open to improvement that they will tolerate experimentation, especially within those components that have a highly technical orientation. Further, there is no reason to publicize your leadership style and management approach until it is clearly demonstrated to produce superior benefits to the organization.

A second significant point is worthy of consideration before we enumerate the possible leadership styles: "A single pattern of leadership behavior used without discretion is unlikely to be successful in a wide variety of managerial situations" (KAZMI69,254). The application of this principle to different people within the organization component is quite straightforward. The leadership style applied to a peer would be different from that of a subordinate or superior. Obviously, the approach to a clerical support person is quite different from that applied to a team leader. This is not a matter of unfairness; it is a function of individual maturity.

The "multiple pattern" approach also applies over time. This was introduced above as we mentioned the potential deterioration of leadership. At that point the failure of the new manager to adapt was cited as a potential problem. This concept should be extended as we recognize that it is a pervasive and perpetual problem. Although time is a parameter, it is not the cause. Power and maturity are the cause. This brings us to the summary principle of this chapter:

> As a manager gets more powerful, more and more control can be exerted through leadership as opposed to formal authority.

In fact, this is the sole reason for power building. The ability to make the transition from managerial authority to effective leadership is critical to the manager's survival through the transition to technimanagement.

Several categories for management styles could be proposed depending on the aspect being considered: (1) reward approach: negative or positive, (2) openness: centralized or decentralized, and (3) distribution: authoritarian, democratic, free-rein. By now the reader must understand that the transition to technimanagement is in the direction of the positive-reward, decentralized, free-rein approach. These are fairly self-explanatory, with the possible exception of the last. The concept of *free rein* might appear to be one of anarchy. However, this is not at all the case. Instead, it fits the ideal technimanagement goals quite well. The name originates from a style of horseback riding. Its full value can be appreciated by advice often given to young riders when crossing a stream: "Give the horse its head." That is, let the horse decide how best to cross the stream. Is this because the horse is smarter than the rider? Certainly not. But in knowing how to best negotiate the slippery rocks and holes, the feet of the horse are far more intelligent than the eyes of the rider. The goals of horse and rider are integrated, and *in this activity the horse knows best*. It is not our intention to denigrate human activity, but the analogy is nearly perfect. It breaks down only as we recognize that the superiority of distributed intelligence of our subordinates over our own is the rule rather than the exception.

There might be some questions as to the reason that free-rein is superior to democratic distribution, or its closest approximation, consensus. Decision making by consensus is merely the basing of decisions on what is most popular with the majority at the time. This is not effective leadership. The principle of distributed decision making within technimanagement requires that the best person or group be established, charged, and empowered to make each decision. This is absolutely not decision making by consensus (however, we can understand the confusion). If this is in any way unclear, we urge the reader to restudy Section 21.2, where the meaning of "walk behind them" is thoroughly clarified.

As the organization matures, it will be necessary to make changes in the leadership styles that are employed. With the possible exception of setbacks, this will generally proceed from the negative, centralized, authoritarian Theory X type of leadership, which we have shown to be quite ineffective, to the positive, decentralized, free-rein approach which characterizes Theory Y.

Before getting into the important chapter on transition, two other concept chapters are necessary to round out our discussion of technimanagement. The first of these has to do with the limitations of experts, a subject for which we have hopefully matured as we recognize that, being engineers and scientists, we are the experts. The second is a review of communication concepts as they relate to technimanagement, which attempts to remedy one of the major limitations of experts.

22

Limitations of Experts

22.1 A Notion Revisited

For the context of this chapter we define an *expert* to be a technically trained specialist. Very few, if any, members of the technical organization would fail to qualify by this general definition. This definition is relatively useless therefore, unless it is viewed as a relative, not an absolute quality. It will help to define the term *degree of specialization* to assess qualitatively the relative "expertness" of a given person. Experts with a very high degree of specialization would be very narrowly educated and experienced, since this might be essential to achieve the depth that is required for them to function in their jobs. At the other extreme is the generalist, the jack of all trades–master of none, who is very flexible but is not able to perform any of the technical tasks that are necessary to accomplish the goals of technical organizations. (This is not a derogatory statement; they still perform many of the necessary administrative, management, marketing, and other support functions that are essential to enabling the specialists to accomplish organizational goals.)

The selection of the degree of specialization that people adopt for their personal career goals is a very complicated optimization problem. It is usually not very well modeled or analyzed, especially by young college students, but instead, is found by a complex default process that is determined largely by personal inclination and ability, opportunity, perceived rewards, and peer and role model influence. While these influences are not perfectly responsive to job market demands, there is a fair correlation, and the high demand for a particular type of expertise does not remain a secret for very long. In any event, the net effect of this process, coupled with the experience that people accumulate by continuous education and on-the-job training and experience, generates a total workforce (membership) of the organization with a fairly continuous range of degree of specializations from the generalist to the highly expert.

With this in mind, let us revisit a counterintuitive principle that we introduced with regard to the accumulation of personal power in Chapter 19:

> That which makes you good makes you bad.

Now, it must be understood that the technical organization cannot accomplish its goals without experts. That which makes most members of the technical organization of value to the organization itself is their high degree of specialization. The purpose of this chapter is to demonstrate that this is precisely the commodity that makes them "bad," where the degree of "badness" is measured by the degree to which this commodity prevents organizational goals from being realized. (Note that there is no intent to use the terms *good* and *bad* in a moralistic sense. We return to this at the end of this chapter.)

There are a number of purposes for going through this exercise: (1) to generate a better understanding of limitations caused by our own degree of specialization, (2) to understand the problems in dealing with experts so that better team balance and motivation might be attained, and (3) to optimize the personal decisions with regard to degree of specialization such that all members of the organization might set a course toward greater productivity. We first analyze the problem further and then attempt to suggest some solutions based on the ramifications of this analysis.

22.2   The Problem

Although we might be tempted to believe that the problem with experts is a phenomenon of the recent advances in our technology, most of the thoughts of this chapter were developed during the 1920s and are attributed to or summarized by Laski (LASKI30). No doubt the early Roman and Egyptian empires had their experts as well. Indeed, the expertise that it must have taken to deal with the remote provinces, or, more technically, to build a pyramid, must have been recognized as a tremendous asset to their leadership. Can there be any doubt that people possessing this specialized expertise would be given great positions of power? Thus the problem that we are approaching is one that is ageless.

In Chapter 19 we discussed the personal power that accrues from possessing scarce, highly specialized expertise which is necessary to the accomplishment of organizational goals. This power will generally manifest itself in the ability of the expert to influence the organizational decision-making process. Although this might seem to be advantageous from a personal point of view, it can be quite counterproductive to the organization. Laski suggested a number of underlying problems that experts have in being effective decision makers beyond the realm of their expertise, which are presented in the display below. Like many analogous lists, this one is neither all-inclusive nor are its elements mutually exclusive. The remainder of this section is dedicated to qualifying it adequately so that we can formulate a model to begin to address this class of problems.

Elaboration of Laski's Causes for Limitations of Experts

- *Sacrifice of common-sense.* We define *common sense* by its normal usage, recognizing that this is one of the most uncommon commodities in the world. Perhaps this is due to the sophistication of our educational system and their tendency to impart faith in the process of test passing as opposed to analytical thinking. So often when an untrained person comes up with a brilliant solution to a problem, its very genius manifests such simplicity that the experts write it off as being "just common sense." Our response is dazed wonder as to the inconsistency between common sense and common practice.

- *Aversion to new ideas.* The expert is the person who we would least expect to be vulnerable to this problem. However, they might well be more susceptible, because the expert is expected to be the one solving the problem, not the one listening to the suggestions of others. This expectation is both internal and external to the expert, which complicates the problem further.

- *Restricted perspective.* Experts tend to see solutions only in terms of their own specialization. Throughout their careers, experts are so focused on acquiring depth of expertise that they do not have time to acquire breadth. The result is a flawed decision-making process due primarily to the lack of a systems view.

- *Feeling of superiority.* When a refined sense of expertise causes one to develop a feeling of superiority over others, this strongly affects the moral realm. Technical managers can easily find themselves in the unhappy position of dealing with several supercilious people, each of whom might possess expertise critical to a given project.

> **Elaboration of Laski's Causes for Limitations
>   of Experts (Continued)**
>
> - *Identification with fellow specialists.* Everyone must have a peer group, and it is only natural for people with common specialties to have close informal relationships with one another. This process has been formalized for most disciplines. While identification with fellow specialists is neither unexpected nor inherently counterproductive, it becomes a negative force to project group cohesion in those interdisciplinary activities that are required in the vast majority of projects today. This interdependence will only grow as projects continue to become more and more technologically sophisticated.
>
> - *Confuses knowledge with wisdom. Knowledge* in this context involves the information and techniques of the person's specialization. We define *wisdom* for this context to be the ability to combine technical with other knowledge to realize goals. It is not that experts cannot have wisdom to operate very far outside their areas of expertise. However, the crux of the problem is that when they lack wisdom, *they do not realize it*. Confusing this body of knowledge for wisdom, they tend to plunge forward into areas that are not within their realm of potential wisdom.

An institutionalized example of expert limitations is the integration of the traditional *staff group* (defined in Chapter 14) within the total organization. This provides the ideal model for how things should not (attempt to) be accomplished. When problems arise and greater efficiencies are sought in the line operations, the very worse thing that can be done from a motivational standpoint is the imposition of a solution from outside this group of experts. Even in the unlikely event that they come up with the ultimate perfect solution, there is little reason to expect that someone in the line organization did not think of it long ago. (However, because they did not have the blessing of the staff, they were not given credibility.) In many cases these staff groups impose solutions inferior to those attained by quality deployment techniques, which only increases project group synergism in opposition to the organization.

My friends in the liberal arts college would argue that the list of characteristics given above argues for a more liberal education. Their theory is that if people do not specialize, this produces a superior educational experience, since now the time can be employed to get a broader flavoring of a wide variety of educational experiences. The fact that a large number of their nontechnical students struggle to get jobs is largely ignored as a message from society, and certainly from the technical organizations that drive our contemporary job markets. But an even more fundamental problem arises when liberal arts majors manifest the same sense of superiority because of pride in their educational experience. That is, their breadth becomes their specialization, or at least equivalent to it, and from there they exhibit all the characteristics given above. This, too, is very highly the result of the indoctrination they receive in our institutions of higher education.

A similar phenomenon might be observed with regard to business or management majors. Certainly they have a greater breadth, but they still possess a specialization (e.g., business, management, marketing, management information systems, etc.), which tends to give them the same characteristics as other experts and possibly a lack of sufficient depth for the immediate job at hand.

We can see clearly how many of the characteristics given above are "educated in" at the outset of individual specialization and then reinforced throughout the career. However, we find little chance that our educational institutions as they are currently structured have any hope for "educating out" these characteristics. This is because technical organizations *demand* specialists. Since specialization is inherently difficult for most, a great deal of motivation is required for these people to acquire an adequate educational experience. If they are not conditioned to believe that their particular specialization has an ultimate payoff, their motivation is insufficient to overcome these demands.

For purposes of discussion, let us consider the possibility of a relationship between degree of specialization and the characteristics presented above:

> A person's degree of specialization is highly correlated with the presence of the six characteristics given above.

This conjures up a number of possible visions, not the least of which is Mr. Rogers (of PBS "Mr. Roger's Neighborhood" fame) gathering his visiting children together and asking: "Can you tell me—what is *prejudice*, boys and girls?" Surely, the assignment of an excess of all six negative characteristics to all specialists, or to say that the degree depends on the degree of specialization, is the height of prejudice. We hope that we caught you in this trap. The statement in the box above just cannot be proven. We conclude that, while all experts are susceptible to the six negative characteristics, these problems are more a function of basic human nature than the degree of specialization that a person possesses.

Nevertheless, we cannot deny that these problems exist. We are not trying to avoid the problem; we are trying to avoid misplacing the blame. But let us not forget that the "six characteristics" were invented by one expert and further elaborated by another (the author). In fact, there are very few generalists in technical organizations, and as we mentioned above, even they might not be immune to the "problem of experts." Thus if we are even going to begin to address these problems, the first step is to go to the nearest mirror and take a good look at one of the causes. With these qualifications, let us proceed to some countermeasures for this very real problem.

22.3 Dealing with Expert Limitations

The most effective way to deal with the limitations described above is through leadership. But like any other silver bullet, this one requires virtually perfect aim. Leadership begins with the acquisition of two rare characteristics: (1) the ability to overcome the "six limitations" within oneself, and (2) the ability to refrain from being judgmental of those who have not yet obtained the first characteristic. The first step is to overcome denial, and those who are unable to see all six limitations within themselves will never come to grips with them and overcome them.

Once denial is overcome, the next step is to mature as a leader and manager. A comparison of the traits of successful leaders and managers given in Chapter 21 shows the limitations of experts to be virtually the opposite of those traits, although stated somewhat differently. Maturity requires humility—the recognition that even though you are a manager, you are merely a part of a team (and generally not a key part of it at that). This leads to a formulation of our overall approach to these problems:

> The limitations of experts can only be overcome by combining their strengths with those of other experts in a complementary way to produce cohesive teams.

If this team is merely a combination of experts with the limitations given above, the team will never reach a state of synergism. It will remain a collection of people who will have as much chance of working against each other as they do of working together.

Before going on to discuss the methods for bringing this about, we must further emphasize the necessity for leadership as opposed to manipulation. When it comes to common sense, we might look to those who were forced to grub their life out of the soil with few of the formulas and machines that make things fairly routine for many today. You may have privilege to know some of these people, and if you do not, you would do well to find, meet, and listen to them. In their world, you can find out what you are really worth on your own (i.e., without your team). For example: "If you're so smart, why ain't you rich?" (Bill Koch, small-time chicken farmer, 1956). A simple question. But its full impact cannot be appreciated if we think Bill merely measured intelligence by wealth. It was a put-down; and a quite effective one at that, because we wanted to be rich and we were not. To the rich Bill would ask, "If you're so rich, why ain't you smart?" Indeed, the irrational things that monetary success makes many rich people do give rise to such a question.

Although we have tried to refrain from anecdotal proofs throughout this book, perhaps you will allow us to make another exception. For Bill Koch was an amazing man who managed to survive on nothing other than common sense. He had no education, no assets, and he leased most of the land that he worked. But somehow he survived and prospered, despite the odds against him. One of his most important unlearned abilities was the uncanny ability to inspire the young men that he hired to do some of the toughest, dirtiest jobs imaginable. (I worked for him cleaning out chicken houses when I was 14.)

A friend of mine, the son of a carpenter whose father worked him very hard, prided himself on his endurance. Desiring to impress Bill with him, we invited him to join us for a few days. He readily accepted the challenge, but when Bill put him to work cleaning out the horse stable, he began to have second thoughts. He related the story to us later: "It must have been 110 degrees in there, the place had not been cleaned out for over a year—the manure must have been a foot thick, packed down by horse hooves and interwoven with straw. When you would break into it, it would steam up and the sulfur made breathing next to impossible. After about an hour, I was just about ready to puke. I told myself that when Bill comes through that door, I am going to tell him what he can do with this job. And then Bill came in with a shovel to help . . . in his bare feet! There was no way that I was about to tell Bill at that point that I couldn't handle it."

Did Bill have any idea that he had affected a major change in behavior? Certainly that was not his intention. It was just natural for him not to ask someone to do something that he was not willing to do himself. It was also natural for him to walk around in the summer without shoes. Would this qualify him to be a natural leader? Or was he just "walking behind" his employees?

Another humbling concept is that the converse of our major principle of this chapter does not hold:

> That which makes me good makes me bad; but
> That which make me bad does not make me good.

Many experts recognize their own limitations, but they excuse them because of the myth that experts must have some shortcomings to compensate for their strengths. This myth is extremely counterproductive, since it leads them to do nothing to improve in their areas of greatest deficiency. Pride in recognition is worse than denial, since there might be a conscious effort to persist in the shortcoming if it perceived to be creating another benefit. Although it is true that everyone cannot be expert at everything, there is no reason that everyone cannot have a broad enough view to enable them to function effectively within a team environment.

A final humbling thought to the reformed expert: As is the case with alcoholics, even experts who recognize and take steps to remedy their limitations never fully overcome them. They are always in recovery, but never fully recovered. Thus as is the case with all personal characteristics, the possession of the six negative characteristics is a matter of degree and direction. We all have limited perspectives, and while they can be broadened, they can never truly put us into another person's shoes. Each of us is only one person, and that brings with it inherent limitations, no matter how broad or narrow our education or occupational experiences.

Approaching the limitations of experts in oneself is a major step toward being able to apply the leadership by which these problems can be overcome in subordinates. Although individuals have complete control over recognizing and controlling their own limitations, the establishment of a personal control system to govern their own actions is still extremely difficult. The establishment of such a mechanism within others is not within the direct power of the manager, and thus it is an order of magnitude more difficult.

This is not to say that we should give up. The answer, as we stated above, is leadership that starts with and depends most heavily on self-control. This leadership does not require that every member of the organization totally overcome the six limitations of experts discussed above. If this were required, no projects would ever be successful. Since the key to overcoming the technical limitations of experts is the participation of balanced teams, team cohesion and cooperation are essential. However, these can be upset easily by the nontechnical limitations.

While the leadership required to motivate experts to work together is much more complex than that required to motivate a person to clean a stable, the basic principles are the same. They relate to the underlying motivations of the people involved. Although the manager cannot totally overcome the limitations of experts, knowing that they exist enables their consequences to be mitigated.

Since the primary problem is one of inspiring teamwork, the limitations that become most difficult to overcome are primarily the fourth and fifth: *feeling of superiority* and *failure to relate to other disciplines*. These are closely interrelated, the first possibly causing the second. It is important to recognize this as a problem of individual optimization. Indeed, pride is one of the major motivators for product quality, and in this regard it can and should be inspired. However, the problem arises when the line is crossed where personal pride affects organizational harmony. This optimal balance will exist at different points for different people. For example, most groups are more apt to tolerate a strongly independent technical expert whose expertise is deemed to be essential to the success of the project. However, strong informal leaders become such because they do not alienate others with their strong feelings of superiority.

Can the same sense of pride that leads to a desire for product quality be used to motivate individual self-improvement and team cohesion? The answer is a qualified yes, qualified by goal integration. It is essential that the person who is motivated to produce a high-quality product recognize that the total team is essential to this product, regardless of the varying degrees and types of skills involved. This will integrate team and personal goals. The need for self-improvement is more difficult to realize on a personal level. However, if personal and team goals are integrated, this can be accomplished by making self-improvement a team goal. Once it becomes socially acceptable to admit shortcomings and work together to overcome them, the process of self-improvement is greatly facilitated.

As a model, we might initiate this process by considering the possibility that everyone possesses an equivalent body of knowledge and skill; however, they are just of varying degrees of breadth, specialization, and maturity. Job market forces recognize that there are additional rewards for the more difficultly attained skills. Those who are willing to pay the price in education and experience attain these rewards. Others who for personal reasons were either unwilling or unable to acquire the more demanded skills fail to be so rewarded. However, their skills are still essential to enable the more skilled the freedom to practice their specialty, and often the efforts that they put forth are of superior intensity to the rest of the team.

In essence, the solution is the never-ending task of promoting an appreciation for the efforts of others, regardless of the skill and specialization level involved. While members of the organization take their cues in this regard from their immediate supervisors, the influence of the informal organization cannot be overestimated. A group attains a personality of its own, either respecting the efforts of each other, or fragmenting themselves into conflicting informal groups. Although conflict resolution will necessarily be required from time to time, the very existence of a conflict is evidence of the failure to anticipate proactively conflicts that result naturally from expert limitations.

One additional consequence of expert limitations is its tremendous deterrents to effective communication. At the same time, the creation of effective communication mechanism between the members of an organization tends to be the most effective method for building the relationships that overcome these limitations. For this reason we devote the next chapter to the subject of communication.

23

Principles of Communication

23.1 Definition and Objective of Successful Communication

We saw in Chapter 22 how the very fact of being an expert can tend to hamper effective personal communications. (This is not pointing the finger elsewhere, since we can all be classified as some type of expert.) Since communication breakdowns account for the vast majority of the problems in technical organizations, it is appropriate that we dedicate this final chapter to it before getting into the details of transition. The very act of grouping specialists together on teams tends to create communication problems, which is a principal cause of increased organizational entropy. However, if the necessity for maintaining channels of communication is appreciated, and if the deterrents to effective communication are understood, countermeasures can be taken to minimize these effects. In this chapter we first define the process of communication and crystallize its basic functions and objectives. With this in mind, the various deterrents to effective communication will be addressed.

We begin with a definition from Kazmier: "The success of the communicative effort is based on the extent of new information or understanding achieved by the receiver" (KAZMI69,178). With this (rather intuitive) definition in mind, we can construct a model of communication. For simplicity, let us assume that only two people are involved: a sender and a receiver. Communication originates with a thought in the mind of the sender, which the sender feels needs to be communicated with the receiver, either because the receiver is perceived to be ignorant of the information or because the sender wants to reinforce a known concept or produce better understanding. The sender then has the task of translating the thought into words that are going to communicate this thought effectively. Then the sender must articulate these words either orally or in writing. Finally, the receiver must interpret these words to mean the same thing as the sender had in mind.

This process is rarely 100 percent successful according to Kazmier's definition, where complete success would require that the exact thought be reproduced in the mind of the receiver. Exceptions would occur in very simple communications. At the other extreme, however, complex thoughts are rarely communicated with a single message. Figure 23.1 presents an analysis of the steps involved to determine the cause for this. With this host of vulnerable deficiency points, we might wonder how any effective communication ever takes place. The answer is: with considerable effort. If this effort is not put forward, communications will inevitably break down. Further, effort alone is not sufficient; it must be directed toward strengthening the points of weakness identified in the model given above. Thus one approach toward improvement is for communicators to go mentally through the steps above and make sure that adequate attention has been given to each.

The first three of the steps are within the control of the sender, and adequate preparation and contingency planning can overcome these pitfalls. Although the fourth is not, the following should always be kept in mind:

> In any attempt to communicate, the burden of proof is on the sender.

This being the case, the sender must assume responsibility for being misunderstood (excepting cases of obvious malice, of course). Since attitude is the most critical aspect of effective reception, most of the countermeasures to the deterrents of successful communication given in the next section deal with this. In general, however, the sender should seek verification of the message sent by means of *feedback*. This might simply take the form of requesting the receiver to repeat the concept in her or his own words. It might require an immediate demonstration, or in more complex cases, it might require close supervision over an extended period of time. This is a simple step, but it is often omitted because it is not totally necessary for all communications. On the other hand, no communication can be verified to be effective without it.

As a final major point on the ultimate goals of communication, it must be recognized that communication is a means to an end, not an end in itself. The purpose of communication is almost always behavioral modification on the part of the sender. Thus we can go back to step 1 and declare that for effective communication to be formulated there must be a clear understanding of the behavioral modification objectives on the part

1. *Sender's conception of the thought.* This is by far the most vulnerable area for communication breakdown, since rarely are complex thoughts totally understood on the part of the sender. Communications suffer to the degree that messages are used to further formulate thoughts, as opposed to transmitting thoughts that are already formulated and resolved. Thus the first responsibility of senders is to be sure that the purpose of the specific communication is known definitively to themselves. Clarify to the receiver whether the communication is just "thinking out loud" or serious communication.

2. *Senders translation of thought.* This is an articulation within the mind of the sender. It is a preparatory step. Effectively it is self-communication, or the replacing of abstract thought with words. It is analogous to loading and aiming of the gun before pulling the trigger. In formal presentations or written reports, it corresponds to the rough draft. Unfortunately, it rarely has an analog in informal oral communications. Instead, words are often poured forth under the assumption that the receiver can and should be able to read the mind of the sender.

3. *Articulation to the receiver.* Any number of distractions can occur in the actual articulation, especially if the sender is not in full control of the environment at the time that this is done. The most significant is an unanticipated reaction from the receiver. Indeed, most communications are not controlled formal presentations but conversations of which control is shared between the parties. These disruptions cause the sender to alter the course from the prepared "script." They might even cause a reformulation of step 1. Although this is quite healthy in brainstorming interaction, it can be quite counterproductive if the goal is to focus on communicating a single concept.

4. *Interpretation of the words received.* Since there might be considerable distortion introduced in the first three steps, communications will be approximate even with perfection here. This step depends most heavily on the attitude of the receiver. Distractions, emotions, unforeseen inferences, preconceived mindsets, and the feeling toward the sender virtually always override the logical thrust of the message. Thus if the wrong words are chosen, this can only compound these problems.

Figure 23.1 Communication Breakdowns by Step

of the sender. If the cause–effect mechanisms are not understood in this regard, the most effective communications might lead to counterproductive changes in behavior. We have all experienced what is commonly called putting one's foot in one's mouth—a consequence of all too successful communication without adequate thought given to the behavioral modifications (i.e., reactions) that result.

23.2 Deterrents to Successful Communication

Given the basic model of communication presented above, the objective of this section will be to approach the subject from the perspective of particular deterrents that have been suggested by Kazmier and Kast, including stereotyping, poor symbology, selective perception, informal communication deficiencies, and a failure to modularize. Each deterrent will be related to the model given above to further unify the discussion of this most important aspect of management.

23.2.1 Stereotyping

In terms of communication, Kazmier defines *stereotyping* to be "a sender's tendency to categorize rigidly on the basis of personal characteristics . . . . " (KAZMI69,182). *Stereotyping* is a pleasant word for *prejudice*; by being more politically correct, it may assist in avoiding denial. Although this might be a noble attempt to organize thought, we should recognize that this tendency creates misperceptions of reality that cannot contribute to effective decision making.

We mentioned prejudice in regard to the possible discrimination against experts that we might have triggered in Chapter 22. We tried to overcome this toward the end of Section 22.2 but may well have failed. Our defense is that we include ourselves (both readers and the author) in the group to which we attributed the potential characteristics, and we qualified the potential by stating quite clearly that the characteristics exist in varying degrees as opposed to being absolute. However, we recognize that just because one is included in the group does not give him or her the right to be prejudiced toward it. (This presented a vexing problem for us, but the only alternative was either to eliminate Chapter 21 or to so water it down as to make it useless. We hope the reader will understand that the very motive of instilling humility within our own numbers creates the potential that these same words will be used to create stereotypes. This is a good example of the paradoxical nature of some communications and the critical requirement for a proper attitude on the part of the receiver.)

Other than the moral aspects of stereotyping, its major problem is that it creates a simplistic but false sense of reality on the part of the manager. Even if the hasty generalization is correct in 99 percent of the stereotyped subset, sooner or later the other 1 percent will be done injustice by it. But the number of exceptions to most stereotyping is usually so significant that the stereotype is generally misleading and thus counterproductive. Although we can all come to this conclusion, we all have a tendency to extrapolate our limited experiences (most of which are indirect) into the formation of classifications in an attempt to apply them in the future. Stereotyping is a very common human characteristic that we all possess, and it takes considerable effort for us to overcome the extremely negative excess baggage that it carries with it.

Stereotyping can affect both the sender and the receiver. On the sender's side, a misperception of the background, ability, or inclinations of the receiver will lead to erroneous assumptions with regard to the communication process. (This includes both assumptions about the receiver's attributes as well as the receiver's belief system, including stereotypes.) This will affect the way that the sender translates thought into words and thus the particular words used and the approach and attitude of the sender toward the receiver. It can lead the sender to apply the same approach to everyone within a given classification, not recognizing that the significant individual differences generally require considerably different approaches. It also cuts off the feedback mechanism, especially in cases where there is a loss of respect on the part of the receiver.

On the receiver's side, the problem might be even worse. We have discussed this characteristic throughout the book, for example, the inability of certain people to communicate because of prejudice against the management positions they hold. This works both ways. If receivers attribute validity based on the messenger rather than the message, enslavement to the leader will follow. This is consensual manipulation at its worst, although it is usually disguised as effective leadership. There is a fine line between convincing someone of the validity of a position and convincing them to follow blindly the champion of that position. However, this is the primary difference between continuous improvement and the management fad of the year.

The solution to this problem is fairly easy to implement on a personal, individualized basis: Judge persons and concepts only by their individual performance as opposed to any other classifications and maintain an accurate perception of each person's performance, potential, attitude, and ability to communicate. Although this can be controlled in oneself, it is impossible to impose this attitude on others. However, as discussed in Chapter 22, leadership and peer pressure can heavily influence these types of attitudes in either direction.

23.2.2 Poor Symbology

By *symbology* we are referring primarily to words, either written or verbally delivered (and secondarily, the diagrams that are usually a part of most technical documentation). However, this also includes the tone and body language that accompanies speech. These can be much more important than the words chosen. Indeed, some people tend to offend by their very tone and mannerisms, whereas others have a conciliating effect. It might seem that these are uncontrollable characteristics, but they can be controlled by those who properly value negative feedback (see Chapter 20). The remainder of this subsection is limited to choices of words as opposed to these highly personal characteristics of communication.

In our model above, this deterrent would emanate from a deficiency in the third step: articulation of the thought to the receiver. This step is often taken for granted and not given consideration before the actual communication session takes place. One way to assure this is to formulate the thoughts in a memorandum first which is either transmitted before the meeting or else taken to the meeting, allowing the receiver to read it at that time. Many written communications by technically trained personnel assume far too much knowledge on the part of the reader. It is imperative that the writer be able to read the communication in an objective and self-critical way. Well-written communications will require several rewrites before they actually say what is intended. When this is not done, the literal language usually says something quite different from what is intended. Most readers can read between the lines; however, this should not be necessary or expected. If you do not revise your writings several times, chances are that they are not accurate representations of your intentions, which can cause serious communication problems.

Unfortunately, we do not have the luxury of being able to review spoken communications. However, we do have immediate feedback via conversation which enables corrections to be made on-line. Still, considerable thought should be given beforehand as to the proper approach and attitude to be exhibited to cause the desired behavioral changes. It is unfortunate how many arguments seem to be due purely to *semantics*. That is, they involve only use of the proper word, not issues involving fundamental differences in philosophy. This is totally counterproductive, and if a change in terminology can accomplish agreement without compromising principle, this is a trivial price to pay.

Certain terminology is loaded with emotional connotations. The use of this rather than logical statements in argumentation presents a calculated risk. If it does not produce the intended behavioral change, the power loss can be tremendous. The resort to emotional rather than logical arguments demonstrates that there may be no logical support for the position. On the other hand, supporting a position with facts demands that those who disagree refute those facts. Thus, although emotional language is not totally ruled out, it demands considerable attention to its ultimate intent (i.e., the behavioral change that it will bring about). Unfortunately, such words are often uttered in the heat of battle, and the result that they bring about is most often negative to the speaker.

Finally, concrete, understandable terms should be used rather than abstract concepts. This is unfortunate since many concepts cannot be expressed in simplistic terms, including many given in this book. Nevertheless, to the extent possible the terminology chosen should clarify and reveal as opposed to obscure. Different levels of terminology might be required with different audiences, or even to communicate different topics to the same audience.

23.2.3 Selective Perception

Did you ever wonder about the large variety of beliefs that exist with regard to politics, economics, and religion? How is it that when two intelligent people with comparable backgrounds are presented with the same basic body of evidence, they come to completely different conclusions? We might be encountering a case of *selective perception*, or the tendency to believe what suits one's purpose as opposed to what is logically justifiable. To answer our question briefly: People generally believe what they want to believe.

We would really have it no other way—this is free will, and the exercise of free will is essential to most of the principles of technimanagement. But selective perception might not seem reasonable to many technically trained people. Its importance is expressed by Kast: "The concept of selective perception is important because voluminous information is received and processed. Individuals select information which is supportive and satisfying, they tend to ignore information which might be disturbing" (KAST70,217). Although this is clearly a problem related to the fourth step in our model of communication (interpretation of the words received), the sender should not just write off those who practice selective perception as being unreasonable. It can probably be proven that we all resort to it from time to time. The important thing is to overcome selective perception.

To overcome selective perception within others it is necessary to demonstrate reasons that it is in their best interests to broaden their views. If people believe what they want to believe, the way to get them to accept reality is to demonstrate to them the benefits of enlarging their perspective. This is a straightforward application of the principle of rewards if, in fact, you are arguing from a position of strength and reality.

Let us illustrate the above by a very vexing example. Suppose that you have a very valuable employee who reports directly to one of your subordinates. This employee has been coming to you about once a month for the last three months with complaints about his supervisor which appear to be quite reasonable. However, you have taken some actions to establish mechanisms that you believe will solve the problem over the intermediate term. In the meantime, your subordinate (this person's supervisor) is performing some critical work for you, and you do not wish to intervene further in this matter for fear of interrupting this very important project. For that matter, intervention would be counter to your style, since you are not interested in micromanaging this situation. Thus, to get everyone around the table and hash it out is not viewed as a feasible alternative (although it might be required later).

To this point you have been a release valve, utilizing primarily a nondirective approach. The few suggestions that you have made to emphasize the authority of your subordinate have been selectively ignored, and the comments regarding changes that have been put in place have been interpreted to mean that you take the complainer's "side." You give concerted thought to this problem and resolve an approach, which you rehearse mentally (steps 1 and 2). On the next incidence of a complaint you inform the employee that you feel he has a false sense of reality which if continued will create a very counterproductive situation for him. You then explain how valuable his supervisor is to your operation, and how you have delegated total responsibility and authority to this person. Mechanisms have been set in motion, but will take time, and it will be up to him to assist and be patient in this regard. If he feels that this is impossible, you will be glad to do all they you can to help him relocate.

In essence, you threatened to fire him if he could not find a way to live with the situation. But it was not done in a way that you did not take sides and act in what he would perceive to be an unreasonable way. This would be the case if you just threatened (e.g., either get along or get out!). This would have brought about a change in perception—he would have decided that you were both equally unreasonable, and you would probably have lost a good person. In this case, you have demonstrated that the original selective perception was counterproductive and that if he continues to persist in it, no benefit would accrue to him, at least not within the context of your organizational component.

23.2.4 Breakdown of Effective Informal Communications

Because we gave an entire chapter to the topic of informal leadership and communications (Chapter 12), we will not elaborate in great detail here. However, the failure of the informal organizations to promote communications is a major problem within many organizations. According to Sayles: "The modern organization depends on lateral relationships precisely because there are so many specialized points of view and so many required contacts that no single manager could handle the communication flow alone" (SAYLE66,424). We can think of a number of reasons that could account for a breakdown in this regard:

1. *Management greediness with communications.* When managers think that all communications must go through them, catastrophe is not far behind. Not only can "no single manager . . . handle the communication flow alone," but no manager can control informal communications. They will continue, but if prohibited, you will no longer know about them.

2. *Management overreaction.* Closely related to the problem given above, this is a stifling of informal communications due to an error caused by informal communications. It should be recognized that the informal organization will cause behavioral modifications that will appear to indicate a lack of control. We have made quite clear that the worst approach to remedy this is any attempt to disable the entire informal communication system. Of course, this would not succeed, but the consequences of the attempt would be devastating.

3. *Alienation of the informal organization.* When the proper working relationships are not maintained with the informal organizations, their communications can become counterproductive to organizational goals.

Informal communications are essential to all organizations. Methods were given in Chapter 12 for fostering, maintaining, and interacting with them. It bears reiteration that the informal organization is generally a much more effective communication mechanism than is the formal management structure for exceptional situations.

23.2.5 Failure to Modularize

Traditionally, the number seven (plus or minus two) has been accepted as the number of *thought units* that the typical human being can assimilate simultaneously. Although this has been based on research, we question the exact number because of the undefined term *thought unit*. However, the principle that there are an optimal number of facts that should be presented to people simultaneously is an important one even if we do not know exactly what this optimal number is. In most cases when the receivers start to get edgy, look around, fall asleep, yawn, and so on, we have gone far enough.

This demonstrates the importance of "packaging" thoughts for communication. The endless bombardment of facts to most upper-management personnel will inevitably lead to the classical question: What's the bottom line? As usual, the audience, their interests, and the topic of communication itself strongly affects the definition of a thought unit and the number of these that can be communicated effectively.

This principle can be proven in written communications. Noisy diagrams of large systems tend to overwhelm and obscure. However, when they are converted to a hierarchy, an effective *hiding* of details takes place, and the diagrams become usable. The highest level provides a table of contents and a concept of the overall system structure. Details can be revealed successively as we go down the hierarchy. The higher levels enable abstraction of the details at the lower levels, and they can be discussed and manipulated without going into their particulars.

This same concept affects step 3 of our model of communication. Words should describe facts, and facts should be organized into presentable units that facilitate communication without leading to confusion. Overwhelming the receiver with more details than can be assimilated effectively communicates only one thing: that the sender does not know how to communicate.

23.3 General Conclusions

Whereas this chapter has summarized some of the major issues with regard to communication, this entire book, management in general, and technical management in particular are totally dependent on effective communications. It is assumed that the principles of technimanagement have furnished you with a more precise sense of reality and the mechanisms by which the true power of organizations can be unleashed. If this is the case, it goes without saying that your ability to bring this about in your organization will depend most on your ability to communicate.

This chapter concludes our presentation of basic principles of technimanagement. The final chapter is dedicated to the process of evolving from the traditional organization to one optimized with respect to these principles.

24

Performing the Transition

24.1 Essential Requirement

The earlier chapters have demonstrated that although the principles of technimanagement are not new, they do not seem to be applied to their optimal degree in most organizations. This might lead some to believe that they have been tried and found wanting. To the extent that they have been applied and allowed to evolve properly, this has not been the case. However, in many attempted implementations there has not been allowance for the time necessary to enable these principles to take root and grow on their own. This is what we are calling *the transition*. The transition is more than a way of getting there, it is an end within itself to adopting the philosophy of constant improvement forever. Any alternative to the transition treats technimanagement as a fad, and its effects will last only until the next trend appears on the horizon.

Technimanagement is more than a major modification in management philosophy; it is a complete reversal of the underlying principles upon which most current management practice is based. Evidence of this is the initial resistance to the principles: "They can never work here," "We tried that before," "This organization is different," "We have an ingrained culture that cannot be overcome," and so on. There is a strong component of truth to these statements. However, inherent in each is the silver-bullet approach, which as we have shown, typifies traditional management. This must be overcome if these major changes are to be made. So as a primary basis we reword the *principle of transition* introduced in Chapter 7:

> The principles of technimanagement cannot be implemented on a crash mandated basis in most existing organizations, due to the cultural momentum of current management practice and experience. The only way that it can be implemented in these organizations is by establishing mechanisms by which the implementation of these principles can evolve.

We say *most* recognizing that there might be a few exceptions. Certainly, one exception is the brand-new organization which breaks out with one of its major objectives being the implementation of technimanagement philosophy. Even this is not without its problems, however, since all new organizations are composed of members who tend to lean on their experience with their old organizations. Few of these have direct experience with this type of implementation. Another exception might be organizations that have had a successful history in applying Theory Y, openness, and distributed management. In these we expect the easiest reception to continuing a transition that has already begun.

The abandonment of pure Theory X in all technical organizations gives rise to its bankruptcy as a driving management philosophy. If the principles of technimanagement were not being applied to some extent in technical organizations they would fail to continue to be successful. Some organizations practiced them much more successfully in their infancy, but as the organization grew, so did the bureaucracy, the division of labor, and all of the other characteristics symptomatic of regression toward traditional management practices. To the extent that an organization already enjoys momentum in the direction of a matured implementation of technimanagement principles, they have a headstart and might be able to skip over some of the initial transitional steps.

The principle of transition is quite emphatic for the vast majority of organizations. We predict that most current attempts to implement TQM (or other "quality movement" initiatives) which are clearly attempts to move in a direction consistent with technimanagement principles will fail because the transition principle is not recognized. Although we say this based on observation, experience is not needed to validate the need for transition. In Figure 24.1 we present three compelling reasons to support this premise. Thus, whereas the transition leads to a solution of many problems, the transition is the solution to the problem of implementation. To stay focused: The transition is an end within itself.

Since it is counterproductive to try to implement technimanagement with Theory X edicts, the organization is forced to fertilize and grow those technimanagement principles that are currently being practiced. Depending on the current state of the organization, the evolution might take from five to ten years to mature to the point where the vast majority of organizational members understand and are totally supportive of it. This does not mean that some of its benefits cannot be obtained immediately. To the extent that the transition progresses, its benefits will be realized by the organization.

- *Universal ramification.* The transition requires a major change in behavior on the part of everyone in the organization. If it were just a management behavioral change, it would be considerably simpler, since the dimensionality of the problem would be greatly reduced. The goal is to liberate the attitudes and energies of the lowest levels of the organization so that they are motivated toward serving their defined customers. This reorientation cannot be attained overnight. (In Section 24.3 we address this problem further and present a basis for evolving an entirely new customer orientation within the organization.)

- *Inability to mandate.* To mandate the principles of technimanagement top down is an oxymoron (give it some thought). Those who cannot internalize this are wasting their time starting the transition. They are viewing the transition process with their old Theory X paradigms. *You cannot implement technimanagement with Theory X edicts.* Customer orientation must be true dedication to the customer alone, not secondary to the manager as the primary customer. We might mandate changes in behavior in a very short term, but surface behavioral changes are not the goal. This brings home the true cultural change that cannot occur in just a few weeks or months.

- *The process will never be completed.* At any point along the transition there will be an optimal mix of traditional practices that are continuing side by side with the conversion to technimanagement principles. The most important thing is the direction, and the second is the speed with which the evolution is progressing. Obviously, an implementation that moves too slowly can itself become counterproductive. With each step in the right direction, more and more of the productivity gains will be realized, and once a "critical mass" is achieved, the entire organization will get behind and accelerate the evolution. However, the optimal mix today will not be the optimal mix tomorrow. As attitudes change, so does the outside world to which we must adapt. The transition that is set in motion cannot be viewed as temporary. "Continuous improvement forever" requires that it be designed to be self-adapting and to last forever.

Figure 24.1 Essential Nature of the Transition

In Section 24.2 we further define the transition in terms of the reorientation that must take place toward the customer throughout the organization. After this, we propose alternative approaches to the transition, depending on your level of management: top down, bottom up, or middle out.

(A major qualifier is in order: No rigid set of steps will work in all organizations. The ones given below are illustrative of models that should be modified to meet the maturity and local culture of the target organization. If technimanagement were as simple as following a straight-line procedure, we would be able to accomplish it with computers rather than human managers.)

24.2 Evolution of a Customer Reorientation

Although much lip-service has been given to "turning the organization on its head," the resistance that we anticipate to the transition will be ample evidence that this has not occurred in most organizations (at least not to the extent required). To illustrate, ask the lowest-level subordinates in an organization who their customers are, and some interesting results will follow. Generally, they will tell you who the organization's customers are. When pressed (e.g., but are these your customers?), they will ultimately admit reality: Their primary customer is the person to whom they report. For example, if a showdown comes between a customer on the phone and a request from the boss, guess who gets priority? Subordinates shake their heads and know better; but they do what the boss says, filing their thoughts for "I told you so" time. This is customer orientation, but to the wrong customer. The major problem is that managers do not perceive this to be a problem, since they are generally served quite well. In fact, loyal employees who put their managers first are very highly valued, and managers' power is often mistakenly assessed by how well they can control their subordinates as opposed to how well the organization's customers are served.

In the traditional organization there is a de facto customer orientation right up the organization. That is, people recognize that primarily the determination of their success is the assessment given by the managers to whom they report. Thus their primary allegiance is to the chain of command. This would work beautifully if two things would occur with reasonable effectiveness: (1) the CEO of the organization has perfect knowledge of customer needs, and (2) the CEO is able to communicate these effectively all the way down through the organization. In reality, neither of these is satisfied. Although the CEO might have considerable intelligence regarding the broad movements of the markets, there is no way that one person can possibly be cognizant of the particular demands being made by all customers.

This is fully recognized, and edicts are passed down in an attempt to address such concerns, not the least of which is the slogan to "put the customer first." But, in fact, the customer comes first only if it is perceived to be in the best interests of the lowest-level subordinate—which does happen at times. However, as likely as not, this is not the perception. Ask yourself the next time that you need to get service but the employees are busy with some other trivial tasks (such as talking among themselves or on the telephone): Why should this be? It is quite simple: They believe that (1) the organization will survive their inconsiderate actions, and (2) so will their jobs. The first is due to the perceived momentum, especially of large organizations. The second is because they continue to please their primary customer: their managers (who at this moment are out of sight).

This brings us to the second necessary prerequisite for top-down organizations to thrive: The communication channels must function effectively to communicate customer needs from the top down. Even if the CEO knows the customers needs (in general, let's say), the two previous chapters should convince us clearly that these will not be conveyed effectively down to the lowest-level subordinates. This serves to explain many of the problems in customer dealings with traditional organizations.

The customer reorientation to which we refer is commonly called "turning the organization on it's head." Theoretically, the model includes the aspects given in Figure 24.2. Thus, while we have turned the organization on its head, the process is done by an evolution that results in the comparable relief to "lighten" the organization so that the head is not crushed under the weight of the demands of an army of tyrant subordinates.

As the reorientation evolves, so do the benefits. The first benefit to the organization comes from the very first step, as problems begin to be approached from a completely new direction. That is, the lowest-level subordinates begin to become totally involved, as opposed to the isolated manager defining and prioritizing customers and their problems (a process that is largely ignored in the firefighting of day-to-day management). Since these are the people who best know and deal with customer needs, they are in the best positions to provide the data required to meet those needs. In addition, when they create the solutions, they are far more motivated to implement them.

The recognition of the lower-level (subordinate) responsibility in this process is essential. As long as they do not attain their own customer orientation, the next-higher level must define and prioritize customers and problems and then translate this into subordinate activities that will address these needs. This might not seem to be too problematic; indeed, until the

> - *Customer prioritization and problem definition.* Lowest-level subordinates effectively define and prioritize their customers in collaboration with their management counterparts, to assure consistency of purpose. This is followed by an ongoing series of work sessions to determine customer problems and needs and to resolve approaches toward better meeting customer demands. (See Section 15.6 for the basics; this is discussed further below.)
>
> - *Subordinates become customers.* As this customer orientation evolves, and to the degree that it exists, first-level managers no longer need to define customer needs (e.g., define work activities) for their subordinates. To the extent that this is implemented, they consider their subordinates as their customers, since now the facilitation of subordinate activity is the most effective way to increase service to the customer.
>
> - *Replication "up the organization."* At each successively higher level of management this process is repeated. However, it generally results in a flattening of the organization, since some of the layers had a primary function of assuring customer orientation at the lower levels. Since this is being handled directly at the lower levels, these as well as many staff resources can be moved in that direction. Hence, as the transition evolves, many middle managers might find themselves in positions of advisors and facilitators to groups at lower levels. While they no longer have direct organizational authority, their recognized expertise and experience produces de facto authority and influence within these groups (project groups, work teams, and task forces).

Figure 24.2 Turning the Organization on Its Head

organization matures, good managers will serve in this capacity as they have in the past. We contend, however, that even at the lowest level, this is an inefficient process that requires a super first-line manager for its accomplishment. This person might get promoted quickly or otherwise be "eaten" by the organization (recall the Peter principle). When there is a breakdown at this level, the next level must perform the process—a process that itself which becomes institutionalized in rules and standard procedures. Thus the problem snowballs (or "dominoes") to the highest levels of management, where each person must look beyond those at the next-lower level and define their customers for them. However, each higher rung of the organization is further removed from direct knowledge of customer needs and wishes. Thus not only are the input data more apt to be flawed, but so is the communication system through which they must travel.

Resistance to reorientation by both managers and their respective subordinates should be expected and is further cause for evolution rather than imposition. For example, subordinates will complain that they are being given tasks that have traditionally been performed by managers in an attempt to make the manager's life easier ("that's not my job"). As another example, managers will complain that the subordinates are "taking over." If these types of attitudes are to be overcome, they must be shown to be counterproductive to the people who hold them—from the perspective of organizational efficiency, not just the preferences of some key managers. The most effective way to counter this resistance is a combination of full dedication and education. These elements will be integrated into the transition approaches that follow.

24.3 The Transition

Section 24.2 helps by giving us a target model, but it is still theory. The question remains: How do you take an organization in its current form and initiate a process of evolution that will transform the organization to sufficient maturity that the process essentially continues on its own impetus? The answer depends on two things: (1) its current state and (2) the management level from which the evolution is to emanate. Generally, in the three subsections that follow we assume the worst-case scenario (i.e., that the organizational culture is very traditionally oriented and therefore that no transition steps have yet been taken). As far as the management level is concerned, we cover the ground by considering the very highest, the very lowest, and middle management, respectively.

We can further subdivide alterations of the basic transition approach according to the maturity of the organization. Clearly, modifications in the overall approach will also be necessary depending on the size, geographic distribution, and general organization type. For example, academic institutions that have enjoyed some degree of distributed management would require a significantly different transition mechanism than that of militarily oriented aerospace organizations. We present a mainline approach targeted toward the three alternative organizational levels and encourage adaptions to be made from that point.

It will become clear that our belief is that the optimal approach is the middle-out transition. We say "our belief" to indicate that studies and sufficient experience are not available (and possibly never will be) to determine which approach is superior. This is largely irrelevant, since it is not a controllable factor. If you are the technimanagement champion, you must move forward regardless of your level. We preface the following sections with this statement so that our presentation approach can be better understood. Both the top-down and bottom-up approaches are designed to be preliminary evolutions to a point where ultimately a middle-out evolution will take place. That is, regardless of your position, the general approach will be to direct your influence over the organization to the point where those who are in the middle-management positions will ultimately run with the ball.

24.3.1 Top-Down

The top-down approach is usually favored in the literature where the assumption is made that the target audience is the CEO and that the CEO has total and ultimate control over the organization. The first of these assumptions rarely applies; the second never applies. While the CEO will have to get totally involved in the implementation before the process is completed, the total change required in every member of the organization makes the CEO only one, and possibly not the most critical, of the many targets. Clearly, we do not share Deming's view, since it is documented that he would not even agree to visit a company without first being convinced that the CEO was totally dedicated to TQM. Even with total CEO dedication, the transition may not succeed. For the CEO's powers are limited to those generated by the success of their subordinates, which depends totally on the subordinates' willingness to change their orientation.

This does not mean that CEOs acting alone cannot effectively implement technimanagement in their companies. They are probably in the best possible power position to initiate the transition. However, if not handled properly, this can cause far more problems than it apparently solves. It should be clear from the evidence presented above that strong leadership, as opposed to strong dictatorship, is totally essential to this effort.

If you are a CEO, the following stepwise procedure is recommended to establish the mechanisms to evolve technimanagement within your organization:

1. Select a key set of middle managers to form what we will call the quality management council (QMC). [Name it to suit your needs, but by all means give it a distinctive name (and do not call it a committee).] The membership of this steering group is critical since it will be the focal point of the transition. This should not be viewed as an experiment—it is the initiation of an evolutionary process that will completely transform the management structure of the organization forever. Characteris-

tics that are essential to membership include:

    a. A cooperative personality, and collectively, the ability to work together.

    b. A systems view and broad experience with the ability to present a wide range of interests from varying viewpoints.

    c. Vertical and horizontal integration; ideally all levels of management and all functions within the organization should be represented. (Some members should be able to represent several areas to reduce the size of the council.)

These are the people who will form the nucleus of the evolution of the organization's management philosophy. They should be recognized leaders—able to follow your lead in influencing their peers and subordinates alike in accomplishing the transition. One major by-product of the council are the relationships that will be developed between its members. *They should have fun working together.*

The size of the QMC is critical but highly dependent on the organization. An optimal balance must be sought between the chaos of an extremely large work team and the need to get a broad representation of levels and views. It will be recognized at the outset that the QMC will function both as a whole and as subgroups are formed and charged by the council as a whole. This argues strongly for a larger group, but only under the rarest of circumstances should everyone within a given level be included (e.g., all department heads or all district managers).

2. Train the QMC in the principles of technimanagement. This should not be rushed. Since you have selected your best people for the QMC, their time is valuable and it is important that the transition not disrupt current activities. Thus this training might initially require two to three hours of intensive lecture, discussion, and exercises per week over several weeks to accomplish. This might be reduced to one hour per week where training can proceed concurrently with some of the early organizational activities, which might include the *pilot* establishment of a local functional councils in one (or more) division(s) of the organization. The QMC will become the means by which technimanagement philosophy is spread to the remainder of the organization. Thus it is essential that everyone thoroughly understand the principles before taking the next step. Reservations about the applicability to their particular organizational components should be discussed thoroughly and resolved by the entire group.

3. Formalize a contract with the QMC for the transition. This will effectively define their charge. In essence, you will agree to share decision-making responsibilities with the QMC to the extent that they do likewise *within their respective organizational components*. This establishes a control system for evolving distributed management in the subset of the organization defined by the QMC. Part of the functioning of the QMC at this point will be a development of their charge based on the principles they have learned. This charge should include the following:

   a. The establishment of pilot work teams and task forces to begin to address system-wide continuous improvement and special problems, respectively.
   b. The generation of a transition plan to extend the pilots to the rest of the organization.
   c. The investigation and resolution of problems within the transition through the use of subgroups of the QMC, which might be extended to include members outside the QMC.

4. Execute the QMC plans for extending the transition throughout the organization. This should include actions to extend the pilot to the entire organization, beginning with training and extending to the creation of work teams and task forces. *In this regard, the procedures applied at the lowest levels for establishing a bottom-up implementation given in Section 24.3.2 will be useful.* It should also include a systematic method for rotating QMC membership.

5. Measure the degree to which technimanagement is being implemented in the pilot subset of the organization. This can be determined by obtaining information on the functioning of coordinating councils that are being established at the lower levels and the work teams and task forces that are being established by these councils. Do not attempt to rush the process. The rate of progress will be a function of the educational efforts, employee acceptance, and the size and diversity of the effort. The process might take months or years to show significant change. The important aspect of measurement is to assure that a positive direction has been set and that progress in that direction is being made. When a given component is deficient in this regard, inquiry should be made to assure that corrective action is forthcoming.

6. Once the QMC is established to this point, it can proceed with a middle-out transition, as discussed in Section 24.3.3. It is expected that the QMC members will become champions of technimanagement and use their

departments to lead their sister components. This will greatly accelerate the middle-out process described below, which (for purposes of exposition) is assumed to start with just a single middle manager. Also, since the highest level of management is already convinced of the validity of technimanagement, the process of proving its value is greatly simplified.

The QMC will continue to meet on a weekly basis to discuss problems encountered by the transition.

As a final comment, note that we have used the principles of technimanagement to transition to technimanagement, since any departure from these principles will be detrimental. Consistent with this, mandates were used as little as possible. They might be required to create and initiate the QMC and to assure that there is adherence to the training regimen. However, these are not seen to require coercion. Once they are accomplished, the QMC essentially becomes a self-directed work team. You as a CEO are no longer solving day-to-day decisions; you are now managing the evolution of a process by which problem-solving mechanisms are generated by the organization itself. (Scary? Be courageous and have confidence in your people. If they cannot accomplish this, chances are they will not be able to accomplish any other meaningful organizational goals.)

24.3.2 Bottom-Up

At the extreme opposite end of the spectrum (from top down) is the bottom-up approach in which the lowest-level managers attempt to implement technimanagement with neither the support nor the overt resistance of their immediate supervisors. In the case that middle management is fully supportive of and understands the transition process, the lower-level manager can move immediately to being supportive of the middle-out approach described in the next section. On the other hand, if there is resistance to change from the traditional approaches, further evaluation is required.

In order for you to begin the evolution to technimanagement, it is essential that you initiate the first step, which is to *fully implement all the principles of technimanagement in your own organizational component.* You might ask how this can be done without the support of your immediate supervisor. If this is the case, you might consider the issues presented in the display on page 356. The best approach might be: Simply don't ask. Evolve technimanagement in your organizational component in the name of accepted good management practice. We do not pretend that the transition will not be noticed, but by the time it is noticed, the effects of higher productivity, greater cooperation, and better morale will be obvious. The result will be an increase in your personal and organizational component's power which will be too formidable for any middle manager to wish to oppose. In fact, anticipate the greatest resistance from your peer managers who might be suspicious of your success. This too should be dealt with proactively, which we will discuss shortly.

What If Your Boss Does not Agree?

OPTIONS

- Accept the current structure.
- Continue attempts to influence your immediate supervisor to allow you to be a prototype.
- Go ahead and do it anyway and let the chips fall where they may.

  Although we certainly do not want to jeopardize anyone's job security or career, the first option is a no-win situation for both yourself and the organization. The third option is certainly the riskiest personally, but it also has the greatest potential for advancement. The middle option might sound like the conservative approach, but it places your fate totally in the hands of another person from whom you have already received negative signals. We emphasize: Weigh the trade-offs; the decision is yours—your fate is totally in your hands.

- *Forgiveness is easier to attain than permission.* If you believe that your supervisor will not take the time to understand and will not bless your efforts even if understanding is forthcoming, avoid the direct request. Surely in an advancing technical organization like yours, innovations are expected. Surely they want you to apply the most advanced management techniques. Surely they want you to do everything in your power to increase your organizational component's efficiency. That is what you are getting paid for, so there should be no objection to you doing just that.

- *Maintain total openness.* You have nothing to hide—total openness to subordinates, peer managers, and upper management is essential. If there is a problem with your management approach, the burden of proof is on them. However, there is a major difference between openness and advertising.

What If Your Boss Does not Agree? (Continued)

- *Communicate effectively.* To attempt to communicate what you are doing to those who do not understand the concepts of technimanagement will result in no communication at all. It can create more problems than it solves. In the final analysis, it will be the performance of your component that will tell the story, not talking about it, especially before the fact. Communication in the upward direction might be counterproductive at this point, but effective communication to your subordinates is essential.

The procedure presented below assumes that there is not a threat to your position in initiating a technimanagement transition within your organizational component. Typically, the organization has tried several management gimmicks in the past, and there will be a tendency for your subordinates to suspect that there is really nothing new here. In this regard, your actions will speak much louder than words. In all cases, total frankness, admission of past wrongs, the need to change on your part, and openness are required. There is no reason that they cannot know all of the theory and purposes of technimanagement. Will you expect greater productivity?—absolutely! Are you trying to manipulate them into it?— absolutely not! Application of the principles of technimanagement empowers everyone in the organization. (Give them the book!) Since they are open for everyone to understand and apply, this is not something that you are making up (or bringing in) just to get them to work harder. On the contrary, productivity and job satisfaction cannot exist separate from each other.

Given that you (as a first-line manager) commit to implement technimanagement in your component, the following is the suggested model procedure for a bottom-up transition which should be modified to meet local needs:

1. Hold a general extended meeting with your subordinates in which you explain the basic principles of technimanagement and explain the steps (suggested below) that you propose for the transition of your component. Explain the evolutionary process and the fact that they, as your ultimate customers, will be involved in determining the rate of this transition by their assumption of a customer orientation. Assure them that to the extent that they assume a customer orientation, they will become your customers. Emphasize the cooperative nature of resolving the mechanisms and groups for optimal decision making. If this is being done in an "hostile" environment (i.e., as an isolated component), impress upon them your

confidence in them and the organizational risks that this entails.

2. Initiate an ongoing training program (minimum: one hour per week) on the principles of technimanagement, involving all your subordinates. This training will emphasize the techniques, empowerment and responsibilities of distributed management, including heavy emphasis on the ramification of the customer orientation. Since some will be assuming leadership of (and serving on) groups for the first time, it should include sessions on teaming and effective meeting conduct. Some of the technical aspects (e.g., statistical quality control) should be covered to the extent and at the time that it is required to provide for better interaction between the groups and their facilitators. Many of the steps that follow can be integrated as exercises into the training program.

   Comment: Steps 3 to 7 present a problem-driven approach in which the members of the component themselves will establish work teams and task forces to address customer needs. It is important that this goal be kept in mind constantly by all participants as they work together so that these steps do not become bogged down and turn into ends within themselves. We say this from the point of view of the elimination of controversies that can arise over a variety of issues. In all of these discussions, considerable time can be saved and controversy avoided by keeping the ultimate goal in mind (e.g., "How is the resolution of this point going to affect our establishing work teams and task forces?")

3. Have a brainstorming session in which your subordinates define their customers, both internal and external to the organization. (In large, diverse organizational components, this might be performed in several separate sessions.) The training sessions should prepare the group(s) by giving them the definition of what a "customer" is (both internal and external) and by having them prepare for the brainstorming session by giving thought to identifying their customers. Output: a complete list of customers.

   (If you are on the list, go back to Chapter 1. On the other hand, other organizational components might need to be included even though they would not be considered as customer components in a mature organization. Because of the bottom-up transition, you might need to view your component as a separate organization in an otherwise hostile environment. Thus all components within the organization which receive information or services from your component must be considered as customers at this point.)

4. Have the group prioritize their customers using a group consensus tech-

nique (Chapter 14). Segment the customers into priority categories, such as most critical, vital few, and important many. There might not be significant differences among customers within a given category, but there will always be significant differences in those in different categories. Output: a prioritized list of customers.

5. Have a brainstorming session that takes the most critical customers and defines their needs and your component's problems in meeting these needs. (Terminology: In many cases *desires* and *problems* are better words than *needs*. We will use the word *needs* to include all of these.) Individual and subgroup assignments might be made at this point to contact customers and survey their needs. Output: a complete list of customer needs.

6. Have the group prioritize these needs within each customer (or customer grouping where several individual customers have identical needs) using a similar group consensus technique as was used in step 4. Output: a prioritized list of customer needs/problems.

7. Transform the list of prioritized needs into a prioritized *action-requirements list* by determining those customer needs and problems that can best be addressed by immediate team action. Whereas steps 3 to 6 are largely routine brainstorming and prioritizing, this transformation step is much more complex. The action requirements will form the basis for the charge of the task forces and work teams that will be established. The reason for this step is that an additional factor must be considered before creating and making assignments to work teams and task forces: the potential that these groups will have to address the problems as defined. The highest priority for study, then, might not be the highest-priority need identified, although we would expect a strong correlation between the two. Output: a prioritized action-requirements list.

Comment. Many suggestions will be made during the course of brainstorming and prioritization which are *so obvious that they can be implemented immediately.* In these clear cases, waiting for the actions of task forces or work teams is a regression to bureaucratic methods. Immediate actions generated from the open feedback mechanisms that naturally result from initiating this process are a major benefit of this process. Not only that but they serve to convince your subordinates further of your commitment to open and distributed management. However, these factors should be seen only as by-products, not as attainment of the primary goal. As stated in the comment before step 2, the goal of these steps is the establishment of the new mechanisms to address continuous improvement in meeting customer needs.

8. Establish, by consensus, work teams and task forces to address the issues at the top of the action-requirements list. Depending on the maturity of the members and the size of the organizational component, this might be accomplished by first establishing a task force for the establishment of other task forces and work teams. Alternatively, the entire component would be involved. Group members should not be appointed by the manager, although as part of the total team, the manager should influence their selection. The manager and the rest of the members of the organizational component should work together to assure that each group's membership is balanced and effective in working together. Needs identified solely within a project group will generally be addressed by a subset of that project group (possibly augmented by management or staff facilitators). Output: a work team and task force definition: names, membership, and charge.

There are a number of issues that affect the approaches applied in carrying out this step, including number, size, content, facilitation, and charge. These are addressed in the below display.

Issues Affecting Group Selection and Operation

- *Number of groups.* An optimal balance must be reached to address the most critical issues while not overburdening your subordinates with activities that detract from their central production activities. Those problems with the highest potential for group solution (or improvement) should be approached first to build momentum and convince your subordinates that you are serious about the transition. Under no circumstances should the number become so excessive that it hinders the efforts of the project groups. However, expending one hour per week to develop better and more efficient methods will return far more in terms of increased productivity.

- *Size.* As with project groups, the size of any given work team or task force should seek for an optimal balance between the factors that argue for enlargement (the need for information, involvement, and representation) and factors that argue for smaller team size (communication efficiency and cost of lost alternative activities). See Chapter 14. (Recognize that the optimal size of a team is sometimes 1.)

Issues Affecting Group Selection and Operation (Continued)

- *Membership.* Those group members from within the organizational component should serve on a voluntary basis. Usually, there will be a champion of a particular cause who will willingly accept the leadership of a given effort. The successful participation of customers, both internal and external, should be sought, and most will be anxious to participate if it means that your component will be more responsive to their needs. Where appropriate, team membership might be extended to upper-management staff personnel.

- *Facilitation.* Some groups require the aid of expertise that is not possessed by team members or customers. Requests of this nature for upper-level staff assistance are not inconsistent with traditional management practice. However, communication to the particular staff person assigned will be necessary to explain the approach being applied. For example, if you appeal for help in the area of statistical and quality control expertise, the staff person(s) assigned should recognize that they are not going independently to create a control system that will be implemented by management at your level. Rather, the development of the system itself will be done by their participation on the work team, and ultimately this work team will implement it in conjunction with all involved members of the organization.

- *Charge.* It is important that the groups be self-directed to the extent that maturity allows. In a sense, the issues on the action-requirements list and the name of the group will begin to define its charge. As manager you should refrain as much as possible from setting the charge—this can kill the morale of a group before it even gets started. Empower the new group to draft their charge, subject to review by the entire group.

Inherent in all these considerations is the necessity that the transition be open to peer managers as well as upper management. This enables the value of these new approaches to be communicated outside the organizational component in a formal way. (They will get communicated informally in any event.) The subtle changes in the fragile relationships that exist with higher-level staff personnel form one of the most critical aspects of communicating the benefits of technimanagement. Success in implementing improvements with their facilitation can open a major conduit for promoting this approach to upper management. On the other hand, alienation of these staff personnel can spell disaster (e.g., "You ought to see what they are trying to do down in the XYZ department . . . ").

9. *Reiterate, reiterate, reiterate.* This is to emphasize that the process of transition is never complete, even within an organizational component. The particular steps that need reiteration depend on the time elapsed since they were last performed. Step 1, extended meetings with your subordinates, should be repeated at least annually to step back and assess the overall progress of the transition and to self-generate overall redirection. Steps 3 to 6 should be reiterated whenever the underlying assumptions of the previous brainstorming and prioritizing sessions no longer hold. This, in turn, will lead to updates in the action-requirements list. In any event, once task forces satisfy their charges, others can be established that address the next needs as given on the action-requirements list, which is a reiteration of step 8. At the same time, the work teams should go through constant self-evaluation. This process should eventually evolve an optimal level of studies and formal cooperative efforts while stimulating the informal organizations to support organizational goals to a greater degree.

Comment: Although it is not your job to dominate these groups in any way, it is your job to see that they do not get bogged down in meaningless red tape and evolve into a bureaucracy. Organizational entropy will assure this if the meta-management process is neglected, and this is a real possibility within any structure. The countermeasure is diligence on your part to see that the transition is moving ahead as measured by the productivity of all groups. Those that are not productive should be evaluated within and corrective measures taken with your facilitation.

10. *Represent technimanagement to the rest of the organization.* Although you can implement and gain the productivity advantages of technimanagement in your organizational component, the full benefits to the component and to the organization cannot be realized until it is

extended organization-wide. This should focus on leading eventually to the middle-out transformation described in the next section. We discuss the promotion of technimanagement further below. However, the natural extension is for your manager to perceive its value and provide leadership in implementing the transition in your sister components.

This stepwise approach to the bottom-up transformation must be viewed as a mere guide. The tremendous danger in formulating such procedural guides is that this will be interpreted as being technimanagement, as opposed to the true intent, which is only to provide guidance with the initiation of the transition. *No single stepwise procedure can capture the essence of the principles of technimanagement that are distributed throughout this book.* This is the reason that the very first step emphasized the need to understand and begin to communicate this philosophy, and the second involved intensive training in this regard.

An example of this is the recognition from the principles covered in Chapter 2 that the stepwise procedure given above is a meta-management reestablishment of the fundamental control systems that determine the organizational component's behavior. The primary symptom of successful implementation is that you, as the manager, are *not nearly as concerned with the specific decisions which are made as you are that they are being made by the best possible group.* You are not as concerned with "running" your operation as much as you are in seeing that it runs itself properly. You do not blame deficiencies on people but on the failure to establish mechanisms to overcome your shortcomings. And this—to establish effective management mechanisms (e.g., project groups, task forces, work teams, and most important, informal relationships)—is now your primary responsibility. But it is a responsibility that you accomplish by means of group activity. Essential to the success of this process is your internal realization of your recognition that such decisions are far superior to those that you alone could generate. (If you do not believe this, please do not call what you are doing technimanagement.)

Despite being multiply redundant, we repeat for emphasis: This does not come overnight; it requires growth. This is the reason that we call the process a never-ending transition and liken the organization to an alcoholic in recovery. As your subordinates pick up and begin to recognize that you are really serious about this distributed management stuff (or whatever you choose to call it), they will begin to play the game under a completely different set of rules. It is crucial that they do not begin using this to manipulate you. *Collective maturity* is essential. As they mature, you will mature to a point where you will be able to empower them more and more because their effectiveness is empowering you. Thus the decision-making mechanisms grow to an optimal level, and they constantly move the organizational component toward greater quality and productivity.

Step 10, the promotion of technimanagement throughout your organization, is not a separate sequential step. It is required at the outset, and this requires considerable further elaboration. The transition within one organizational component is a start, but it alone cannot save an organization. In fact, if the transition does not grow to other organizational components, its chance of survival at all is quite small. The reason for this is one mentioned above: The greatest resistance is anticipated from your peer managers. This will occur for several reasons. First, your productivity ratings will go through the roof as your people become more dedicated to the customer. This will receive particular favor from your internal customers, and you can expect this to be communicated throughout the organization. Looking for an easy excuse for this, the peculiarities of your opportunities, or the competency of your people might be cited as reasons for this. *Do not disagree*! Above all, do not try to claim credit for what is happening. If you are doing what you should, you will not have to.

A second problem might arise in reaction to the major morale increase that occurs in your operation. This will not remain a secret. As your people talk with their colleagues in other departments, complaints might be met with what will seem to be very strange responses. For example, a complaint about a work assignment in another department might receive a reaction that in "our department we worked out our own assignments." This will get back to your peer management group, and when it does you can expect criticism for your management style.

Anticipation of and proactive approaches toward these attitudes is critical to the growth of technimanagement throughout the organization. It will take a very strong second-level manager to counter the flak if you are outnumbered by 10 to 1. However, if the proper actions on your part are initiated at step 1, these problems can be controlled. The adjacent display presents approaches to mitigate these problems (nothing short of full transition can eliminate them totally). In summary, positive efforts must be made to keep from alienating those within the organization who are not moving toward transition as quickly as you are.

Proactive Approaches Toward Peer Resistance

- *Communicate.* Although not announcing what you are doing, do not try to hide it from your peers. Generally, their support and alliances are essential to your component's success. Share with them technimanagement concepts, and provide written information on it to anyone who is interested. The chances of their taking this very seriously at this point is relatively low, since most of them will view it as a mere gimmick.

- *Demonstrate intentions.* Convince your peers that you are moving your component of the organization in this direction because it is sound management practice (cite our sources). This should also establish (implicitly) that the objective is not to get a competitive advantage over them or to create bad morale in their components. It is to empower them, especially if they are customer components—this is not a zero-sum game.

- *Be positive.* Do not criticize any manager or component for not joining in the transition. Your assumption is that their components will also be transitioned at the appropriate time. Technimanagement requires that each organizational component be transitioned at its own rate and time. Remember, problems that might arise between organizational components are symptomatic of the lack of proper mechanisms for resolving these problems rather than being the fault of your peer managers.

- *Team up.* If you share enthusiasm for technimanagement with one or more of your peer managers, initiate your transitions simultaneously and compare notes as the transitions progress. Communications with your common manager on the subject of management philosophy (if they take place at all) should concentrate on the positive aspects of the transition in terms of the established broader principles involved rather than anything new.

Since in this section we are assuming that the transition is being made from the bottom up (and possibly by only one component), we must suspect that there is not a large amount of support for change in this direction from your supervisor. This presents a timing problem—it is advantageous that the benefits of technimanagement become apparent to your manager before major organizational resistance surfaces. The endorsement of your own manager depends on the clear demonstration of benefits to him or her—greater cooperation, higher productivity, better morale. Mere words or explanations may not suffice because of the counterintuitive nature of many of the principles, some of which are believed to have been "tried" before. If your manager objects to the transition itself, appeal for time to serve as a pilot.

Never get defensive. Technimanagement must prove itself, and if you are introducing it to your organization, you should be willing to accept a fair test of its capabilities. Openly state that if it cannot be shown to produce superior results, it should be abandoned and you will gladly lead the way. This will diffuse any direct attacks. After all, they can hardly prevent you from attempting to improve your operation. Significant benefits should be realized from the outset as far as increased morale and cooperative effort are concerned. Bottom-line benefits will accrue within one year. However, the full benefits will not be realized until the entire organization matures.

The bottom-up transition cannot occur in isolation from the rest of the organization. It must be accompanied by a promotional effort—to your subordinates, peer managers, upper-level staff, your immediate supervisor, to middle management in general, and ultimately to upper management. We have discussed above the first four of these, to which you have direct access. Yet if the entire organization is to be transitioned, middle management will have to be included, since the optimal transition is one that proceeds middle-out.

There are two routes to middle management, both of which require that your operation be recognized for its superior performance. The most effective is through your immediate supervisor, who will hopefully not only communicate this to his or her peers but also encourage your peer managers to initiate the transition in their organizational components. However, a lack of initiative might make an indirect route to middle management necessary. Patience is of considerable value, and we strongly recommend that you not get overanxious. When your organizational component matures to the point where it is realizing half of the benefits of technimanagement, the organization will take notice. You can expect to be contacted by upper management seeking to determine what is occurring and how this might be extended. Be sure to give your immediate supervisor credit at this point for allowing (encouraging?) you to continue with the prototype.

Being the (possibly sole) champion of technimanagement within your organization may involve considerable risk, but once it has proven successful you might be in very high demand. Keeping a firm hold on reality is essential to the extension of the transition to the rest of the organization. All must recognize that the transition within each organizational component must proceed at its own self-determined pace, depending on the current state of that component's current management mechanisms. Above all, middle and upper management must be made aware of the dangers of trying to impose top-down that which was so successful in your component mainly because *it was allowed to evolve from within.* Instead, they should be referred to the top-down approach of Section 24.3.1 as well as the middle-out approach discussed in the next section.

24.3.3 Middle-Out

The primary reason that we believe that middle-out is the optimal transition approach is that the top-down approach does not lend itself to lowest-level self-direction, and the bottom-up approach may take years to spread to the entire organization unless it is occurring simultaneously in several components. Thus both the top-down and bottom-up approaches have been formulated to lead to the middle-out approach, which is initiated by middle management. By *middle management* we mean all the management levels which also have managers reporting to them, with the exception of the CEO and those who report directly to the CEO (who are usually classified as upper management).

To further our understanding, it helps to elaborate the ideal model. This would occur where the majority of middle managers readily accept the principles of technimanagement and have convinced upper management to provide support by facilitating the transition process. This avoids the pitfalls of a top-down-mandated implementation. However, middle managers need to recognize that the same problems will apply at the levels immediately below them if they impose the transition. *Thus it is contingent on them to provide education in technimanagement principles to the first-level managers and allow them the latitude to evolve the new philosophy as best fits their particular operations.* The role of middle managers is to assure that all first-line managers are proceeding in the right direction and in a fairly consistent pace and manner. However, rules and edicts along these lines should be avoided.

This is the ideal scenario, which requires that enough middle managers are "aboard" to get it kicked off. As illustrated in the preceding two sections, this can occur by leadership from the top down, or it can proceed as the momentum of success builds from the bottom up. However, if the technimanagement champion is in middle management, it might emanate from within this level, and that is what will be assumed here. Regardless of how we got to this point, the procedure presented in this section will go the rest of the way toward getting the entire organization into the transition process.

At this point we assume that as a middle manager, you have made the decision to implement technimanagement in the components of the organization that are under your jurisdiction. The following steps provide a guide for initiating the transition:

1. Select a set of first-line managers and nonmanagerial personnel to form a quality steering task force (QSTF). This will be a pilot group, and their selection is critical, since the success of the transition depends totally on their cooperative efforts. The characteristics of these people is much the same as that give for the QMC in Section 24.3.1 (it will be noticed that they are analogous in most aspects even though the QSTF is at a lower level). To recap, these include a cooperative personality, a systems view, and broad experience. Vertical and horizontal integration should be sought even if this creates a fairly large group. Significant formal and informal leaders should be included. Above all, QSTF members must be committed to the principles of technimanagement to the point that they will attain a working familiarity with them and be able to represent them throughout your part of the organization.

2. Train the QSTF in the principles of technimanagement. The considerations for training given in Section 24.3.1 apply equally well here. Once training is completed the QSTF should develop its charge, which should include the following:

    a. The establishment of work teams and task forces to address continuous improvement and special problems, respectively, which transcend the purview of any first-line manager
    b. The generation of a transition plan to implement technimanagement throughout the organizational components within your scope
    c. The investigation and resolution of problems within the transition through the use of subgroups of the QSTF, which might be extended to include outside members

    The plan in item b should be formulated after study of the top-down, bottom-up, and middle-out procedures suggested in this chapter.

3. Train all first-line managers in the principles of technimanagement. This training will not be as intensive as that given the QSTF. Time commitments of first-line managers might prevent this. However, training must be adequate to give them an excellent working knowledge of it so that they can begin to develop plans for its implementation in their respective organizational components.

Sec. 24.3 The Transition

4. Encourage all first-line managers to initiate the transition to technimanagement within their respective organizational components patterned after the applicable steps of the procedure detailed in Section 24.3.2. Provide all of the support for this, and promote the QSTF, which can be used as a resource in the following ways: (a) members of the QSTF might be facilitators of lowest-level work teams and task forces, and (b) issues that arise in the transition might be handled by the QSTF rather than by their line managers.

   Comment. It is important to recognize throughout this process that great pains are being taken to provide an environment whereby technimanagement can grow from the bottom up without being (and thus perceived as being) imposed by upper management. The QSTF serves this function well, and first-line managers must recognize it as a mechanism for facilitating the transition rather than one for eroding their authority.

5. Execute the QSTF plan for the transition throughout your components of the organization, especially as it involves intercomponent coordination and cooperation. This should include a clear vision of the role that the QSTF will continue to play as the transition progresses, which will generally include a systematic method for rotating QSTF membership. The QSTF will continue to meet on a weekly basis to discuss problems encountered by the transition.

6. Measure the degree to which technimanagement is being implemented at the lowest levels of the organization. This can be determined by monitoring the variety of activities that are necessary to this process (Section 24.3.2). It should be expected that progress within the various components will proceed at different rates. Recall that the various rates of progress will be a function of the educational efforts, employee acceptance, and the size and diversity of the effort. The process might take several months to show significant change. However, it is essential that corrective action be taken from within those components that fail to demonstrate progress.

   Comment. This establishes a control system for evolving technimanagement in the subset of the organization over which you have formal authority. The most critical element of this control system is the correction element. In the spirit of power preservation discussed in Chapter 19, this should take the form of encouragement and leadership as opposed to coercion. A regression to Theory X top-down intimidation at this point can set back the effort by many months, if not years.

   As was true in all others, the key to this transition is that the principles of technimanagement are employed immediately, especially with regard to bringing the transition into existence. The stepwise procedure given above did not assume any interaction with peer managers or upper management. Needless to say, this transition cannot possibly go unnoticed by either of these constituencies. Nor would this be desirable. Although we do not envision complete understanding by upper levels, we do anticipate relatively passive cooperation [i.e., what is commonly called the "wait and see" attitude (you can't blame them)]. In most cases, only success at your level will convince them, and once that happens, they will probably want to impose it on the rest of the organization. This is where things again get critical, and they should be referred to the top-down procedures given in Section 24.3.1.

Resistance from your peer managers could well precede any clear demonstrations of increased productivity. This problem is much the same as discussed above in conjunction with the resistance of the first-line manager's peer group in the pure bottom-up approach (Section 24.3.2); thus the same proactive approaches for anticipating and dealing with these problems are recommended. Needless to say, the larger the number of middle managers who are replicating your transition approach simultaneously, the better.

Unfortunately, some of the worst resistance might come from your own subordinates, i.e., the first-line managers, who see this as an erosion of their authority and an attempt by you to micro-manage their operations. This is paradoxical, in that you are attempting to accomplish just the opposite. Unless you are dealing with totally unreasonable people, this attitude should be seen as a flaw on your part in the training and communication process. However, those who complained about everything just for the sake of complaining will not necessarily be cured by technimanagement. These we deem to be unreasonable, and we assert that they are in most cases motivated by the fear that technimanagement might just succeed. Hopefully they are not in the majority, and ultimately peer pressure will convince most of them to cooperate.

We advise against any radical action on your part, especially during the early stages of the transition, for the following reasons: (1) those who resist are probably quite useful to organizational goals or they would not be in their current positions (another case of "that which makes you good makes you bad"); (2) severe action will be (and thus will be perceived as) a Theory X reaction, justifying their very assertions; and (3) technimanagement will convince them on its own merits. This said, we add that people come in all shapes, sizes, and mindsets, and we anticipate that quite creative solutions will be required on your part to counter those who are counterproductive to a transition that has but one goal: self-improvement. We urge you, as the initiating middle manger, to give considerable thought and study to methods of motivation and conflict resolution that may apply here. In most cases the solution is within your grasp if you will take the time, get all the facts, get advice, and then take definitive action. Regression to coercion is a sure sign that you have not applied the discipline that this requires.

To conclude on the positive side, most first-line managers will welcome technimanagement, and many might be way out in front of you as far as the transition is concerned. If this is the case, use their components as pilots and allow the other managers to follow their lead, as opposed to having the principles dictated from above. This is the primary reason that the QSTF was established early—to enable the informal organization to "spread the word" independent of the formal organization. This is not manipulation, in that identically the same principles are being espoused by both the formal and informal organizations. Indeed, they can and should be clearly documented so that there can be no argument in this regard. In all three approaches to the transition, *this openness is the most critical ingredient*. There is nothing to hide—we are all on the same team, have the same goals, and should be able to agree on the best ways for their accomplishment, which is always open to better mechanisms for its own improvement.

24.4 Postscript on Training

The suggested model transition steps given above require a considerable amount of training. This training begins with the principles presented in this book. However, it is not limited to this. In particular, methodological training is required in the areas of problem identification, measurement, evaluation, and corrective action formulation. Specific methods will depend heavily on the nature and goals of the organization. More important, *teaming concepts* are required independent of the product of the organization. A few examples of these were introduced in Chapter 14. However, we tried to impress that they only scratched the surface of the methods and techniques that could be applied.

Entire books have been written on each of these topics. Most of them have some excellent techniques that can serve the technical manager well. We were tempted to include many more of these techniques than the very few that were presented. However, we saw a grave danger in doubling the size of this book for this purpose (which this would require). The primary danger is that the principles of technimanagement, which tend to be counterintuitive, would get lost among these techniques. These techniques are not technimanagement. Although they will greatly assist in enabling the transition to proceed, they cannot be substituted for the principles that will drive the transition.

When members of the organization understand the principles of technimanagement, they will be motivated to seek out and acquire the specific training which they need to move the organization forward. However, these techniques are dead within themselves, since alone they provide no inherent motivation for continuous improvement. It was our fear that technique would be substituted for principle that led us to place minimum emphasis on these techniques. For this reason, we hope the methodology of this chapter will not be confused for the essence of technimanagement, which could be implemented in any number of transition approaches.

A second reason for excluding techniques is that these are documented adequately elsewhere. Managers need to secure books, videos, short courses, and any other means by which their groups can become aware of the techniques they need. However, this should be done in recognition that no technique can make up for a lack of commitment to sound principles. The most effective techniques will vary considerably between industries and organizational personalities, but the underlying principles remain the same.

24.5 Where Will All This Take Us?

This brings us to the last section of a rather prolonged effort. Our goal has been to present time-proven principles that apply to the management of technical organizations. We have tried to present them in a logical order, beginning with the very fundamental concepts and ending with those that are the most difficult to accept on an intuitive basis. This chapter on transition approaches is the closest to art, and we urge that great liberties and flexibility be used in its application. If you started reading the last chapter first because you thought you already knew the principles of technimanagement, we urge you to consider the other chapters carefully before proceeding further, for without an understanding of the basic principles, the transition will become nothing more than the *management gimmick of the year*.

In this regard, we have made every attempt to use the principles of technimanagement to affect the transition toward technimanagement. It bears restatement that any departure from this approach is a contradiction in terms and the height of oxymoronic hypocrisy. If you, the manager, are truly convinced of the principles of technimanagement, your actions with respect to your subordinates will quickly reveal this reality. Feel free to modify these transition steps to your operation in accordance with the general principles revealed in the first 23 chapters. The results will come with persistence and patience, and ultimately, you will succeed. On the other hand, if you see technimanagement as another method for manipulation, your perception of the message of this book, and (we suppose) your perception of reality in general, is quite flawed. This will ultimately work against you.

The heading of this final section is a fair question. Yet it is one of uncharted ground. Many organizations have begun to make the transition, but obviously none have arrived, since the target destination is continual change for improvement. However, we can project the benefits which have come as a result within those organizations that have begun the transition.

As for the impact within any given organization, we can see a time when the reward systems are self-administered, where task forces and work teams are indeed self-directed and empowered, and where the entire organization is fully integrated into the establishment of its goals. However, this will occur only after the pendulum swings several times. With each swing, it will go past the optimal point of distributed management for the organization's current state of maturity, and there will be a reaction that will bring it past this optimal point again in the other direction. The astute manager will recognize, however, that the pendulum itself is on a moving platform, the platform of organization maturity. It is the direction of this platform, not the swing of the pendulum, that will determine the fate of the organization.

If we can imagine a situation in which the power of each person and each informal organization is utilized fully and harmoniously to further the goals of the organization, it is easy to visualize an order-of-magnitude increase in productivity. But that is just the material end of it. We also see entire organizations cooperating with others in the same way that the components of a single organization cooperate, each providing for society what they can provide best (as opposed to running each other out of business or gobbling one another up). Further, we see an extension of these concepts to our government, which is nothing more than a reflection of the society that it represents.

The vision above is quite idealistic, and if it should come about in any measure, it will take decades of evolution. But then, if we had started this 20 or 30 years ago, we would be well on the way by now—so let's get on with it!

Glossary

The definitions that follow are not an attempt to reflect current usage or to imply that such should become a standard. Our objective in providing this list is merely to increase the readers' understanding of techni-management by clarifying the use of the terminology within the context of this book. Note that terms appearing in *italics* in the definitions are also entries in this glossary and are not included in the see and see also entries.

arbitrator as applied to conflict resolution, a person who, after hearing both sides, formulates resolutions that the parties agree beforehand to accept.

authority the ability to realize *goals*. See *formal authority, functional authority, informal authority*.

belief system the perceptions that the members of the organization hold toward it.

binary thinking the tendency to either accept or reject a notion wholesale, as opposed to determining the parts of it that apply and those that should be discarded.

brainstorming a group technique for stimulating creative thought in which the only thing disallowed is criticism of another person's suggestion.

bureaucracy an organization that has evolved improperly to a point where it is rich in procedure but virtually devoid of principle.

bureaucratic of or pertaining to a *bureaucracy*.

Glossary

buy in bywords of common *management* practice, the idea that *subordinates* or colleagues must be sold a policy that is already devised by management. (We note that the term could be used in a positive sense if it implies full participation; however, this is rarely the case, hence we always use it with a negative connotation.)

CEO chief executive officer; also commonly referenced by other acronyms, this is the person in *traditional organizations* who has ultimate legal *authority* over and responsibility for all aspects of the organization.

chain of command collective reference to the lines of *authority* and responsibility that are established in an organization; primarily of military origin, the term and its implications are of very little value within *technimanagement*.

charismatic leaders those who assume *authority* by virtue of their inherent (not necessarily natural) *leadership* abilities, usually appealing to a cause or principle greater than themselves.

clinical approach an approach to *management* that is analogous to the general practice of medicine (e.g., incomplete knowledge, diagnosis, treatment selection, evaluation, and adjustment if the expected response is not attained). Essential to this process is complete objectivity and the ability to accept criticism.

coercive authority the extreme point along the continuum between itself and *free-will authority* at which the manager causes action by making any alternative to undesirable behavior less desirable than the behavior itself.

committee one traditional approach to creating temporary *organizational structures* outside the line organization when additional interactions between line organizations are required.

committee chairperson (or chair) a person recognized and charged with oversight of a committee. The chair does not automatically endow any line position or authority.

component see *organizational component*.

compression corollary to Parkinson's law work is compressed into the time allowed for it.

continuous job enrichment a countermeasure to the *Peter problem* which calls for a seamless alternative to *discrete promotion*.

control a state of a system that cannot be attained without the elements of a unified set of *goals, measurements* against those goals, and a *correction* capability to restore the system when measurements show unacceptable deviation from the goals.

correction as an *element of control,* the process of restoring a system to an improved course.

countercontrol a state where attempts at *correction* actually cause the system to deviate further from the *goal*.

creative incompetence negatively perceived behavior that would disqualify one for consideration for promotion, but not so extreme as to hurt one's intended career.

cross-group representation the same people appointed intentionally to participate on several relevant groups to provide a *formal communication* link.

defense mechanisms internalized rationalizations that tend to preserve a person's feeling of self-worth; these are counterproductive when they create a false sense of reality.

delegation of responsibility (authority) the transfer of responsibility (*authority*) for some function to a lower-level position (*subordinate*).

deliverable a tangible, recognizable product that signals the attainment of a *milestone*.

discrete promotion the act of plunging employees who are totally skilled and absorbed in one function into another for which they have no background, thus invoking the *Peter principle*. See also *continuous job enrichment*.

distributed decision-making principle for each decision that must be made to attain the *goals* of an organization, there is a unique person or organizational support group who/which is in the best position to make that decision.

effective leadership the ability to generate cooperative behavior in others without the use of *coercive authority*. See *free-will authority*. See also *coercive authority*.

effective management those *management* activities that will foster productivity, organizational longevity, and the assurance that the organization will meet its *goals*.

Glossary

elements of control see *control, goals, measurements, correction.*

empowerment the process of moving an entire organization toward a positive, functional working relationship between the *formal* and *informal organizations.* (Because of current misuse of this word, it is essential to recognize that within *technimanagement*, it must be viewed in the context of *optimization*; what is sought is a degree of the *distributed decision-making principle* which is optimal for the particular organization under consideration at this particular time.) See *ineffective empowerment, shared power principle.*

entropy in thermodynamics, the measure of disarray or disorganization within a physical system. If unattended, physical systems tend to evolve toward states of greater entropy. For applications, see *organizational entropy, second law of thermodynamics.*

equifinality the proposition that equivalent results can be attained by a variety of applied methods.

evaluation the process of translating *control* measurements into useful information for *correction.* See also *elements of control.*

favored motivator principle it is easier to motivate most people with dissatisfaction than with satisfaction.

first-line management the lowest level of *management*; traditionally, those responsible for the supervision of those who are not considered to be part of management.

first-order needs in *Maslow's hierarchy of needs*, basic physiological needs.

formal authority the *authority* that the formal organization legally attributes to a given manager. Due to the general nature of authority, however, this often does not reflect reality.

formal communications those meetings, memos, and other interactions necessitated by the *formal organization.*

formal leaders *management* personnel who are usually given titles by those higher up in the organization itself.

formal organization that organization which legally exists based on the (actual or implied) contractual relationships between the corporate whole and its membership.

free-will authority the extreme point along the continuum between itself and *coercive authority* at which action proceeds because it is accurately perceived to be in the best long-term interests of the *sub-*

ordinates, independent of the need for management persuasion to that effect. *Effective leadership.* See *leadership.*

frustration a condition within a person that results from an inability to satisfy a *need* because of either *motivational conflict* or external factors.

functional authority *formal authority* assigned to a person for a specific function outside the *chain of command*. (*Note*: Although the formalized chain of command appears quite neat on paper, it rarely reflects reality, or even what is really formally desired; *functional authority* is an example of an attempt to overcome some of these shortcomings.)

functional conflict a significant difference of opinion that arises between individuals or *organizational components* when behavior that is perceived to be *rewarded* by one component of the organization is perceived to be frustrated by demands made by another component or person. See also *personal conflicts.*

goal integration the process of uniting personal *goals* with those of the *organizational component.*

goal refinement the process of refining a *goal* that analyzes it into accomplishable *objectives* and in the process might alter it due to pragmatic considerations.

goals the articulation of the highest-level ideals of the organization.

golden parachute a large financial incentive given to a key person within an organization who has been fired. The purpose is to minimize the disincentives that such a relief of duty might have on this person's potential successors.

groups a generic term for *project groups, task forces*, and *work teams.*

hasty generalization the logical flaw that leads one to believe that if a premise applies to a small subset, it should apply to the entire population.

Hawthorne effect a phenomenon discovered by early studies at the Hawthorne works by Mayo and others. Generally misinterpreted as the simplistic conclusion that workers respond to any change, the cause of this behavioral change was later shown to be improved human relations.

healthy friction that state of relationship between the *formal and informal organizations*, and also between individual *subordinates*, in which constructive criticism is expected and encouraged to flow in both directions. Healthy friction is essential to synergism.

hierarchy of needs a concept that *needs* can be arranged in a theoretical hierarchy, according to the priority in which most human beings seem to value their attainment; once one need is satisfied, it is no longer perceived to be a need, and the fact that it continues to be attained does not provide the same satisfaction (the intent of a *reward*) that it did when first attained. See also *first-order needs, higher-order needs*.

higher-order needs in *Maslow's hierarchy of needs*, safety and security, belonging and social activity, esteem and status, and self-realization and fulfillment.

horizontal integration the membership and involvement of people representative of a cross-section of line organizations.

Icarus paradox (the *technimanagement* version) that which makes you good makes you bad.

individual power struggle a conflict between people in which the *power* differential between the parties is inadequate to predetermine the outcome. See also *organizational power struggle*.

ineffective empowerment an attempt at *empowerment* that is motivated by the desire for *manipulation*.

ineffective leadership *leadership* by people who are issue- rather than principle-driven, which by its very nature strives for *optimization* over the short term.

informal authority the *authority* that any member of the organization (manager or not) possesses inherently; it is a function of *formal authority*, but its more powerful attributes accrue from *informal leadership*.

informal leaders those who exercise de facto *leadership* within *informal organizations*.

informal organizations organizations that generate themselves spontaneously within all *formal organizations* to meet sociological needs that cannot be met by the formal organization.

leadership the ability to generate collective actions in others which it is not likely that they would perform independently.

learning organization the essence of evolving the organizational culture toward *technimanagement*, an organization with *control* structures that enable it to continue to improve not only its product but also the very *management structure* of the organization itself.

line management managers with direct *authority* and responsibility for day-to-day operation and production.

logical flaws flaws in the logical reasoning process caused by a misperception (intentional or not) of reality. See *binary thinking, hasty generalization, proof by anecdote, tunnel vision.*

maintenance factors factors, including company policy and administration, supervision, salary, interpersonal relations, and working conditions, which are related to environment or context of job. See also *motivational factors* and *motivational maintenance.*

management the activities inherent in being responsible for the productivity of a system of people, machines, and/or materials. Management requires giving up the satisfaction of "doing it yourself" for the increased productivity attained by group effort. The word is also often used to refer to the collective class of managers in a given organization. See also *effective management, first-line management, line management, middle management, staff management, upper management.*

management by exception (MBE) a clear definition of those areas over which the *subordinate* has flexibility to exercise judgment and make direct decisions; implicit in this definition is that the subordinate must seek upper-level approval for resolving any decisions that must be made outside these areas. See *management by negotiated exception.*

management by negotiated exception (MBNE) the guidelines for *subordinate* decision-making limits are never viewed as fixed; instead, they are determined by recurring negotiations, which have as their objective the establishment of the optimal distribution of *authority* and responsibility for that point in the organization's *transition.* See also *management by exception.*

management by objectives (MBO) a *management* technique that requires definitive written objectives for specific periods of time and assessments of their completion at the end of that time span.

management mechanisms see *management structures.*

management structures the *meta-control* system by which an organization can either progress toward the ideal of *technimanagement* or regress to a *bureaucracy.*

manipulation the attempt to implement a *control* system on people without their full understanding of it.

Maslow's hierarchy of needs see *hierarchy of needs*.

McGregor's theories see *Theory X* and *Theory Y*.

measurements as an *element of control*, the ability to assess the degree to which *objectives* (and hence *goals*) are being met.

mechanisms see *management structures*.

mediator see *negotiator*.

meltdown a degree of *organizational entropy* at which the organization's own *management structures* are working at a degree of *countercontrol* that is irreversible.

meta a prefix that seeks to imply the next higher level of abstraction. Examples: meta-thought is thought about thought, meta-management is *management* of the management process, and so on.

meta-control the design, implementation, and operation (i.e., *control*) of control systems.

meta-decision question who determines which decisions are made by the manager and which are made by the *subordinates*? Answer: It is determined collectively within a healthy organization.

middle management collectively, those managers who are above the rank of first-line managers but who are not in *upper management*.

milestone a particular recognizable point (instant in time) in the life of a project.

mindset habitual routine of thinking which is extremely difficult to change.

motivation factors factors including achievement, recognition, the sense of accomplishment of the work itself (especially its quality), responsibility, and advancement. These are the items that lead to job satisfaction without which the workplace is at best tolerable for *professionals*. See also *maintenance factors* and *motivational maintenance*.

motivational conflict a condition within a person in which two or more *motives* of relatively equal strength cannot be realized simultaneously.

motivational maintenance the balancing of *motivational factors* and *maintenance factors* to maintain *goal integration*.

motive alignment the process of internalizing a peaceful coexistence between individual *motives* that are at conflict.

motives (as distinguished from *rewards*) that which produces behavior internal to the individual. See also *theory of rewards*.

need anything that a person perceives at the moment to be essential to his or her well-being or happiness (not restricted to food, shelter, and clothing).

negative feedback all criticism, constructive and destructive.

negative feedback principle the concept that your greatest asset is negative feedback about yourself, regardless of who it comes from.

negotiators as applied to conflict resolution, a person who observes and possibly comments to suggest potential resolutions.

objective the articulation of one activity (usually among many) in the accomplishment of a *goal*.

objective function in the process of *optimization*, a real or imaginary mathematical function of the objective (e.g., cost, benefit, profit, etc.) which it is the goal to either minimize or maximize.

optimization the process by which trade-offs are made between various decision parameters with a view toward obtaining the best possible overall solution. See also *optimum solution, objective function*.

optimum solution the set of decision parameters that yields the best value of the *objective function* in a process of *optimization*.

organizational component the manager's span of *authority* and responsibility: that part of the organization for which the manager has been assigned responsibility.

organizational entropy a measure of an organization's actual (as opposed to perceived) disorder, decay, and proximity of *meltdown*.

organizational power the organization's potential (or that of a *component*) to accomplish its *goals*.

organizational power struggle an *individual power struggle* that becomes further generalized to involve entire *organizational components*.

organizational structures see *management structures*.

organizational support groups those *groups* that are designated specifically to improve organizational decision making; namely, *task forces* and *work teams*.

Glossary

paradigm a model, pattern, example, or template.

paradigm problem once thought is well established within a given *paradigm, optimization* tends to take place within this *mindset*; it is exceedingly difficult for most people even to conceive of another paradigm, and once one is suggested, it is difficult to gain general acceptance of it.

Parkinson's law the amount of time that it takes to perform a task expands to fill the time available to do it.

personal conflicts other than *functional conflicts*, all differences of opinion between two members of the organization that cause a decrease in organizational productivity.

Peter principle in any large organization, the most competent people at a given level will inevitably get promoted to jobs of higher and different (often more complex but less technical) responsibilities, until ultimately they reach a level where they are both nonproductive to the organization and miserable in their job satisfaction.

Peter problem the encapsulation of all the problems that accrue from all of the negative ramifications of the validity of the Peter principle.

positive environment the collective psychological state of an organization in which individual action is based on trust and honesty as opposed to *manipulation*.

power the capacity possessed by a person, an organization, or an *organizational component* to bring about its desired effects either within or outside the organization.

principle of authority the proposition that *authority* exists only to the point where *subordinates* will allow it to exist.

principle of consensus to build consensus effectively requires an attitude on the part of every member of the organization that they would rather yield themselves to the group than to "have their own way" if this is against the will of the group.

principle of fear fear within *management* is the sole cause of management by fear.

principle of free-will authority *technimanagement* provides *rewards* by means of the *distributed decision-making principle*; in the meantime, the *free-will authority* of managers increases to the extent that organizational *goals* are fulfilled through decisions made by their *subordinates*.

principle of functional conflict all healthy organizations will inherently produce *functional conflicts*.

principle of organizational entropy the *second law of thermodynamics* applies to organizations.

principle of synergism *informal leaders* will ultimately determine the *synergism* of groups both internally and with respect to each other. *Formal leaders* have a very high capacity to destroy synergism, but they have little capacity to create it directly.

principle of transition The principles of *technimanagement* cannot be implemented on a crash-mandated basis in most existing organizations, due to the cultural momentum of current *management* practice and experience. The only way that they can be implemented in these organizations is by establishing *mechanisms* by which the implementation of these principles can evolve.

principle of volunteerism the membership of all *work teams* and *task forces* should come only from the subset of the organization who have enough confidence in *empowerment* to be fully motivated to serve enthusiastically in the group.

professional a person who manifests a professional attitude (as opposed to an earned degree); the characteristics of this attitude include the following: (1) self-starting mental capacity and inclination, (2) a love for the work itself, and (3) satisfaction from seeing high-quality work accomplished separate and distinct from that obtained from the monetary *reward* system. Although not degree dependent, the discipline and indoctrination required by most notable universities usually guarantees that those graduating with technical degrees have the basic characteristics, if not the fully developed character, to assume the role of professionals.

professional organization an organization consisting predominantly of *professionals*; see *technical organization*.

project group an organizational unit at the lowest line level that works in direct pursuit of organizational *goals*.

project group leader a person at the lowest level of *management* placed in charge of a *project group*.

promoting managers the set of managers who have the *authority* to promote a given subset of members of the organization.

proof by anecdote the logical flaw of *hasty generalization* which accepts an anecdote as being accurate and then generalizes it to apply to most similar situations.

Glossary

Protestant ethic a value system derived or evolved from puritanism which emphasizes hard work as ample (ethical and spiritual) justification for the accumulation of wealth.

punishments physical and psychological deterrents that are intended to affect negatively the *motives* of the individual. Note that withholding of a *reward* is, by definition, a punishment.

rational-legal authority the final step in the evolution of an organization in which *bureaucracy* develops.

reality truism unless managers can honestly face the fact that they are the cause of resentment, there is absolutely no way that they will be able to devise measures to eliminate it.

rewards physical and psychological inducements that are intended to affect the *motives* of the individual. See also *punishments*.

role conflict a person's internal problem caused by incompatible role expectations.

second law of thermodynamics left to themselves, all physical systems tend to increase in *entropy*. Application: The second law of thermodynamics applies to organizations and individuals.

selective perception the tendency to accept those communications that tend to reinforce one's beliefs as opposed to that which is logically correct (reality).

shared power principle effective *empowerment* increases the *power* of managers in the same proportion that it increases the power to their *subordinates*. Alternatively: Effective empowerment increases the power of *informal organizations* in the same proportion as it increases the power of *management*. See *ineffective empowerment*.

simpatico the sense of being able to understand another person's true feelings (apart from overt actions).

social Darwinism a value system modeled after Darwin's theory of evolution which emphasizes that those who rise to the top of the social hierarchy do so through a natural process that *rewards* those most capable and resourceful.

spiritual needs nonphysical *needs* omitted from *Maslow's hierarchy of needs* which in certain people have the capability of displacing any of the other needs in the hierarchy.

stability of authority a principle which states that stability is a function of the degree to which *subordinates* see decisions made to be consistent with their long-term interests. See also *coercive authority, free-will authority*.

staff group a traditional alternative to creating a permanent organizational structure outside the line organization; this is established in the *traditional organization* to deal with issues that are generally common to many, if not all, of the line *components* of the organization.

staff management people who report to line managers (at various levels, depending on the organization) for particular specialized activities but who are not directly responsible for implementing policy. [*Note*: A quote from Kazmier (KAZMI69,31) provides insight: "The relationship between line and staff activities in most organizations is quite complex. Contrary to what one might expect, for example, the staff is not invariably subordinate to the line organization."]

suboptimal any decision set that is not optimal with respect to the system despite the fact that some of the *components* might be optimized.

subordinate a member of an organization who holds a position below another in the formal *chain of command*. Use of this word is not intended to imply that one person is morally superior to another or that one person has greater intrinsic value than another. (On the contrary, the principles of *technimanagement* demand that neither the *superior* nor the subordinate hold this attitude.)

superior a member of an organization who holds a position above another in the formal *chain of command*. The use of this word is not intended to imply that one person is morally superior to another or that one person has greater intrinsic value than another. (On the contrary, the principles of *technimanagement* demand that neither the superior nor the *subordinate* hold this attitude.)

synergism an encapsulation of the concept that a properly functioning whole is an order of magnitude more productive than the individual efforts of the functioning parts. It goes beyond collective productivity to the root cause of this productivity—collective enthusiasm. See also *principle of synergism*.

systems view the notion that the application of *optimization* within each *component* of a system will generally not produce *optimization* at the full system level.

task force equivalent in all respects to the *work team*; however, this group is expected to have a finite life due to its charge, which will be to address a specific problem as opposed to the constant improvement of a process.

technical organization organization of scientists, engineers, or other *professionals* who are performing creative design and development activities. This term is used synonymously with *professional organization*.

technimanagement an open-ended philosophy of *management* directed at organizations composed primarily of technical *professionals* (i.e., *technical organizations*). It is based on established principles of human nature and human interactions which have been validated in the last half-century.

theory of rewards a principle of experimental psychology which states that activities that are *rewarded* tend to be repeated and emulated by others; activities that are punished tend to be avoided and not emulated.

Theory X a *paradigm of management* put forth by McGregor which is generally applied in most *traditional organizations*—that the function of management is to establish *control* systems to assure that the workers do what they were charged to do. See *Theory Y*.

Theory Y a *paradigm of management* put forth by McGregor which generally reflects that which should be applied in *technical organizations*—that the function of management is to establish *meta-control* systems to assure that all members of the organization participate in its decision-making processes. See *Theory X*.

total quality management (TQM) a philosophy that has as its intent to provide guidance and motivation for organizations to realize fully the benefits of McGregor's *Theory Y*.

traditional authority/leadership that which typically comes after *charismatic leadership* where lower levels recognize the necessity for additional structure and allow formal positions to be established to satisfy this requirement.

traditional management those *management* approaches that are generally applied to production-oriented organizations rather than *technical organizations*.

traditional organizations predominantly production-oriented organizations which have a well-developed and tested set of products, with the major objective of producing them with a competitive quality and price.

transition a process of evolution that is essential if the full benefits of *technimanagement* are to be realized; this is due to the requirement for maturing on the part of all members of the organization. See *principle of transition*.

tunnel vision the inability to see the global ramifications of a position.

unwin to accept what is best for the organization (or *organizational component*) without claiming victory.

upper management collectively, the *CEO* and all line managers who report to the CEO.

upward mobility with reference to *Maslow's hierarchy of needs*, the perception and reality of freedom on the part of the person to move to the next higher *need* when a lower need is satisfied.

vertical integration the membership of all pertinent work group and managerial levels necessary to accomplish the group's *objectives* assembled together on an equal basis.

work team a nontraditional structure that has a number of *empowerment* characteristics, to distinguish it from a *committee* or *staff group*.

References

BARKE91 Barker, J. "The Business of Paradigms" (videotape and discussion guide), Charthouse Learning Corporation, Burnsville, MN, 1991.

BELLM57 Bellman, R., *Dynamic Programming*, Princeton University Press, Princeton, NJ, 1957.

BLAU62 Blau, P. M., and Scott, W. R., *Formal Organizations*, Chandler Publishing Company, San Francisco, 1962. Obtained from KAST70.

BROWN76 Brown, D. B., *Systems Analysis and Design for Safety*, Prentice Hall, Englewood Cliffs, NJ, 1976.

CLELA68 Cleland, D. I., and King, W. R., *Systems Analysis and Project Management*, McGraw-Hill, New York, 1968. Obtained from KAST70.

DALE65 Dale, E. (ed.), *Readings in Management: Landmarks and New Frontiers*, McGraw-Hill, New York, 1965.

DAVIS69 Davis, K., and Scott, W. G. (ed.), *Human Relations and Organizational Behaviors: Readings and Comments*, McGraw-Hill, New York, 1969.

DEMAR87 Demarco, T., and Lister, T., *Peopleware: Productive Projects and Teams*, Dorset House Publishing Co., New York, 1987.

DEMIN85 Deming, W. E., "Transformation of Western Style of Management," *Interfaces*, Vol. 15, No. 3, pp. 6–11, 1985.

DRUCK58 Drucker, P. F., *Technology, Management and Society*, Harper & Row, New York, 1958 (1977).

HERZB59 Herzberg, F., Mausner, B., and Snyderman, B., *The Motivation to Work*, Wiley, New York, 1959. Obtained from KAZMI69.

KAST70 Kast, F. E., and Rosenzweig, J. E., *Organization and Management: A Systems Approach*, McGraw-Hill, New York, 1970.

KATZ66   Katz, D., and Kahn, R. L., *The Social Psychology of Organizations*, Wiley, New York, 1966. Obtained from KAST70.

KAZMI69   Kazmier, L. J., *The Principles of Management: A Program for Self-Instruction*, McGraw-Hill, New York, 1969.

KEITH71   Keith, L. A., and Gubellini, C. E., *Introduction to Business Enterprise*, McGraw-Hill, New York, 1971.

LASKI30   Laski, H. J., "The Limitations of the Expert," *Harper's*, Vol. 162, pp. 102–106, Dec. 1930. Obtained from DALE65.

LEAVI65   Leavitt, H. J., "Applied Organizational Change in Industry: Structural, Technological and Humanistic Approaches," in J. G. March (ed.), *Handbook of Organizations*, Rand McNally, Chicago, 1965. Obtained from KAST70.

MASLO43   Maslow, A. H., "A Dynamic Theory of Human Motivation," *Psychological Review*, Vol. 50, pp. 370–396, 1943. Obtained from STACE65.

MASLO54   Maslow, A. H., *Motivation and Personality*, Harper & Row, New York, 1954. Obtained from KAZMI69.

MAYO45   Mayo, E., *The Social Problems of an Industrial Civilization*, Division of Research, Harvard Graduate School of Business Administration, Boston, 1945.

MCGRE57   McGregor, D. M., "The Human Side of Enterprise," *Proceedings of the 5th Anniversary Convocation of the School of Industrial Management*, MIT, Cambridge, MA, 1957. Obtained from DAVIS69.

MCGRE60   McGregor, D. M., *The Human Side of Enterprise*, McGraw-Hill, New York, 1960.

MILLE90   Miller, D., *The Icarus Paradox*, HarperBusiness, New York, 1990.

MILLE65   Miller, J. G., "Living Systems: Basic Concepts," *Behavior Science*, July 1965. Obtained from KAST70.

PARKI57   Parkinson, C. N., *Parkinson's Law and Other Studies in Administration*, Riverside Press, Cambridge, MA, 1957.

PETER69   Peter, L., *The Peter Principle*, Morrow, New York, 1969.

SAYLE66   Sayles, L. R., and Strauss, G., *Human Behavior in Organizations*, Prentice Hall, Englewood Cliffs, NJ, 1966. Obtained from KAST70.

SCHOE57   Schoen, D. R., "Human Relations: Boon or Bogle," *Harvard Business Review*, pp. 91–97, March–April, 1957. Obtained form DAVIS69.

SCOTT67   Scott, W. G., *Organization Theory*, Richard D. Irwin, Homewood, IL, 1967. Obtained from KAST70.

STACE65   Stacey, C. L., and DeMartino, M. F. (eds.), *Understanding Human Motivation*, World Publishing, New York, 1965.

References

TOWNS70 Townsend, R., *Up the Organization*, Fawcett Publications, Greenwich, CT, 1970.

URWIC56 Urwick, L. F., *The Pattern of Management*, University of Minnesota Press, Minneapolis, MN, 1956. Obtained from DALE65.

WHITE36 Whitehead, T. N., *Leadership in a Free Society*, Harvard University Press, Cambridge, MA, 1936. Obtained from DALE65.

YNTEM60 Yntema, T. O., "The Transferable Skills of a Manager," *Journal of the Academy of Management*, pp. 79–80, August 1960. Obtained from KEITH71.

Index

'Aggression, 33
Arbitration, 270-72, 274
Authority, 59-73
    coercive, 60-61, 66
    delegation, 158
    empowerment, 155-62, 181-84, 213, 364
    exception principles, 74-78
    formal service, 296
    leadership, 60, 63-65, 313-14, 320-24
    organizational evolution, 61-63, 68 (*See also* transition)
    personal power, 279-80, 287, 290
    principle of, 59
    responsibility, 65-68
    stability of, 61-63

Barker, J., 128
Belief system, 26-29, 123-24, 338-39
Bellman, R., 101
Binary thinking, 91
Blame, 56, 220-21, 267, 271, 274
Blau, P. M., 41
Brainstorming, 204-5
Bureaucracy, 14, 79, 104-9, 120, 359, 362
Buy in, 16, 54, 69, 157, 174

CEO, 182, 214, 288, 348-49, 352-55, 367
Chain of command, 238-39, 260, 348
Chairperson, 179
Charismatic authority, 62
Cleland, D. I., 287, 289
Clinical approach, 242-51, 254, 265, 284, 292-93, 303
Coercive authority, 60-61, 65-68

Committee, 16, 69, 179-81, 184
    chairperson (*See* chairperson)
    problems, 184, 352
Communication, 335-44
    deterrents, 338-44
    model, 335-38
    modularized, 343-44
Complexity (*See* human complexity)
Component (*See* organizational component)
Compression corollary to Parkinson's law, 106-8
Compromise, 32
Conflict, functional (*See* functional conflict)
Conflict, personal (*See* personal conflict)
Consensus building, 200-9
Continuous job enrichment, 121-22, 126-27
Control, 8-17
    analysis of, 9-11
    correction, 14-15
    effect of rewards, 18-48, 223
    elements, 8-17
    empowerment, 153-62, 364
    for process improvement, 218-31
    goals and objectives, 11-13
    individual entropy, 280-85
    influence, 290-92
    leadership, 319, 322-24
    loss, 166-70, 300
    measurement and evaluation, 13-14
    of professionals, 15-17
    purpose, 8-9
    self, 15, 200, 249-51
    statistical process, 234-35

Control *(cont.)*
    transition, 354, 363, 369
    under MBNE, 77-78
Control systems, 8-17
    functional conflict resolution, 258
    in total quality management, 223
    personal objectivity, 251
    (See also control)
Correction, 14-15
    control element, 9-11, 14-15
    in clinical approach, 251
    in total quality management, 219-20, 233-34, 234-35
    with empowerment, 153-62
    with goal integration, 34-36
    within decision making, 72-73
Countercontrol, 11, 47
Counterintuitive, 45
Creative incompetence, 120-22
Criticism:
    acceptance, 220-21
    administering, 38-40
    brainstorming, 204-5
    clinical approach, 246-48
    during prioritization, 208
    ego, 313-14
    external reaction, 303-8
    group formation, 173
    group harmony, 196-97, 199
    in meta-control, 47
    in transition, 364-66
    internal reaction, 302-3
    negative feedback principle, 300-10
    personal conflicts, 264
    reaction, 302-8
Cross-group representation, 192
Customer:
    definition, 236-37
    involvement, 238
    orientation:
        during transition, 347-51, 357-58
        in total quality management, 236-39
        leadership, 313
        organization for, 238-39
        resolving conflicts, 260-61, 264
        TQM infrastructure, 236-39
    reorientation, 348-51

Dale, E., 389
Davis, K., 389
Decision maker selection, 69-70, 75-78
Decision making, 59-73
    agreement, 246
    authority relationship, 59-73

barriers, 250, 338
cultural affects on, 91-95
distributed, 67-70, 181, 313
empowerment, 155-62, 181-84, 192-93, 364
expert limitations, 326-27
future reality, 248-49
goal integration, 174
improvement, 149-50, 173, 192-93, 222, 236
informal organizations, 145-48, 173
leadership, 313, 320-24
mechanisms, 73
optimization, 95-100
process, 69-73
responsibility, 69-70
systems view, 101-2
transition, 354, 357-58, 363
Defense mechanisms, 30-34, 135
Defining the customer, 236-37
Delegation:
    of authority, 158-62
    of responsibility, 158-62
Deliverable, 83-84
Delphi technique, 207
Demarco, T., 5, 57, 64, 197, 316
Deming, W. E., 85-86, 132, 210-12, 214-16, 232, 352
Deming's obligations of management *(See* obligations of management)
Discrete promotion, 121
Distributed decision-making principle, 183
Drucker, P. F., 389

Effective:
    empowerment, 156-62, 364
    leadership, 60, 64, 311-24
        creating, 149-50
        need for, 77-78, 81
    management, 1, 317-19, 355-67
Ego:
    clinical approach, 244, 249-51
    group harmony, 197
    growth, 249-51
    Hawthorne effect, 42
    informal organization, 142
    negative feedback, 300, 303
    of the leader, 313-14
    organizational entropy, 284, 292-94, 297-99
    protection, 249-51
    resentment, 78-80
    theory Y, 55, 58

Index

Elements of control, 8-17
    in empowerment mechanisms, 159
    (*See also* control, goals, measurements, correction)

Emotions:
    clinical approach, 246-47, 249-51
    complexity of, 44-45
    group synergism, 194
    in communication, 337, 340
    in personal conflicts, 269, 274

Employee involvement, 233-36

Empowerment:
    effective, 155-62, 364
    in MBNE, 76
    in total quality management, 213, 224
    in transition, 355-67
    individual, 170, 280, 289-99
    informal organizations, 158-62
    mechanisms, 161
    of groups, 181-84

Entropy:
    effect of communication, 335
    ego effects, 249-51
    in groups, 194, 201
    in total quality management, 216, 219
    in transition, 362
    individual, 279-99
    informal communication, 162-63
    negative feedback, 303
    organizational, 165-77
    personal conflicts, 262
    (*See* organizational entropy, second law of thermodynamics)

Environment (*See* positive environment)

Equifinality, 100-1

Evaluation:
    element of control, 13-14
    in decision making, 72
    in transition, 355, 362

Evolution:
    Darwinian, 28
    organizational (*See* organizational evolution)

Expert limitations, 325-34

Favored motivator principle, 171

First-order needs, 23 (*See also* hierarchy of needs)

Formal authority, 59, 61-63, 65, 294, 314, 320-24

Formal communications, 148-49

Formal leaders (*See* formal organization)

Formal organization:
    authority, 65
    entropy, 166-70, 174-77
    group dynamics, 178-81
    in functional conflicts, 253, 255, 258-59
    in personal conflict, 265-58
    in role conflict, 276-78
    in total quality management, 223-24
    in transition, 371
    informal interaction, 139-62, 249
    leadership, 320
    personal power, 279-80

Free-will authority, 60-61

Friction (*See* healthy friction)

Frustration, 30-34

Frustration from motivational conflict, 29-31

Functional conflict, 252-61, 265
    benefits of, 255-58
    causes, 252-55
    in goal refinement, 12
    resolution of, 258-61

Generalization (*See* hasty generalization)

Goal conflicts, 12

Goal integration:
    expert limitations, 333
    group dynamics, 181
    in clinical approach, 247
    in conflict resolution, 259-60
    in total quality management, 216, 227, 236
    organizational entropy, 174-77
    personal power, 288
    process definition, 22-23
    productivity, 34-36
    responsibility and authority, 65-68
    rewards, 23, 34-37
    theory X, 50-52

Goal refinement, 12

Goals:
    as control element, 9, 11-13
    as rewards, 21-24, 34-36, 45-48
    group dynamics, 183
    in informal organizations, 145-53
    in MBO, 83-88
    in role conflict, 276-78
    in theory X and Y, 50-53
    in total quality management, 216, 223, 226-27, 230-31, 236
    of communication, 335-44
    of functional conflict resolution, 260
    of leadership, 314-17, 322-24
    of personal power, 280-99
    organizational entropy, 174-77
    related to paradigm problem, 130

Goals *(cont.)*
    systems view, 101-2
    within transition, 354, 359, 371, 373
Golden parachute, 220
Group:
    dynamics, 178-209, 232
    empowerment, 181-84, 364 *(See also* empowerment)
    formation:
        management response, 173-74
        organizational entropy, 168, 170-74
    harmony, 187, 193-200
    morale, 199-200
    selection, 184-89, 197-99, 360-61
    sizing, 189-93
    synergism, 194-97, 232, 272, 288, 328
Groups, 179-80, 350

Harmony (*See* group harmony)
Hasty generalization, 91, 338-39
Hawthorne effect, 40-42, 55, 200, 213
Healthy friction, 196
Herzberg, F., 37
Hierarchy of needs, 23-26, 29, 36, 52-53, 225, 231
Higher-order needs, 23-26, 37-38, 56-57, 231
Horizontal integration, 70, 180, 192, 229, 232, 234, 236, 368-69
Human complexity, 44-45
Human factor, 38-45
Humanism, 42-44, 49 *(See also* industrial humanism)

Icarus paradox, 299-300
Individual entropy, 279-99
Individual power struggle, 279
Industrial humanism, 42-44, 49
Ineffective empowerment, 157-59
Ineffective leadership, 64 *(See* effective leadership)
Influence spectrum, 290-92
Informal authority *(See* informal organizations)
Informal communication:
    breakdown, 342-43
    group dynamics, 190, 193
    leadership and organizations, 148-49
    value of, 162-64, 342-43, 371
Informal leaders:
    expert limitations, 333
    group dynamics, 186, 195, 200
    group formation, 170-74
    in transition, 368, 371

    (See also informal leadership)
Informal leadership, 138-64
    communications, 342-43
    in MBNE, 81
    leadership, 319-22
Informal organizations, 138-64
    benefits, 145-53
    downside, 152-53
    empowerment, 153-62
    identification methods, 139-41
    in the clinical approach, 249
    influence, 33
    paradigm problem, 137
    problems of current approaches, 141-45
    systems view, 102
Integration *(See* goal, horizontal, vertical integration)

Kast, F. E., 28, 42, 62, 100, 102, 134, 184, 278, 290, 338, 341
Katz, D., 279, 282-83, 290
Kazmier, L. J., 9, 21-22, 29-30, 32, 34, 37, 50, 52-53, 139, 197, 276, 294, 297, 315, 322, 335-36, 338
Keith, L. A., 38, 65, 134, 145, 163
Killing the messenger, 301-2

Laski, H. J., 326-27
Leadership, 311-24
    authority, 63-65
    decision making, 60, 67-68, 73
    during transition, 352-54, 358
    effective *(See* effective leadership)
    evolution, 61-65, 68
    group dynamics, 179-80, 185-86, 195, 200
    in MBNE, 77-78, 81
    in meta-control, 17
    in total quality management, 221
    informal, 65, 138-64
    negative feedback, 303-4, 308-10
    personal power, 285, 289-91
    styles, 322-24
    traits, 314-17
    versus management, 319-22
Learning organization, 81, 136-37
Leavitt, H. J., 182
Limitations of experts, 325-34
Line management *(See* Glossary)
Logical flaws, 91-92

Maintenance factors, 37-38
Management:
    by exception, 74-80
    by negotiated exception, 75-78, 85

Index

by objectives, 83-89, 106
mechanisms, 125-27, 141, 319, 357, 363 (*See also* management structures)
structures, 118, 230
traits, 317-19
Manipulation, 16, 19-20, 32, 157
defense mechanisms, 32-34
during transition, 363-64
motivational conflict, 29-31
Maslow, A. H., 23
Maslow's hierarchy of needs (*See* hierarchy of needs)
Mayo, E., 2, 5, 40-43, 49, 172
MBE (*See* management by exception)
MBNE (*See* management by negotiated exception)
MBO (*See* management by objectives)
McGregor, D. M., 4, 49-59, 211-12
McGregor's theories, 6, 49-58 (*See also* Theory X and Theory Y)
Measurement:
as element of control, 13-14
in clinical approach, 251
in total quality management, 230-31, 234-35
Mechanisms:
defense, 32-34
meta-control, 46-48
of correction, 14-15
of decision making, 72-73
of informal leadership, 141, 147, 149-50, 161
of negative feedback, 309-10
of personal power, 285, 289, 297-98
of theory Y, 56
of total quality management, 224, 224-28, 229
(*See also* management structures)
Mediation, 260, 270-72, 274 (*See also* negotiation)
Meeting conduct, 201-4, 358
Meltdown (*See* organizational meltdown)
Meta-control:
and process improvement, 218
and rewards, 45-48
control basis, 10, 12, 15, 17
in decision making, 68, 70
in group selection, 186
in MBNE, 77-78, 80
in theory of rewards, 45-48
in Theory X and Y, 51-53, 56
organizational entropy, 167
Meta-decision, 69, 77, 94
Meta-decision question, 69
Middle management, 351-52, 367-71

Milestone, 83-85
Miller, D., 299
Miller, J. G., 165
Mind, internal complexity, 44-45 (*See* emotion)
Mindset, 128-29, 239-41
Modularized communication, 343-44
Morale
group (*See* group morale)
individual, 38-40, 150-52
Motivation factors, 37-38, 332
Motivational conflict, 29-34
Motivational maintenance, 37-38
Motive alignment, 30-31
Motives:
clinical approach, 248
criticism, 305
informal communications, 162-64, 342-43
personal conflicts, 265-66
rewards, 20-23, 29-31
versus rewards, 20-23

Needs, sociological, 151-52 (*See also* hierarchy of needs)
Negative feedback, 251, 300-10
Negative feedback principle, 300-10
Negotiation, 75-78
NFL draft method, 208

Objective function, 95
Objectives, within control, 9, 12
Objectivity, 3, 242, 247, 249-51, 265-67
and control systems, 251
personal, 251
Obligations of management, 212-13, 215-33
Optimization, 95-100
effective empowerment, 153-62, 364
expert limitations, 325, 333
general description, 95-102
group harmony, 195
paradigm problem, 128-30
Parkinson's law, 107-8
personal power, 287-88
short-term gains, 214-15
Optimum solution, 95
Organizational:
component, 21 (*See also* Glossary)
entropy, 165-77
clinical approach, 249
communication, 335
group formation, 170-73
in total quality management, 216, 219
in transition, 362

Organizational *(cont.)*
    personal power, 280-81
    principle of, 166
    (*See also* entropy)
  evolution, 61-63, 68 (*See also* transition)
  meltdown, 169, 176
  power, 170, 280-83, 297
  structures, 179-80, 215, 243, 251, 284
    (*See* also management structures)
  support groups, 180

Paradigm problem, 128-37, 153-55, 168, 217, 239
  solutions to, 134-37
  validity of, 130-34
Parkinson, C. N., 103
Parkinson's law, 103-9, 120
  applications, 108-9
  compression corollary, 106-8
  general applicability, 103-6
Perception:
  of current reality, 244-45
  of future reality, 248-49
Personal conflicts, 200, 248, 255, 262-79, 303, 318-19
  involving manager, 272-76
  involving subordinates, 264-72
  prevention, 263-64
  resolution of, 262-78
Personal power, 279-99
  group dynamics, 200
  limitations of experts, 326-27
  negative feedback, 306-7
  role conflict, 276-78
Peter, L., 110
Peter principle, 110-27, 144, 168, 249-50, 350
  shortcomings of, 117-19
  validity of, 112-16
Peter problem, 119, 168
  individual countermeasures, 120-22
  organizational countermeasures, 122-27
  solutions to, 119-27
Polarization, 151, 170
Positive environment, 32-34
Power; *see* personal power, organizational power
Principle:
  distributed decision-making, 183
  of authority, 59-73
  of consensus, 206
  of fear, 81
  of free-will authority, 67
  of functional conflict, 253
  of MBNE, 76
  of organizational entropy, 166
  of resonance, 47
  of synergism, 195
  of transition, 88, 345
  of volunteerism, 188
Principles of communication, 335-44
Prioritization techniques, 205-9
Prioritizing the customers, 237-38
Problem individual, 55
Productivity through goal integration, 34-36, 181
Professional, 1, 3, 58
Professional organization (*See* Glossary)
Professionals, control of, 15-17
Project group:
group dynamics, 178-209
in total quality management, 234-35
in transition, 359-66
leadership, 179-80, 326-27
Promoting managers, 111
Promotion, discrete, 121
Proof by anecdote, 91
Protestant ethic, 26-29
Punishment, 18-19, 60, 66, 213

Quality function deployment, 233-34
Quality management approaches, 233-35

Rational-legal authority, 62
Resentment, 37, 78-80, 226
  as a deterrent, 78
  truism, 79
Resonance, 47
Responsibility:
  and authority, 65-68
  delegation of, 158-59, 178-81
Rewards:
  theory of, 18-48
  versus motives, 20-23
Role conflict, 144, 276-78

Sayles, L. R., 343
Schoen, D. R., 242, 249
Scott, W. G., 43-44
Second law of thermodynamics, 165-67, 280-85
  individuals, 280-85
  organizations, 165-68
Selection (*See* group selection)
Selective perception, 341-42
Self-direction, 100
Semantics, 340
Shared power principle, 156-57

Index

Short-term gains, 214-15, 316
Simpatico, 138
Sizing (*See* group sizing)
Social Darwinism, 26-29
Sources of power, 285-97
Spiritual needs, 23, 26-29
Spontaneous group decomposition, 174-77
Stability of authority, 60-63, 68
Stacey, C. L., 390
Staff group, 179, 218, 233-34, 328
Staff management (*See* Glossary)
Statistical process control, 234-35
Stereotyping, 338-39
Suboptimal, 129
Symbology, 340-41
Synergism, 194-200
    expert limitations, 328, 330
    in functional conflicts, 261
    in personal conflicts, 262, 272
    in total quality management, 216, 226, 232
    leadership, 314, 318-19
    negative feedback, 310
    personal power of, 280-82, 285, 288
    (*See also* group synergism)
Systems view, 101-2, 327, 353

Task force, 180, 184, 260, 359-66, 368
Technimanagement, 1
Theory of rewards, 18-48, 224, 228-29
Theory X, 50-52
    authority, 69-71
    group dynamics, 185
    in functional conflicts, 257
    in personal conflicts, 265-67
    in total quality management, 218, 222, 224-28, 230-31, 232, 239
    in transition, 346-47, 370
    leadership styles, 322-24
    personal power, 296-97
Theory Y, 52-53
    authority, 59-61, 69-73
    fate of, 53-57
    in functional conflicts, 252
    in total quality management, 211-12, 224, 239-40
    in transition, 346
    leadership styles, 322-24
    technimanagement comparison, 57-58
Time value of power, 286-90
Total quality management, 85, 210-41, 289, 346, 252
Townsend, R., 5, 293, 302, 311
TQM (*See* total quality management)
TQM and technimanagement, 239-41
Traditional authority, 62, 65-66
Traditional management (See traditional organizations)
Traditional organizations, 1, 3, 5, 57, 80, 228, 348-49
Training, during transition, 221-23, 231-32, 371-72
Transition, 345-73
    bottom-up, 355-67
    middle-out, 367-71
    need for, 7, 23, 43-44, 47, 49, 68, 70, 74, 77-78, 86, 88-89, 119, 122, 127, 139, 161-62, 176, 184, 215, 230, 236, 239-40, 249, 257, 280, 298, 321-22, 324, 345-48, 363-64
    requirement, 345-48
    top-down, 352-55
    ultimate end, 372-73
Tunnel vision, 92

Unwin, 292-94
Upper management, 78, 113-17, 220, 352-55
Upward mobility, 25-26
Urwick, L. F., 391

Vertical integration, 180, 182, 187, 229

Walker, J. R., 293
Whitehead, T. N., 64, 135, 154, 173
Winning and losing, 292-94
Withdrawal, 33
Work team, 100, 180, 260, 354-55, 359-66

Yntema, T. O. 317, 320